PREFACE

PRACTICAL ANALYSIS OF ELECTRONIC CIRCUITS THROUGH EXPERIMEN-
TATION begins with a review of field effect transistors and then procedes through the
analyzation of multistage hybrid and monolithic circuits, power supplies, RC oscilla-
tors, inverters and switching regulators. It is a continuation book to PRACTICAL
ANALYSIS OF AMPLIFIER CIRCUITS THROUGH EXPERIMENTATION (The
Technical Education Press, 1975).

Although the book is broader in scope with a variety of circuits, the approach to
analyzing the circuitry remains the same; Ohm's Law and common sense are the basic
tools for analysis. Of particular interest will be a new transistorized approach for analyzing field effect transistors and a practical technique for obtaining relatively accurate
FET parameters. Also, in the bulk power supply section of the book, there are several
new approaches to simplify the analysis (and possible design) of unregulated power
supplies.

All the circuits in this text-laboratory book have been thoroughly tested and they work
with excellent results. However, the circuits can be modified for optimization once a
thorough understanding of the circuit is obtained. And optimization is encouraged, once
the initial calculations and measurements have been made, since it is the role of the
technician or working engineer to provide optimization to any circuit or system that he or
she may be working on.

The book is intended for technical people, who have had DC and AC circuit theory,
algebra, some trigonometry, and a solid understanding of transistor circuit analysis. A
knowledge of the calculus is not necessary, however, since it is used only to provide
understanding with regard to the derivation of some formulas.

The number of laboratory hours necessary to complete all of the experiments in the
book is between 45 and 60 hours. This assumes the use of spring boards, proto boards,
EL boards, or any form of breadboarding technique that accommodates discrete and
monolithic devices, while not requiring soldering. This facilitates the initial set-up and
minor circuit changes. The number of laboratory hours can be shortened through selected experiments, if the ideas of several experiments are combined, but this must be done
judiciously to prevent learning loss, because most of the learning takes place when all the
ingredients of calculating, constructing, and measuring are combined.

A debt of gratitude is owed to the hundreds of students who, through feedback and
constructive criticism helped develop this text-laboratory book to its present form. I am
also thankful to Professors Bob Kellejian and Bob Shapiro and to Messrs. John A.
Scullion, Donald Snell, and LeRoy Olson who, collectively provided advise, support,
and inspiration. Especially, I am indebted to my wife Annette, who typed all of the
countless revisions and final copy.

1978   Lorne MacDonald

ELECTRONIC CIRCUIT ANALYSIS SERIES
from The Technical Education Press

DIRECT CURRENT CIRCUIT ANALYSIS THORUGH EXPERIMENTATION 3/e
by Kenneth A. Fiske and James H. Harter

ALTERNATING CURRENT CIRCUIT ANALYSIS THROUGH EXPERIMENTATION 2/e
by Kenneth A. Fiske and James H. Harter

PRACTICAL ANALYSIS OF AMPLIFIER CIRCUITS THROUGH EXPERIMENTATION
by Lorne MacDonald

PRACTICAL ANALYSIS OF ELECTRONIC CIRCUITS THROUGH EXPERIMENTATION
by Lorne MacDonald

DIGITAL TTL INTEGRATED CIRCUIT LOGIC & DESIGN THROUGH EXPERIMENTATION
by Darrell D. Rose

PRACTICAL ANALYSIS OF ELECTRONIC CIRCUITS THROUGH EXPERIMENTATION

Lorne MacDonald

COLLEGE OF SAN MATEO

THE TECHNICAL EDUCATION PRESS
SEAL BEACH, CALIFORNIA

PRACTICAL ANALYSIS OF ELECTRONIC CIRCUITS THROUGH EXPERIMENTATION

ISBN: 911908-08-0

MANUFACTURED IN THE UNITED STATES OF AMERICA

CONTENTS

DEDICATED
TO
LYNN
LESLIE
LISA
LOUISE
LAURA

1 | FIELD EFFECT TRANSISTORS AND SINGLE STAGE FET AMPLIFIER CIRCUIT ANALYSIS — A REVIEW

GENERAL DISCUSSION

The field effect transistor is a high input impedance, low noise, voltage controlled device that has characteristics similar to those of the vacuum tube. However, it does not have the disadvantages of higher operating voltages, filament power requirements, short life span, and physical size.

The two families of field effect transistors are the junction field effect transistor (JFET) and the metal oxide semiconductor field effect transistor (MOSFET). The MOSFET is also known as the insulated gate field effect transistor (IGFET).

The JFET is a voltage depletion majority carrier type device, where the PN junction of the gate to source is reverse biased in controlling the amount of current flow through the device. In other words, the JFET is normally "on" since there is maximum current flow through the device when no reverse biasing exists across the gate-to-source junction and, as reverse gate-to-source bias voltage is increased, the device approaches minimum current flow. There are both depletion and enhancement type MOSFET devices. The enhancement type device provides just the opposite effect to that of the depletion type device — it is normally "off" and has to be turned on with forward biasing. The enhancement characteristic is similar to that of the bipolar transistor. Both families of FET devices, their characteristics, and their effect on various single stage circuits are analyzed in this review chapter.

DEVICE SYMBOLS

The schematic symbols for the junction field effect transistor, both N-channel and P-channel, are shown in Figure 1-1 and, since the JFET is a depletion.type device, the reverse biasing across the gate to source is shown. The reverse biased condition also provides the high input inpedance and, if the device is foward biased momentarily, the advantage of the high input impedance would be lost and extreme (non-linear) distortion to the processed signals would occur.

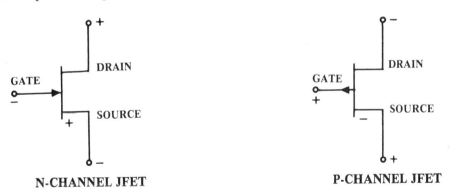

N-CHANNEL JFET   P-CHANNEL JFET

FIGURE 1-1

The schematic symbols for the depletion type MOSFET, both N-channel and P-channel, are shown in Figure 1-2. However, the MOSFET, unlike the JFET, does not require the gate-to-source junction to be reverse biased for a high input impredance. The high input impedance is obtained from the insulating properties of the gate terminal which perform a capacitive coupling effect between the gate electrode and the main body (or channel) of the device. This insulated condition allows the gate voltage to be either negative, positive, or zero volts — with respect to the source terminal — in controlling the current flow through the device.

The schematic symbols for the enhancement type MOSFET, both N-channel and P-channel, are shown in Figure 1-3. The interrupted channel lines denote a normally "off", no current flow, condition. The channel current is turned on by applying "forward bias" similar to that used with transistors and, through increased gate-to-source voltage, increased device current flows.

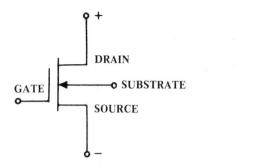
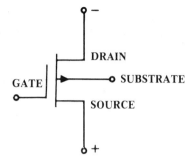

N-CHANNEL DEPLETION TYPE MOSFET **P-CHANNEL DEPLETION TYPE MOSFET**

FIGURE 1-2

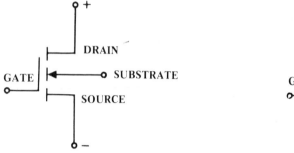
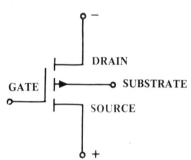

N-CHANNEL ENHANCEMENT TYPE MOSFET **P-CHANNEL ENHANCEMENT TYPE MOSFET**

FIGURE 1-3

DIRECT CURRENT CONSIDERATIONS IN THE LINEAR OPERATION OF JFET'S

1. The gate-to-source junction is reverse biased in controlling the amount of output (drain) current flow through the JFET (source to drain). The power supply connections and the reverse biased gate-to-source junction for both the N-channel and P-channel devices are shown in Figure 1-4.

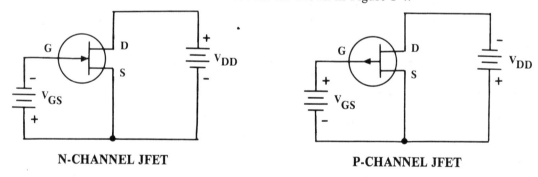

N-CHANNEL JFET   **P-CHANNEL JFET**

FIGURE 1-4

2. Maximum output (drain) current flows through the JFET (source to drain) when zero volts are applied across the gate-to-source junction and when the applied drain-to-source voltage is equal to, or greater than, the pinch off voltage line (V_p), but less than the breakdown voltage region (BVD_{GS}). This is the operating region of the JFET. The static curve for $V_{GS} = 0$ V is shown in Figure 1-5, where the maximum or saturation current, estimated at the knee of the curve (intersection V_p) is labeled I_{DSS}.

3. Therefore, the static curves can be broken into two regions with regard to drain-to-source voltage across the JFET — the operating region beyond the pinch off voltage line (knee of the curve) and the ohmic region before the pinch off voltage lines. FET's biased in the ohmic region can be used as voltage controlled variable resistances and FET's biased in the operating region, primarily, provide amplification functions.

4. The minimum drain current (theoretically) occurs when the reverse gate-to-source voltage equals the

8

pinch off voltage, $V_{GS} = V_p$. Theoretically, the minimum current is 0 mA, but, practically, it is considered at 1% of the maximum current (I_{DSS}) condition. For instance, if $I_{DSS} = 12$ mA, then the current flow at the pinch off voltage could be 0.12 mA.

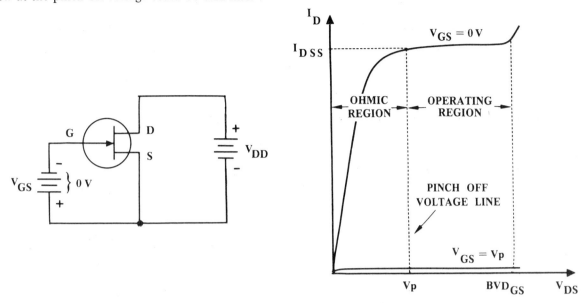

FIGURE 1-5

5.   The pinch off voltage (V_p) for the JFET is the amount of reverse gate-to-source voltage that will reduce the drain current flow to 1% of the saturation current level (I_{DSS}). However, V_p is also the minimum drain-to-source voltage of the operating region, where $V_p = V_{DS}$. For instance, if the pinch off voltage for the JFET is known at 4 volts, then the minimum V_{DS} value cannot be lower than 4 volts and must be higher to process a signal voltage, since the pinch off voltage line limits (distorts) the processed peak-to-peak signal. Therefore, the value of the pinch off voltage can be estimated at the knee of the curve of the $V_{GS} = 0$ V static characteristic curve (which can be taken off of a curve tracer) or by finding the value of V_{GS} that reduces the drain current (I_D) to approximately 1% of I_{DSS}, or from manufacturer's data sheets.

6.   Theoretically, the FET operates between the extreme current conditions of I_{DSS} and the approximate 0 mA condition at pinch off and the drain current is decreased from the I_{DSS} condition by applying reverse gate-to-source voltages until minimum current is reached. Additionally, the JFET is normally biased somewhere in the operating region between the pinch off voltage line and the breakdown voltage line and, as the reverse gate-to-source voltage is increased from $V_{GS} = -1$V to $V_{GS} = -4$V, the voltage across the drain to source at the pinch off line decreases accordingly. For instance, for $V_{GS} = 0$ V, $V_{DS} = 4$ V, for $V_{GS} = -1$ V, $V_{DS} = 3$ V, for $V_{GS} = -2$ V, $V_{DS} = 2$ V, for $V_{GS} = -3$ V, $V_{DS} = 1$ V, and for $V_{GS} = -4$ V, $V_{DS} = 0$ V. This effectively exponential ($V_p' = V_p - V_{GS}$) pinch line is illustrated in Figure 1-10, where all the characteristic curves for the above V_{GS} conditions are given.

7.   In the following example, the relationship between saturation current (I_{DSS}), drain current (I_D), reverse gate-to-source voltage (V_{GS}), and pinch off voltage (V_p) are found by utilizing the formula, $I_D = I_{DSS} \times (1 - V_{GS}/V_p)^2$ and solving for the drain current for $V_{GS} = 0$ V, -1 V, -2 V, -3 V, and -4 V. The remaining characteristics of the N-channel device of $I_{DSS} = 8$ mA and $V_p = 4$ V are given.

a.   For a V_{GS} of 0 volts:

$$I_D = I_{DSS}(1 - V_{GS}/V_p)^2 = 8 \text{ mA}(1 - 0 \text{ V}/4 \text{ V})^2 = 8 \text{ mA}(1) = 8 \text{ mA}.$$

b.   For a V_{GS} of -1 Volt:

$$I_D = I_{DSS}(1 - V_{GS}/V_p)^2 = 8 \text{ mA}(1 - 1/4)^2 = 8 \text{ mA}(0.75)^2 = 4.5 \text{ mA}$$

c.   For a V_{GS} of -2 volts:

$$I_D = I_{DSS}(1 - V_{GS}/V_p)^2 = 8 \text{ mA}(1 - 2/4)^2 = 8 \text{ mA}(0.5)^2 = 2 \text{ mA}$$

9

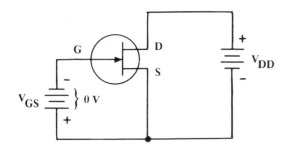

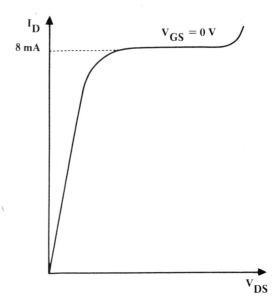

a. For a V_{GS} of 0 volts:
   $I_D = 8$ mA

FIGURE 1-6

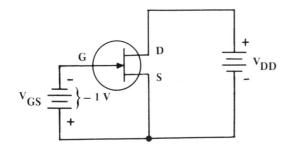

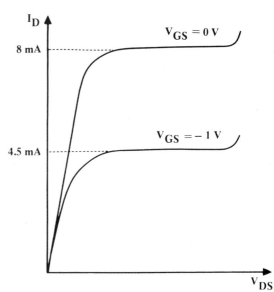

b. For a V_{GS} of — 1 volt:
   $I_D = 4.5$ mA

FIGURE 1-7

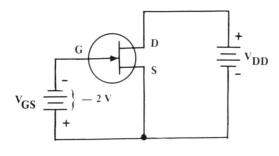

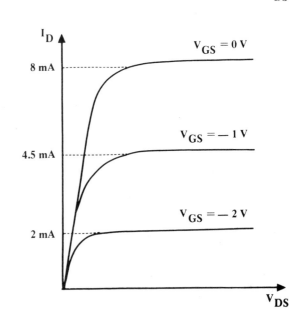

c. For a V_{GS} of —2 volts:
   $I_D = 2$ mA

FIGURE 1-8

10

d. For a V_{GS} of -3 volts:

$$I_D = I_{DSS}(1 - V_{GS}/V_p)^2 = 8 \text{ mA}(1 - 3/4)^2 = 8 \text{ mA}(0.25)^2 = 0.5 \text{ mA}$$

e. For a V_{GS} of -4 volts:

$$I_D = I_{DSS}(1 - V_{GS}/V_p)^2 = 8 \text{ mA}(1 - 4\text{ V}/4\text{ V})^2 = 8 \text{ mA}(0)^2 = 0 \text{ mA}$$

NOTE: The pinch off voltage line is shown in Figure 1-10 to illustrate the effect of diminished drain-to-source voltage with increased reverse biased gate-to-source voltage.

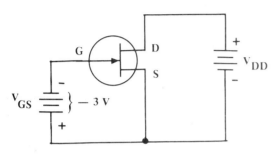

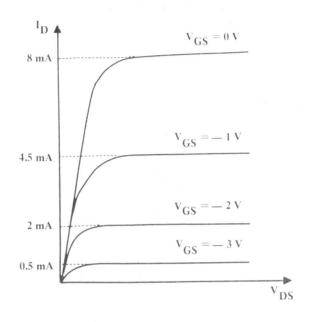

d. **For a V_{GS} of -3 volts:**
   $I_D = 0.5$ mA

FIGURE 1-9

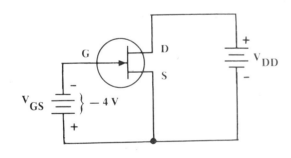

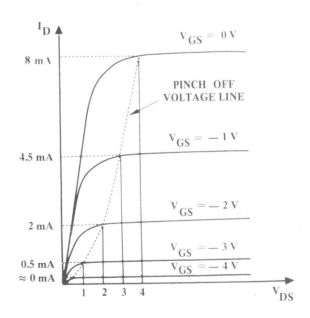

e. **For a V_{GS} of -4 volts:**
   $I_D = 0$ mA

FIGURE 1-10

6. Three techniques used in DC biasing the discrete JFET's are shown in Figure 1-11. The simple self-biasing circuit is shown in Figue 1-11(a), the universal circuit is shown in Figure 1-11(b), and the two power supply circuit is shown in Figure 1-11(c).

11

7. Resistor values for the self-biasing cirucit of Figure 1-11(a) and 1-12(a) can be obtained by utilizing the device characteristics and curves as shown in Figure 1-10. For instance, if the V_{DD} is established at 24 V DC and the operating drain current is chosen at a nominal 2 mA, then (from the curve of Figure 1-10) the gate-to-source voltage is -2 volts. (Remember, the curves were mathematically derived from the device characteristics and the $V_{GS} = -2$ V, $I_D = 2$ mA relationship is computed in Figure 1-8.)

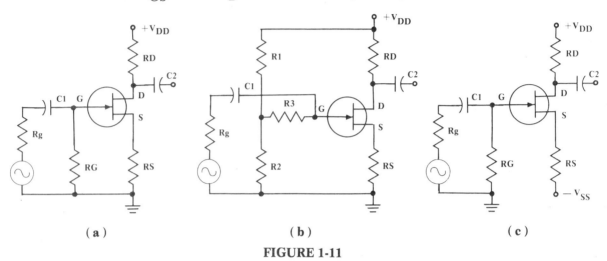

FIGURE 1-11

Therefore, in order to establish a reverse gate-to-source voltage of -2 volts, it is only necessary to establish a voltage drop of 2 volts across the source resistor RS. Since 0 volts exist at the gate terminal and since $I_D = 2$ mA, RS is chosen at $1|k\Omega$. (The zero volts exist at the gate of the JFET because the extremely low gate current of the device does not develop an appreciable voltage drop across the gate resistor RG.) Hence, 0 volts at the gate, and a positive 2 volts at the source, provide a reverse biased condition of 2 volts across the gate-to-source (diode) junction. This is shown in Figure 1-12(a).

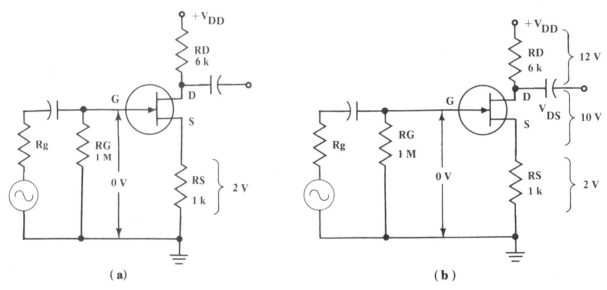

FIGURE 1-12

The remaining voltage drops around the circuit are shown in Figure 1-12(b). Since the 2 mA of drain current flows through the 6 kΩ RD resistor developing 12 volts across it, the remaining (available) voltage of 10 volts is distributed across the drain to source.

8. The 1 MΩ gate resistor is chosen so that if fulfills two requirements: (1) that the voltage across the gate resistor remains at approximately zero volts and (2) that the resistor value be high enough so that the impedance of the circuit is relatively high. For instance, the gate current for the 2N3819 is given in the manufacturer's data sheet as 2 nanoamps (2×10^{-9} A) and, for a 1 MΩ gate resistor, the voltage developed across the gate resistor RG is $10^6 \times 2 \times 10^{-9} = 2$ mV.

Also, the gate-to-source impedance of the device is usually greater than 100 MΩ, and this is seen in parallel with the 1 MΩ gate resistor, where $Z_{in} = R_G \parallel R_{GS}$. Therefore, the value of R_G cannot be too high as that can provide instability, and it cannot be too low as that will negate the high input impedance of the JFET.

9.   The voltage breakdown of the JFET is similar to the bipolar transistor in that the device will destruct if the voltage limits across gate-to-source, gate-to-drain, or drain-to-source are exceeded. The voltage breakdown for the gate-to-drain and drain-to-source is effectively the same, since the breakdown test is performed with the gate-to-source shorted. Therefore, the terms BVD_{SS} and BVD_{GS} are the same, and the latter BVD_{GS} term is the device specification normally given in data sheets. The gate-to-source breakdown voltage is also specified if data sheets and it is generally very close in value to that of BVD_{GS}.

AC CONSIDERATIONS IN THE LINEAR OPERATION OF JFET'S

1.   The transconductance of the FET is given in manufacturers' data sheets as $\underline{g_m}$, $\underline{y_{fs}}$, or $\underline{g_{fs}}$. However, the transconductance can more accurately be calculated from $g_m = 2/V_p\sqrt{I_D I_{DSS}}$, when the DC parameters of V_p, I_D, and I_{DSS} are known. For instance, solving in terms of the DC characteristics of Figure 1-12, where $V_p = 4$ V, $I_D = 2$ mA, and $I_{DSS} = 8$ mA, the transconductance is:

$$g_m = \frac{2}{V_p}\sqrt{I_D I_{DSS}} = \frac{2}{4}\sqrt{2 \text{ mA} \times 8 \text{ mA}} = 0.5 \text{ V} \times 4 \times 10^{-3} \text{ A} = 2000 \ \mu \text{ mho.}$$

2.   INPUT IMPEDANCE — The input impedance of the device, neglecting the effect of the biasing resistors, for both the common source and common drain configurations is nominally 100 MΩ or greater. For the common gate configuration, it is approximately equal to the inverse of the transconductance or $rs' = 1/g_m$.

NOTE: The source resistance rs' is similar to the base emitter diode resistance re (re') of bipolar transisters.

3.   OUTPUT IMPEDANCE: The output impedance for both the common source and common gate configurations, neglecting the effect of the load resistor is nominally 100 kΩ. For the common drain configuration, the output impedance, neglecting the effect of the load resistor, is the inverse of the transconductance or $rs' = 1/g_m$.

4.   DRAIN RESISTANCE: The drain resistance is approximately equal to the drain-to-source device resistance (r_{ds}) and can be calculated from $r_{ds} = \Delta V_{DS}/\Delta I_D$. This is where the $\Delta V_{DS}/\Delta I_D$ can be obtained from the output characteristic curves and where the normal slope in the operating region is relatively "flat" (large ΔV_{DS} to small ΔI_D) insuring both a high r_{ds} and output impedance.

5.   SOURCE RESISTANCE — The source resistance can be best illustrated by analyzing the equivalent circuits of either the common gate or common drain circuits. The common gate is chosen as shown in Figure 1-13(a), and the mid-frequency equivalent (model circuit) is shown in Figure 1-13(b).

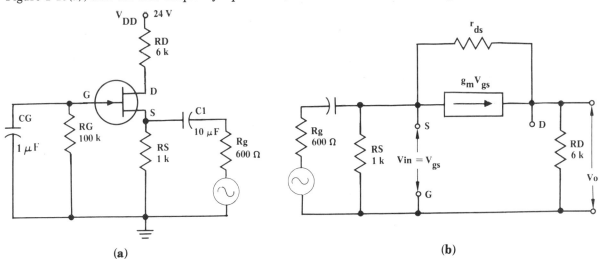

(a)

(b)

FIGURE 1-13

The input impedance of the circuit is the impedance looking into the source in parallel with the biasing resistor RS, but since it is necessary to solve for the input impedance of the source only, RS is eliminated in the calculations. Therefore, referencing the equivalent circuit of Figure 1-13, rs (the impedance looking into the source terminal) can be solved from rin = Vin/Iin, where Vin is the applied input generator voltage across the gate to source. Hence, Vin = V_{gs}.

a. $Iin = g_m \times Vin + \dfrac{Vin - Vo}{r_{ds}}$, where $g_m \times Vin = g_m \times V_{gs} \approx ID$ and $Vin - Vo/r_{ds}$ is the current flow through rds. Therefore, $Iin = ID + I_{rds}$.

b. $Iin = g_m \times Vin + \dfrac{Vin}{r_{ds}} - \dfrac{Vo}{r_{ds}}$

c. $Iin = g_m \times Vin + \dfrac{Vin}{r_{ds}} - \dfrac{Iin \times RD}{r_{ds}}$, where $Iin \approx Io$ for the common base circuit, and $Vo = IoRD \approx IinRD$

d. Collecting terms: $Iin - Iin \times \dfrac{RD}{r_{ds}} = g_m \times Vin + Vin \times \dfrac{1}{r_{ds}}$

e. Combining: $Iin\left(\dfrac{1 - RD}{r_{ds}}\right) = Vin\left(\dfrac{g_m + 1}{r_{ds}}\right)$

f. $\dfrac{Vin}{Iin} = \dfrac{1 - \dfrac{RD}{r_{ds}}}{g_m + \dfrac{1}{r_{ds}}}$, but $rin = RS = \dfrac{Vin}{Iin}$

g. and $\dfrac{1 - \dfrac{RD}{r_{ds}}}{g_m + \dfrac{1}{r_{ds}}} \approx \dfrac{1}{g_m}$, since $RD < r_{ds}$ and $1 << r_{ds}$

h. Therefore: $rs \approx 1/gm$ and $rs' = 1/gm$

i. As an example, if the transconductance is 2000 μmho, as was previously calculated in Step 1, then the approximate source resistance is

$rs' = 1/g_m = 1/2000\,\mu mho = \dfrac{1}{2 \times 10^{-3}} = \dfrac{1000\,\Omega}{2} = 500\,\Omega$

6. SOURCE RESISTANCE (rs') EXAMPLES — Standard FET formulas and the modified "R parameter" approach will be used in the following sample problems to demonstrate the validity of both methods.

a. For the common source, source resistor bypassed circuit at mid-band frequencies:

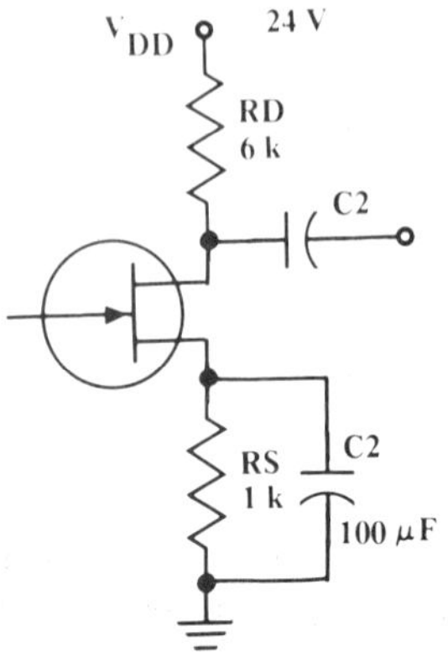

$V_{DD} = 24\,V$ $I_D = 2\,mA$

$RD = 6\,k\Omega$ $RS = 1\,k\Omega$

$I_{DSS} = 8\,mA$ $V_p = 4\,V$

$g_m = 2000\mu mho$ $rs' = 500\,\Omega$

1. $A_v = g_m RD = 2000\,\mu mho \times 6\,k\Omega = 2 \times 10^{-3} \times 6 \times 10^3 = 12$

2. $A_v = \dfrac{RD}{1/g_m} = \dfrac{RD}{rs'} = \dfrac{6\,k\Omega}{500\,\Omega} = 12$

FIGURE 1-14

14

b. For the common source, unbypassed source resistor at mid-band frequencies:

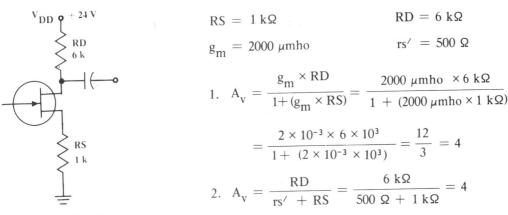

$RS = 1\ k\Omega$ $RD = 6\ k\Omega$

$g_m = 2000\ \mu mho$ $rs' = 500\ \Omega$

1. $A_v = \dfrac{g_m \times RD}{1 + (g_m \times RS)} = \dfrac{2000\ \mu mho \times 6\ k\Omega}{1 + (2000\ \mu mho \times 1\ k\Omega)}$

$= \dfrac{2 \times 10^{-3} \times 6 \times 10^3}{1 + (2 \times 10^{-3} \times 10^3)} = \dfrac{12}{3} = 4$

2. $A_v = \dfrac{RD}{rs' + RS} = \dfrac{6\ k\Omega}{500\ \Omega + 1\ k\Omega} = 4$

FIGURE 1-15

c. For the common drain circuit a mid-band frequencies:

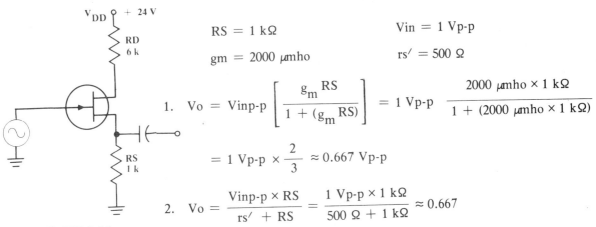

$RS = 1\ k\Omega$ $Vin = 1\ Vp\text{-}p$

$gm = 2000\ \mu mho$ $rs' = 500\ \Omega$

1. $Vo = Vinp\text{-}p \left[\dfrac{g_m\ RS}{1 + (g_m\ RS)} \right] = 1\ Vp\text{-}p\ \dfrac{2000\ \mu mho \times 1\ k\Omega}{1 + (2000\ \mu mho \times 1\ k\Omega)}$

$= 1\ Vp\text{-}p \times \dfrac{2}{3} \approx 0.667\ Vp\text{-}p$

2. $Vo = \dfrac{Vinp\text{-}p \times RS}{rs' + RS} = \dfrac{1\ Vp\text{-}p \times 1\ k\Omega}{500\ \Omega + 1\ k\Omega} \approx 0.667$

FIGURE 1-16

d. For the common gate circuit at mid-band frequencies:

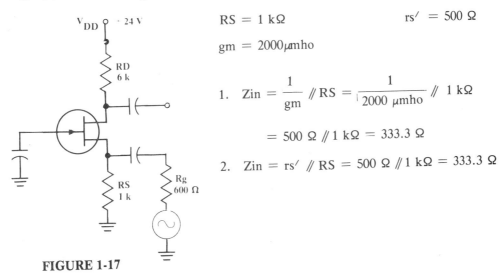

$RS = 1\ k\Omega$ $rs' = 500\ \Omega$

$gm = 2000\mu mho$

1. $Zin = \dfrac{1}{gm}\ /\!/\ RS = \dfrac{1}{2000\ \mu mho}\ /\!/\ 1\ k\Omega$

$= 500\ \Omega\ /\!/\ 1\ k\Omega = 333.3\ \Omega$

2. $Zin = rs'\ /\!/\ RS = 500\ \Omega\ /\!/\ 1\ k\Omega = 333.3\ \Omega$

FIGURE 1-17

7 BIPOLAR BASE-EMITTER (re) EQUIVALENT EXAMPLES — The second solution of all of the
above example problems for the JFET were modified "R parameters" borrowed from bipolar transistor

15

analysis. Therefore, if modified "R parameters" can be used in the analysis of FET's, then the standard FET formulas can be used to solve for bipolar analysis. This is shown in the following examples.

a. Common emitter, bypassed emitter resistor:

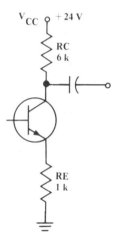

FIGURE 1-18

$V_{CC} = 24$ V $\qquad I_C = 1.3$ mA

$RC = 6$ kΩ $\qquad RE = 1$ kΩ

$$re = \frac{26 \text{ mV}}{IE} = \frac{26 \text{ mV}}{1.3 \text{ mA}} = 20 \ \Omega$$

$g_m = 1/re = 1/20 \ \Omega = 0.05$ mho $= 50$ mmho

$\qquad = 50,000 \ \mu$mho

1. $A_v = \dfrac{RC}{re} = \dfrac{6 \text{ k}\Omega}{20 \ \Omega} = 300$

2. $A_v = g_m RC = 50$ mmho $\times 6$ k$\Omega = 300$

b. Common emitter, unbypassed emitter resistor:

$RC = 6$ kΩ $\qquad RE = 1$ kΩ

$re = 20 \ \Omega$ $\qquad g_m = 50$mmho

1. $A_v = \dfrac{RC}{re + RE} = \dfrac{6 \text{ k}\Omega}{20 \ \Omega + 1 \text{ k}\Omega} \approx 5.882$

2. $A_v = \dfrac{g_m \times RC}{1 + (g_m \times RE)} = \dfrac{50 \text{ mmho} \times 6 \text{ k}\Omega}{1 + (50 \text{ mmho} \times 1 \text{ k}\Omega)}$

$\qquad = \dfrac{50 \times 10^{-3} \times 6 \times 10^3}{1 + (50 \times 10^{-3} \times 10^3)} = \dfrac{300}{1 + 50} \approx 5.882$

FIGURE 1-19

c. Common collector:

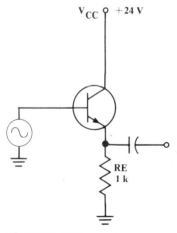

FIGURE 1-20

$RE = 1$ kΩ $\qquad re = 20 \ \Omega$

$g_m = 50$ mmho $\qquad Vin = 1$ Vp-p

1. $Vo = \dfrac{Vinp\text{-}p \times RE}{re + RE} = \dfrac{1 \text{ Vp-p} \times 1000 \ \Omega}{20 \ \Omega + 1000 \ \Omega}$

$\qquad = 0.98039$ Vp-p ≈ 0.98 Vp-p

2. $Vo = Vinp\text{-}p \times \dfrac{g_m RE}{1 + (g_m RE)} = 1 \text{ Vp-p} \times \dfrac{50 \text{ mmho} \times 1 \text{ k}\Omega}{1 + (50 \text{ mmho} \times 1 \text{ k}\Omega)}$

$\qquad = 1 \text{ Vp-p} \times \dfrac{50 \times 10^{-3} \times 10^3}{1 + (50 \times 10^{-3} \times 10^3)} = 1 \text{ Vp-p} \times \dfrac{50}{1 + 50}$

$\qquad = 0.98039$ Vp-p ≈ 0.98 Vp-p

16

d.   Common Base:

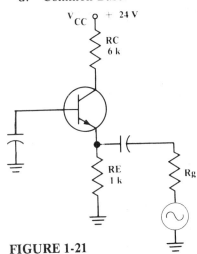

RC = 6 kΩ RE = 1 kΩ

re = 20 Ω g_m = 50 mmho

1.   Zin = re // RE = 20 Ω // 1 kΩ ≈ 19.6 Ω

2.   Zin = $\dfrac{1}{g_m \text{ // RE}}$ = $\dfrac{1}{50 \text{ } \mu\text{mhos // } 1 \text{ k}\Omega}$ = 20 Ω // 1 kΩ ≈ 19.6 Ω

FIGURE 1-21

NOTE: The high transconductance of bipolar devices gives them considerably higher voltage capabilities than JFET's, and the signal voltage loss in bipolar device voltage follower circuits is considerably less because re is much smaller than rs′ (20 Ω as compared to 500 Ω).

8.   The analysis of the maximum peak-to-peak output voltage swing of field effect transistors is not as straight forward as that of bipolar transistors. This is because the peak-to-peak voltage can almost reach the extremities of $I_{C(sat)}$ and V_{CC} for bipolar devices, but for the field effect devices the pinch off voltage knee effects the upper excursion limit. For this reason, if large peak-to-peak voltage swings are to be processed, field effect devices having high pinch off voltages cause the power supply requirements to increase and the non-linear characteristics associated with the device cause increased non-linear distortions. Therefore, comparing the peak-to-peak voltage swing at the output of the bipolar versus field effect transistors:

   a.   The peak-to-peak voltage swing for the bipolar device can swing 12 Vpeak towards V_{CC} but only 11.5 Vpeak towards the $I_{C(sat)}$ because of alpha crowding distortion. Therefore, for the bipolar transistor with a 24 volt DC V_{CC}, the voltage swing at the output is approximately 23 Vp-p without distortion. This is shown in Figure 1-22.

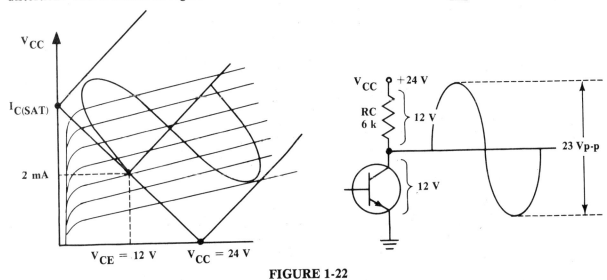

FIGURE 1-22

   b.   The peak-to-peak voltage swing for the JFET can swing 12 Vpeak towards V_{DD} but only 9 Vpeak towards the $I_{D(sat)}$ because of the pinch off voltage intersecting the DC load line at 3 volts. Therefore, with a 24 volt V_{DD}, the voltage swing at the output for the JFET , without distortion because of exceeding the pinch off voltage intersect is only 18 Vp-p. This is shown in Figure 1-23.

17

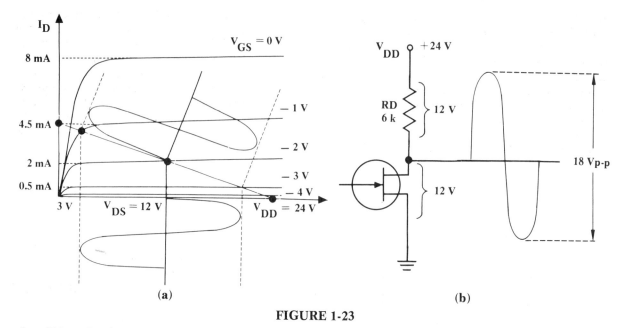

FIGURE 1-23

9. Although a large peak-to-peak voltage swing at the output of field effect transistors is not a positive feature of the device because of the non-linear characteristics of JFETS, it is important, nevertheless, to know the input and output limitations that the device presents to a large processed signal.

a. For the bypassed source resistor circuit of Figure 1-24(a), the peak-to-peak voltage swing is 18 Vp-p. This is shown in the characteristic curve of Figure 1-23(a) where the limiting factor of the output voltage swing is determined by the pinch off line voltage and the DC load line intersection. Since the intersection is at 3 volts, then only a 9 volt positive excursion can be processed and hence there can only be an 18 Vp-p output.   The input voltage peak-to-peak swing is determined by the slope of the DC load line and the $V_{GS} = -1$ V to $V_{GS} = -3$ V swing capability. Theoretically, the best input voltage swing is from a V_{GS} of 0 volts to a V_{GS} equal to V_p or, for the circuit of Figure 1-23(b), -4 volts. However, the slope of the DC load line for the circuit only allows an input peak-to-peak voltage swing of 2 Vp-p from -1 volt to -3 volts, with the operating voltage at a V_{GS} of -2 volts. However, this is more than enough since the input voltage peak-to-peak swing need only be 1.5 Vp-p to process an 18 Vp-p output voltage swing, when the gain of the stage is 12, where Vinp-p = Vop-p/A_v = 18 Vp-p/12 = 1.5 Vp-p.

NOTE: A generalized, non-graphical equation that includes the effect of the pinch-off voltage line for varying degrees of gate-to-source voltage is $2(V_{DS} - V_p + V_{GS})$. Therefore, the maximum peak-to-peak voltage swing, where $V_p = 4$ V and $V_{GS} = -1$ V can be solved from $2(V_{DS} - V_p + V_{GS} = 2(12$ V $-$ 4V $+$ 1 V) = 18 Vp-p. Also, $2I_DR_D = 2(2$ mA $\times 6$ kΩ) = 24 Vp-p, but the smaller of the two is used.

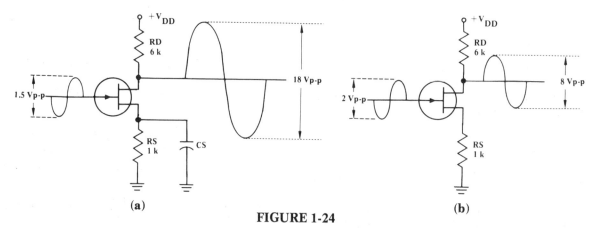

FIGURE 1-24

b. For the unbypassed source resistor circuit of Figure 1-24(b) the peak-to-peak voltage swing is only 8 Vp-p, because the maximum input peak-to-peak voltage swing can only be 2 Vp-p and the gain of the unbypassed source resistor circuit is 4. Therefore, |Vop-p = Vinp-p $\times A_v$ = 2 Vp-p $\times$ 4 = 8 Vp-p.

10. Distortion, temperature drift, and degree of transconductance can be influenced by the operating condition of I_D with respect to I_{DSS} of the device. For instance, low distortion to the processed signal can be obtained if the operating current I_D is greater than 0.5 of I_{DSS}, because in this region the transconductance curve is "relatively straight" with regard to the lower region of the transconductance curve. On the other hand, minimum drift with temperature occurs when low drain current is utilized. However, theoretical "best" current for minimum drift condition can be too low for general operating use and other techniques of biasing and feedback in providing stability must be utilized instead. Maximum transconductance occurs when $I_D = I_{DSS}$. Therefore, if highest possible voltage gain is a requirement, the I_D will have to be operated as close to I_{DSS} as possible. The areas of high transconductance, low distortion, and low temperature drift are illustrated in Figure 1-25.

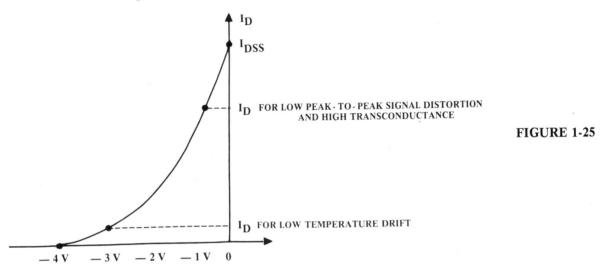

FIGURE 1-25

11. All field effect transistors have internal capacitances inherent to the fabrication process that, when coupled with the circuit resistance, provide high frequency roll off. Of primary consideration are the capacitances between gate and drain and between gate and source as shown in Figure 1-26.

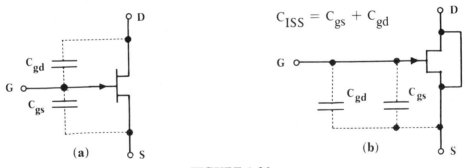

FIGURE 1-26

The gate-to-source capacitance (C_{gs}) and the gate-to-drain capacitance (C_{gd}) are obtained from manufacturers' data sheets. However, sometimes only C_{IS} or C_{ISS} is given and in this case it is necessary to extrapolate the values of C_{gs} and C_{gd}. C_{ISS} is the gate-to-source capacitance with the drain to gate shorted, which means the gate-to-source and gate-to-drain capacitances are in parallel or $C_{ISS} = C_{gs} + C_{gd}$.

For instance, for the 2N3819, C_{ISS} is given at 8 pF maximum and, since the JFET is effectively symmetrical prior to biasing, then $C_{ISS} = C_{gs} + C_{gd}$ or 8 pF ≈ 4 pF + 4 pF. However, remember this is an unbiased capacitance condition and the capacitance can change under varying operating conditions. Hence, $C_{ISS} = 8$ pF could be 5 pF + 3 pF.

NOTE: Additional stray capacitance due to leads, components, and chassis proximity will contribute approximately 30 pF of additional capacitance.

JUNCTION FIELD EFFECT TRANSISTOR CIRCUIT ANALYSIS

The three techniques used in the direct current biasing of JFET's, as shown in Figure 1-11, were the simple biased circuit, the universal circuit, and the two power supply circuit. In this section, a direct current anlysis of these circuits are obtained, and the alternating current analysis for each of the common source, common gate, and common drains are provided for the self-biased and universal circuits. Since the two power supply alternating current anlysis is similar to the self-biased circuit, the alternating current analysis of this circuit is omitted.

THE SELF-BIASED JFET AMPLIFIER — DIRECT CURRENT CIRCUIT ANALYSIS

The DC analysis for the self-biased JFET amplifer of Figure 1-27 is solved in terms of the drain current. However, because of the dependence on the device's characteristics in solving for drain current (I_D), the device's pinch off voltage (V_p) and saturation current (I_{DSS}) must be known.

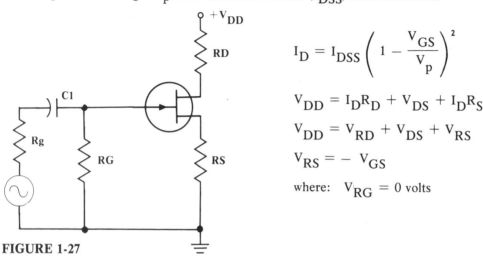

$$I_D = I_{DSS} \left(1 - \frac{V_{GS}}{V_p} \right)^2$$

$$V_{DD} = I_D R_D + V_{DS} + I_D R_S$$

$$V_{DD} = V_{RD} + V_{DS} + V_{RS}$$

$$V_{RS} = - V_{GS}$$

where: $V_{RG} = 0$ volts

FIGURE 1-27

The DC analysis begins by solving for the drain current (I_D) in terms of the device's pinch off voltage (V_p), saturation current (I_{DSS}), and selected gate to source voltage (V_{GS}). Next the voltage drop across the source resistor (V_{RS}) and drain resistor (V_{RD}) is found, and then the voltage drop across the drain to source (V_{DS}) is solved. Since minimal gate current flows (normally $< 10^{-8}$ A), little or no voltage drop is developed across the gate resistor (R_G). Therefore, the voltage at the gate is zero volts and, with respect to the positive voltage at the source, provides a reverse biased condition for the gate to source diode junction.

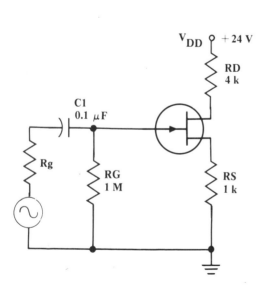

FIGURE 1-28

Given: $V_{DD} = 24$ V $V_p = 6$ V

$V_{GS} = - 3$ V $I_{DSS} = 12$ mA

$$\cdot \; I_D = I_{DSS} \left(1 - \frac{V_{GS}}{V_p} \right)^2 = 12 \text{ mA} \left(1 - \frac{3 \text{ V}}{6 \text{ V}} \right)^2$$

$$= 12 \text{ mA} \times 0.5^2 = 12 \text{ mA} \times 0.25 = 3 \text{ mA}$$

$$V_{RS} = I_D R_S = 3 \text{ mA} \times 1 \text{ k}\Omega = 3 \text{ V}$$

$$V_{RD} = I_D R_D = 3 \text{ mA} \times 4 \text{ k}\Omega = 12 \text{ V}$$

$$V_D = V_{DD} - V_{RD} = 24 \text{ V} - 12 \text{ V} = 12 \text{ V}$$

$$V_S = V_{RS} = 3 \text{ V}$$

$$V_G = V_{RG} = 0 \text{ V}$$

$$V_{DS} = V_D - V_S = 12 \text{ V} - 3 \text{ V} = 9 \text{ V}$$

Figure 1-29 shows the DC voltage distribution for the circuit. It is important to remember that all the calculated and measured voltage drops between V_{DD} and ground must equal the applied V_{DD}.

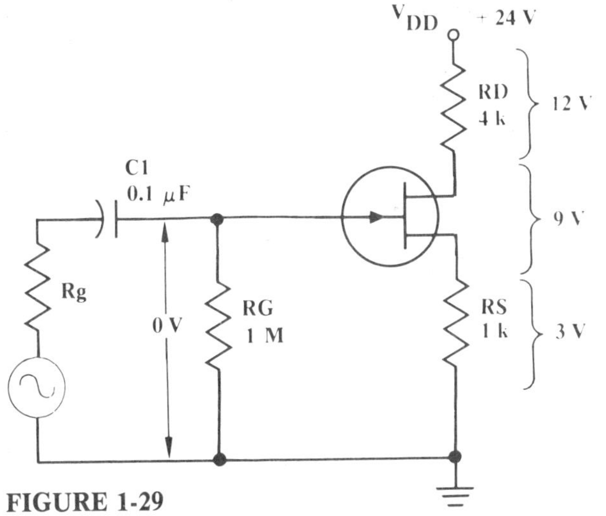

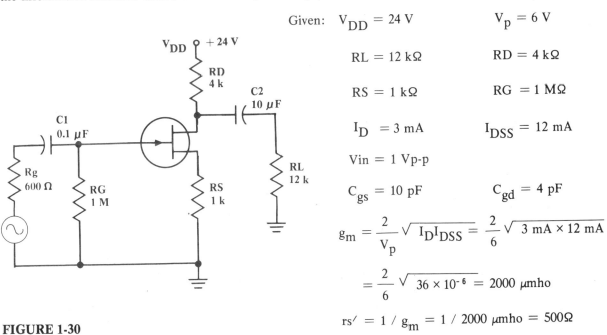

$$V_{DD} = V_{RD} + V_{DS} + V_{RS}$$

$$24\ V = 12\ V + 9\ V + 3\ V$$

$$V_{GS} = V_G - V_S = 0\ V - 3\ V = -3\ V$$

where: $V_S = V_{RS}$

FIGURE 1-29

NOTE: This prior DC analysis is applicable for all of the self-biased JFET circuits. This includes the common source, common gate, and common drain amplifier circuit configurations. Therefore, the DC analysis need only be solved once for all of these circuits.

THE SELF-BIASED JFET AMPLIFIER — AC CIRCUIT ANALYSIS

Once the DC voltage drops are known, the the AC circuit parameters are found. Voltage gain, input impedance, output impedance, power gain, and the peak-to-peak voltage swing at the output must be determined for each of the common source, common gate, and common drain connections.

THE COMMON SOURCE AMPLIFIER CONFIGURATION — ALTERNATING CURRENT CIRCUIT ANALYSIS

The connections for the common source amplifier circuit configuration are shown in Figure 1-30. From the information included with the illustration, the AC parameters will be solved.

Given: $V_{DD} = 24\ V$ $V_p = 6\ V$

$RL = 12\ k\Omega$ $RD = 4\ k\Omega$

$RS = 1\ k\Omega$ $RG = 1\ M\Omega$

$I_D = 3\ mA$ $I_{DSS} = 12\ mA$

$Vin = 1\ Vp\text{-}p$

$C_{gs} = 10\ pF$ $C_{gd} = 4\ pF$

$$g_m = \frac{2}{V_p}\sqrt{I_D I_{DSS}} = \frac{2}{6}\sqrt{3\ mA \times 12\ mA}$$

$$= \frac{2}{6}\sqrt{36 \times 10^{-6}} = 2000\ \mu mho$$

$$rs' = 1 / g_m = 1 / 2000\ \mu mho = 500\Omega$$

FIGURE 1-30

21

VOLTAGE GAIN:

$$A_v = \frac{g_m(RD /\!/ RL)}{1 + g_m RS} = \frac{2000 \ \mu mho(4 \ k\Omega /\!/ 12 \ k\Omega)}{1 + (2000 \ \mu mho \times 1 \ k\Omega)} = \frac{(2 \times 10^{-3})(3 \times 10^3)}{1 + (2 \times 10^{-3})(1 \times 10^3)} = \frac{6}{3} = 2$$

Also: $A_v = \dfrac{RD /\!/ RL}{RS + rs'} = \dfrac{4 \ k\Omega /\!/ 12 \ k\Omega}{1 \ k\Omega + 500 \ \Omega} = \dfrac{3 \ k\Omega}{1500 \ \Omega} = 2$

VOLTAGE OUT:

$$Vop\text{-}p = Vinp\text{-}p \times A_v = 1 \ Vp\text{-}p \times 2 = 2 \ Vp\text{-}p$$

INPUT IMPEDANCE AND OUTPUT IMPEDANCE

$Zin \approx RG = 1 \ M\Omega$ $\qquad\qquad$ $Zo \approx RD = 4 \ k\Omega$

POWER GAIN TO THE LOAD RESISTOR:

$$PG = A_v{}^2 \times \frac{Zin}{RL} = 2^2 \times \frac{1 \ M\Omega}{12 \ k\Omega} \approx 333.3$$

FREQUENCY RESPONSE — HIGH CORNER FREQUENCIES:

a. The first high corner frequency is determined by the input capacitance Cin, and the generator resistance Rg in parallel with the input impedance Zin.

$$f_c(\text{high \#1}) = \frac{1}{2\pi(Rin /\!/ Rg)Cin} = \frac{0.159}{(1 \ M\Omega /\!/ 600\Omega)22 \ pF} = \frac{0.159}{(6 \times 10^2)(22 \times 10^{-12})}$$

$$= \frac{0.159}{132 \times 10^{-10}} = \frac{0.159 \times 10^9}{13.2} = 12.06 \ MHz$$

NOTE: The input capacitance Cin is the miller effect capacitance C(miller) in parallel with the gate-to-source capacitance C_{gs}. The miller effect capacitance for the common cource circuit is the output capacitance $(Co \approx C_{gd} = 4 \ pF)$ amplified and reflected back into the input by the effective voltage gain of the circuits.

$$C(\text{miller}) = (1 + A_v)C_{gd} = (1 + 2)4 \ pF = 12 \ pF$$

$$Cin = C(\text{miller}) + C_{gs} = 12 \ pF + 10 \ pF = 22 \ pF$$

b. The second high corner frequency is determined by the output and stray capacitance $C_{gd} = C_o$ and C_s and the parallel combination of the output resistors RD and RL.

$$f_c(\text{high \#2}) = \frac{1}{2\pi(RD /\!/ RL)(C_{gd} + C_s)} = \frac{0.159}{(4 \ k\Omega /\!/ 12 \ k\Omega)(4 \ pF + 30 \ pF)}$$

$$= \frac{0.159}{3 \ k\Omega \times 34 \ pF} = \frac{0.159}{3 \times 10^3 \times 34 \times 10^{-12}} = \frac{0.159 \times 10^9}{102} \approx 1.56 \ mHz$$

NOTE: Since the high corner frequencies are less than 10 : 1 removed from each other, they will interact. Hence, the roll-off will occur at a lower frequency than the calculated 1.56 MHz. The high corner frequency is therefore "obtained" by adding the RC time constants together underneath the 0.159 (1/2π). Therefore:

$$f_c(\text{high}) = \frac{0.159}{(13.2 \times 10^{-9}) + (102 \times 10^{-9})} = \frac{0.159 \times 10^9}{115.2} \approx 1.38 \ MHz$$

Another technique used in solving for the frequency response of two corner frequencies less than 10 : 1 removed from each other is to solve for the corner frequencies in parallel. Therefore:

$$f_c(\text{high}) = 12.06 \text{ MHz} \parallel 1.56 \text{ MHz} \approx 1.38 \text{ MHz}$$

FREQUENCY RESPONSE — LOW CORNER FREQUENCIES

a.   The first low corner frequency is determined by the input capacitor C1 and the series resistance of Rg and Zin.

$$fc(\text{low \#1}) = \frac{1}{2\pi(R_g + Zin)C1} = \frac{0.159}{(600\ \Omega + 1\ M\Omega)0.1\ \mu F} \approx \frac{0.159}{10^6 \times 10^{-7}} \approx 1.59 \text{ Hz}$$

b.   The second low corner frequency is determined by the output capacitor C2 and the series resistance of RD and RL.

$$fc(\text{low \#2}) = \frac{1}{2\pi(RD + RL)C2} = \frac{0.159}{(4\ k\Omega + 12\ k\Omega)10\ \mu F} = \frac{0.159}{1.6 \times 10^4 \times 10^{-5}} \approx 1 \text{ Hz}$$

NOTE: Since the low corner frequencies interact (and effectively add), the low corner frequency roll-off is approximately 2.6 Hz.

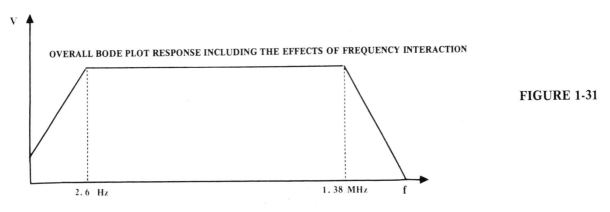

FIGURE 1-31

THE COMMON SOURCE, SOURCE BYPASSED, AMPLIFIER CONFIGURATION ALTERNATING CURRENT CIRCUIT ANALYSIS

The connections for the common source, source bypassed, amplifier circuit configuration are shown in Figure 1-32. From the information included with the illustration, the AC parameters will be solved.

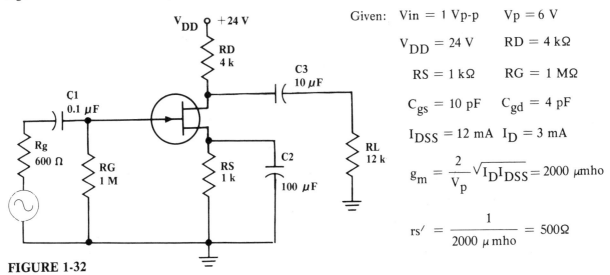

FIGURE 1-32

Given: $Vin = 1\ Vp\text{-}p$ $Vp = 6\ V$

$V_{DD} = 24\ V$ $RD = 4\ k\Omega$

$RS = 1\ k\Omega$ $RG = 1\ M\Omega$

$C_{gs} = 10\ pF$ $C_{gd} = 4\ pF$

$I_{DSS} = 12\ mA$ $I_D = 3\ mA$

$$g_m = \frac{2}{V_p}\sqrt{I_D I_{DSS}} = 2000\ \mu mho$$

$$rs' = \frac{1}{2000\ \mu mho} = 500\Omega$$

23

VOLTAGE GAIN:

$$A_v = g_m(RD \mathbin{/\!/} RL) = 2000 \ \mu\text{mho}(4 \ k\Omega \mathbin{/\!/} 12 \ k\Omega) = 2 \times 10^{-3} \times 3 \times 10^3 = 6$$

Also: $\quad A_v = \dfrac{RD \mathbin{/\!/} RL}{rs'} = \dfrac{4k\Omega \mathbin{/\!/} 12 \ k\Omega}{500\Omega} = \dfrac{3 \ k\Omega}{500 \ \Omega} = 6$

VOLTAGE OUT:

$$Vop\text{-}p = Vinp\text{-}p \times A_v = 1 \ Vp\text{-}p \times 6 = 6 \ Vp\text{-}p$$

INPUT IMPEDANCE:

$$Zin = RG = 1 \ M\Omega$$

OUTPUT IMPEDANCE:

$$Zo = RD = 4 \ k\Omega$$

POWER GAIN TO THE LOAD RESISTOR:

$$PG = A_v{}^2 \times \frac{Zin}{RL} = 6^2 \times \frac{1 \ M\Omega}{12 \ k\Omega} = 3000$$

FREQUENCY RESPONSE — HIGH CORNER FREQUENCIES:

a.   The first high corner frequency is determined by the input capacitance Cin and the generator resistance Rg in parallel with the input impedance Zin = RG.

$$f_c(\text{high \#1}) = \frac{1}{2\pi(Zin \mathbin{/\!/} Rg)Cin} = \frac{0.159}{(1 \ M\Omega \mathbin{/\!/} 600 \ \Omega)38 \ pF} \approx \frac{0.159}{6 \times 10^2 \times 38 \times 10^{-12}} = 6.98 \ \text{MHz}$$

where: Cin = C(miller) + C_{gs} = 28 pF + 10 pF

1.   C(miller) = $(1 + A_v)C_{gd}$ = (1 + 6)4 pF = 28 pF

2. C_{gs} = 10 pF

b.   The second high corner frequency is determined by the output and stray capacitance C_{gd} and C_s and the parallel combination of the output resistors RD and RL.

$$f_c(\text{high \#2}) = \frac{1}{2\pi(RD \mathbin{/\!/} RL)(C_{gd} + C_s)} = \frac{0.159}{(4 \ k\Omega \mathbin{/\!/} 12 \ k\Omega)(4pF + 30 \ pF)} = \frac{0.159}{3 \ k\Omega \times 34 \ pF}$$

$$= \frac{0.159}{3 \times 10^3 \times 34 \times 10^{-12}} = \frac{0.159 \times 10^9}{102} \approx 1.56 \ \text{MHz}$$

NOTE: Since the high corner frequencies are less than 10 : 1 removed from each other, the roll-off will occur at a lower frequency than the lowest calculated at 1.56 MHz.   Therefore:

$$f_c(\text{high}) = \frac{1}{2\pi(102 \times 10^{-9}) + (22.8 \times 10^{-9})} = \frac{0.159 \times 10^9}{124.8} \approx 1.275 \ \text{MHz}$$

or from: $f_c(\text{high}) = f_c(\text{high \#1}) \mathbin{/\!/} f_c(\text{high \#2}) = 6.98 \ \text{MHZ} \mathbin{/\!/} 1.56 \ \text{MHz} \approx 1.275 \ \text{MHz}$

FREQUNCY RESPONSE — LOW CORNER FREQUENCIES:

a.   The first low corner frequency is determined by the input capacitance C1 and the series resistance of

R_g and Zin.

$$f_c(\text{low \#1}) = \frac{1}{2\pi(Rg + Zin)C1} = \frac{0.159}{(600\Omega + 1\ M\Omega)0.1\ \mu F} \approx \frac{0.159}{10^6 \times 10^{-7}} \approx 1.6\ Hz$$

b.   The second low corner frequency is determined by the output capacitor C3 and the series resistance of RD and RL.

$$f_c(\text{low \#2}) = \frac{1}{2\pi(RD + RL)C3} = \frac{0.159}{(4\ k\Omega + 12\ k\Omega)10\ \mu F} = \frac{0.159}{(1.6 \times 10^4)10^{-5}} \approx 1\ Hz$$

c.   The third low corner frequency is determined by the source bypass capacitor C2 and the parallel combination of the source resistor RS and the inverse device transconductance $1/g_m = rs'$. The inverse transconductance $(1/g_m)$ is the "resistance" of the device looking into the source terminal so, for this circuit (Figure 1-32) it is 500 Ω.

$$f_c(\text{low \#3}) = \frac{1}{2\pi(RS\ /\!/\ 1/g_m)C2} = \frac{0.159}{(1\ k\Omega\ /\!/\ 500\ \Omega)100\ \mu F} = \frac{0.159}{333.3\ \Omega \times 10^{-4}} \approx 4.7\ Hz.$$

$$\text{where:}\quad \frac{1}{g_m} = \frac{1}{2000\ \mu mho} = \frac{1}{2 \times 10^{-3}} = 500\Omega$$

NOTE: Since the low corner frequencies are less than 10 : 1 removed from each other, they effectively add. Therefore, the low corner frequency is approximately 7 Hz.

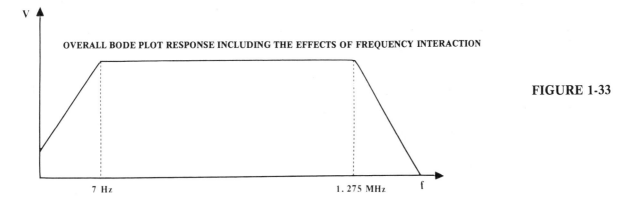

OVERALL BODE PLOT RESPONSE INCLUDING THE EFFECTS OF FREQUENCY INTERACTION

FIGURE 1-33

THE COMMON GATE, SELF-BIASED CIRCUIT CONFIGURATION

The connections for the common gate, self-biased amplifier circuit are shown in Figure 1-34. The DC analysis for the circuit is identical to the preceding common source circuits. The common gate capacitor has no effect on the DC distribution unless it is defective. From the information given with Figure 1-29, the AC parameters will be solved.

VOLTAGE GAIN:

$$A_v = g_m(RL\ /\!/\ RD) = 2000\ \mu mho(12\ k\Omega\ /\!/\ 4\ k\Omega) = 2 \times 10^3 \times 3 \times 10^{-3} = 6$$

$$A_v = \frac{r_L}{rs'} = \frac{RD\ /\!/\ RL}{rs'} = \frac{4\ k\Omega\ /\!/\ 12\ k\Omega}{500\ \Omega} \approx \frac{3\ k\Omega}{500\ \Omega} = 6$$

NOTE: The maximum voltage gain for the CB stages is solved from $g_m \times r_L$ and occurs when the RG resistor is (fully) capacitor bypassed. This is also true for the CS stage when the RS resistor is (fully) capacitor bypassed.

25

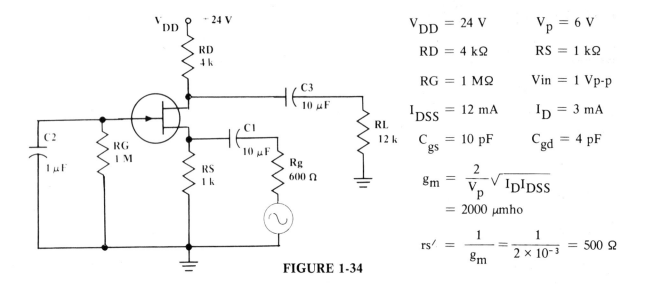

FIGURE 1-34

$V_{DD} = 24$ V $V_p = 6$ V

$RD = 4$ kΩ $RS = 1$ kΩ

$RG = 1$ MΩ $Vin = 1$ Vp-p

$I_{DSS} = 12$ mA $I_D = 3$ mA

$C_{gs} = 10$ pF $C_{gd} = 4$ pF

$$g_m = \frac{2}{V_p}\sqrt{I_D I_{DSS}}$$

$$= 2000 \ \mu mho$$

$$rs' = \frac{1}{g_m} = \frac{1}{2 \times 10^{-3}} = 500 \ \Omega$$

VOLTAGE OUT:

$$Vop\text{-}p = Vinp\text{-}p \times A_v = 1 \ Vp\text{-}p \times 6 = 6 \ Vp\text{-}p$$

INPUT IMPEDANCE:

$$Zin = RS \ /\!/ \ 1/g_m = RS \ /\!/ \ rs' = 1 \ k\Omega \ /\!/ \ 500 \ \Omega = 333.3 \ \Omega$$

POWER GAIN TO THE LOAD RESISTOR:

$$PG = A_v{}^2 \times \frac{Zin}{RL} = 6^2 \times \frac{333.3 \ \Omega}{12 \ k\Omega} \approx 1$$

FREQUENCY RESPONSE — HIGH CORNER FREQUENCIES:

a.   The first high corner frequency is determined by the input capacitance Cin and the parallel generator resistance and input impedance.

$$f_c(high \ \#1) = \frac{1}{2\pi(Zin \ /\!/ \ Rg)Cin} = \frac{0.159}{(333.3 \ \Omega \ /\!/ \ 600\Omega)10 \ pF} = \frac{0.159}{214.2 \times 10^{-11}}$$

$$= \frac{0.159 \times 10^{11}}{214.2} = 74.2 \ MHz, \ where: \quad Cin = C_{gs}$$

b.   The second high corner frequency is determined by the output and stray capacitances ($C_{gd} + C_S$) and the parallel combination of the output resistance RD and the load resistor RL.

$$f_c(high \ \#2) = \frac{1}{2\pi(RD \ /\!/ \ RL) \ (C_{gd} + C_S)} = \frac{0.159}{(4 \ k\Omega \ /\!/ \ 12 \ k\Omega) \ (4 \ pF + 30 \ pF)} = \frac{0.159}{3 \ k\Omega \times 34 \ pF}$$

$$= \frac{0.159}{102 \times 10^{-9}} = \frac{159 \times 10^6}{102} \approx 1.56 \ MHz$$

NOTE: Since no effective interaction of high frequency roll-offs occur, 1.56 MHz is the high corner frequency.

FREQUENCY RESPONSE — LOW CORNER FREQUENCIES:

a. The first low corner frequency is determined by the input capacitor C1 and the series Rg and Zin resistors.

$$f_c(\text{low \#1}) = \frac{1}{2\pi(Rg + Zin)C1} = \frac{0.159}{(600\ \Omega + 333.3\ \Omega)10\mu F} = \frac{0.159}{933.3 \times 10^{-5}} = \frac{15900}{933.3} = 17\ Hz$$

b. The second low corner frequency is determined by the output coupling capacitor C3 and the series RD and RL resistances.

$$f_c(\text{low \#2}) = \frac{1}{2\pi(RD + RL)C3} = \frac{0.159}{(4\ k\Omega + 12\ k\Omega)10\ \mu F} = \frac{0.159}{3 \times 10^3 \times 10^{-5}} = \frac{0.159 \times 10^2}{3} = 0.53\ Hz$$

c. The third low corner frequency is determined by the C2 bypass capacitor and the RG resistor.

$$f_c(\text{low \#3}) = \frac{1}{2\pi R_G\ C2} = \frac{0.159}{1\ M\Omega \times 1\ \mu F} = \frac{0.159}{10^6 \times 10^{-6}} = 0.159\ Hz$$

NOTE: The low corner frequency is determined from $f_c(\text{low \#1})$ or 17 Hz, since no effective interaction from the other two low frequencies occurs.

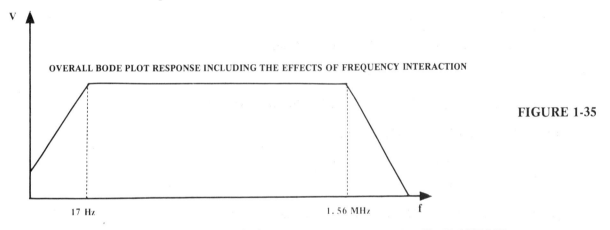

OVERALL BODE PLOT RESPONSE INCLUDING THE EFFECTS OF FREQUENCY INTERACTION

17 Hz   1.56 MHz

FIGURE 1-35

THE COMMON DRAIN, SELF-BIASED CIRCUIT CONFIGURATION

The connections for the common drain, self-biased amplifier circuit are shown in Figure 1-36. The DC analysis for the circuit is similar to the preceding self-biased circuits except for the voltage drop across the drain to source (V_{DS}). This voltage is solved from $V_{DD} - V_{RS} = 24\ V - 3\ V = 21\ V$. The AC parameters are solved from the information included with Figure 1-36.

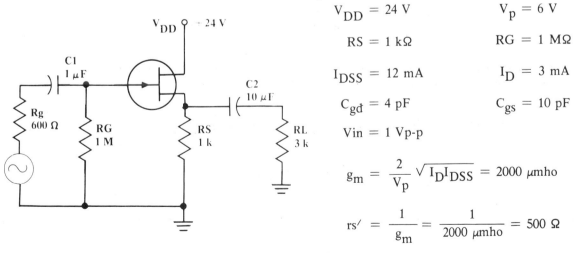

$V_{DD} = 24\ V$ $V_p = 6\ V$

$RS = 1\ k\Omega$ $RG = 1\ M\Omega$

$I_{DSS} = 12\ mA$ $I_D = 3\ mA$

$C_{gd} = 4\ pF$ $C_{gs} = 10\ pF$

$Vin = 1\ V\text{p-p}$

$$g_m = \frac{2}{V_p}\sqrt{I_D I_{DSS}} = 2000\ \mu mho$$

$$rs' = \frac{1}{g_m} = \frac{1}{2000\ \mu mho} = 500\ \Omega$$

FIGURE 1-36

27

VOLTAGE GAIN:

$$A_v = \frac{g_m r_L}{1 + g_m r_L} = \frac{g_m (RS \,/\!/\, RL)}{1 + g_m (RS \,/\!/\, RL)} = \frac{2000 \,\mu mho(1\,k\Omega \,/\!/\, 3\,k\Omega)}{1 + (2000\mu mho\,[1\,k\Omega \,/\!/\, 3\,k\Omega])}$$

$$= \frac{2 \times 10^{-3} \times 750\,\Omega}{1 + (2 \times 10^{-3} \times 750\,\Omega)} = \frac{1.5}{1 + 1.5} = \frac{1.5}{2.5} = 0.6$$

also: $\quad A_v = \dfrac{r_L}{rs' + r_L} = \dfrac{RS \,/\!/\, RL}{rs' + RS \,/\!/\, RL} = \dfrac{1\,k\Omega \,/\!/\, 3\,k\Omega}{500\,\Omega + (1\,k\Omega \,/\!/\, 3\,k\Omega)} = \dfrac{750\,\Omega}{500\,\Omega + 750\,\Omega} = 0.6$

NOTE: The latter technique of solving for voltage gain demonstrates the signal loss across the source resistance as it is processed to the load. The formula is, therefore, the simple voltage divider equation.

INPUT IMPEDANCE:

$$Zin \approx RG = 1\,M\Omega$$

OUTPUT IMPEDANCE:

$$Zo = RS \,/\!/\, rs' = 1\,k\Omega \,/\!/\, 500\Omega = 333.3\,\Omega, \quad \text{where:} \quad rs' = 1/gm$$

VOLTAGE OUT:

$$Vop\text{-}p = Vinp\text{-}p \times A_v = 1\,Vp\text{-}p \times 0.6 = 0.6\,Vp\text{-}p$$

POWER GAIN TO THE LOAD RESISTOR:

$$PG = A_v^2 \times \frac{Zin}{RL} = 0.6^2 \times \frac{1\,M\Omega}{3\,k\Omega} = 0.36 \times 333.3 = 120$$

FREQUENCY RESPONSE — HIGH CORNER FREQUENCIES:

a.   The first high corner frequency is determined by the input capacitance Cin and the generator resistance Rg in parallel with the input impedance (Zin = RG).

$$f_c(high\,\#1) = \frac{1}{2\pi(Zin \,/\!/\, Rg)Cin} = \frac{0.159}{(1\,M\Omega \,/\!/\, 600\,\Omega)8\,pF} = \frac{0.159}{6 \times 10^2 \times 8 \times 10^{-12}}$$

$$= \frac{0.159 \times 10^{10}}{48} = 33.125\,MHz$$

where: $Cin = c_{gd} + (1 - A_v)C_{gs} = 4\,pF + (1 - 0.6)10\,pF = 4\,pF + (4\,pF) = 8\,pF$

b.   The second high corner frequency is determined by the output and stray capacitances and the output impedance in parallel with the load.

$$f_c(high\,\#2) = \frac{1}{2\pi(Zo \,/\!/\, RL)\,(Co + C_S)} = \frac{0.159}{(333.3\,\Omega \,/\!/\, 3\,k\Omega)\,(10\,pF + 30\,pF)}$$

$$= \frac{0.159}{3 \times 10^2 \times 40 \times 10^{-12}} = \frac{0.159 \times 10^{10}}{120} = 13.25\,MHz$$

where: $C_S = 30\,pF \quad$ and $\quad Co \approx C_{gs}$

NOTE: Because of the interaction of the less than 10 : 1 high corner frequencies, the overall high corner frequency is solved from:

28

$$f_c(\text{high}) = \frac{1}{2\pi(RC[\text{hi \#1}] + RC[\text{hi \#2}])} = \frac{0.159}{48 \times 10^{-10} + 120 \times 10^{-10}} \approx 9.46 \text{ MHz}$$

or from: $f_c(\text{high}) = f_c(\text{high \#1}) \mathbin{/\mkern-5mu/} f_c(\text{high \#2}) = 33.125 \text{ MHz} \mathbin{/\mkern-5mu/} 13.25 \text{ MHz} \approx 9.46 \text{ MHz}$

FREQUENCY RESPONSE — LOW CORNER FREQUENCIES:

a. The first low corner frequency is determined by the input capactior C1 and the input impedance Zin = RG in series with the generator resistance Rg.

$$f_c(\text{low \#1}) = \frac{1}{2\pi(Rg + Zin)C1} = \frac{0.159}{(600\,\Omega + 1\,\text{M}\Omega)10^{-6}} \approx \frac{0.159}{10^6 \times 10^{-6}} = 0.159 \text{ Hz}$$

b. The second low corner frequency is determined by the output capacitor C2 and the series output and load resistances.

$$f_c(\text{low \#2}) = \frac{1}{2\pi(Zo + RL)C2} = \frac{0.159}{(333.3\,\Omega + 3\,\text{k}\Omega)10\,\mu\text{F}} = \frac{0.159}{3.333 \times 10^3 \times 10^{-5}}$$

$$= \frac{0.159 \times 10^2}{3.333} = 4.77 \text{ Hz}$$

NOTE: The low corner frequency is determined from $f_c(\text{low \#2})$ or approximately 4.8 Hz, since there is no effective interaction from the other low frequency.

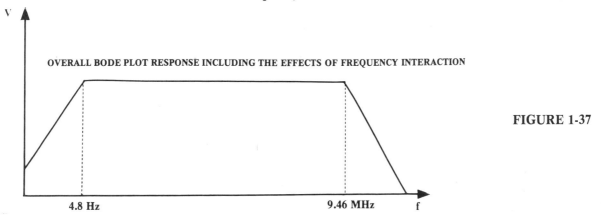

OVERALL BODE PLOT RESPONSE INCLUDING THE EFFECTS OF FREQUENCY INTERACTION

FIGURE 1-37

THE UNIVERSAL JFET AMPLIFIER CIRCUIT — DIRECT CURRENT ANALYSIS

The connections for the universal JFET amplifier are shown in Figure 1-38. Included with the illustration are the basic equations for the direct current analysis.

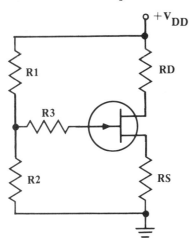

$$V_{R1} = \frac{V_{DD} \times R1}{R1 + R2} \qquad V_{R2} = \frac{V_{DD} \times R2}{R1 + R2}$$

$$V_{R3} = I_G R3 \approx 0 \text{ V}$$

$$I_D = I_{DSS}\left(1 - \frac{V_{GS}}{V_p}\right)^2$$

$$V_{RS} = I_D RS$$

$$V_{RD} = I_D RD \quad \text{and} \quad V_{GS} \approx V_{RS} = V_{R2}$$

FIGURE 1-38

The DC analysis begins by solving for the voltage drops across the R1 and R2 voltage divider. Next, the drain current is solved in terms of the device's pinch-off voltage Vp, the device saturation current I_{DSS}, and the selected gate-to-source voltage V_{GS}. Then, the voltage drop across the source resistor V_{RS} is found, followed by the voltage drop across the drain resistor V_{RD}.

Since the gate current flow is minimal, little or no voltage drop occurs across the R3 resistor. The voltage drop across the drain-to-source junction is found in the standard manner, and the gate-to-source voltage is found by subtracting the voltage developed across R2 from the voltage developed across V_{RS}.

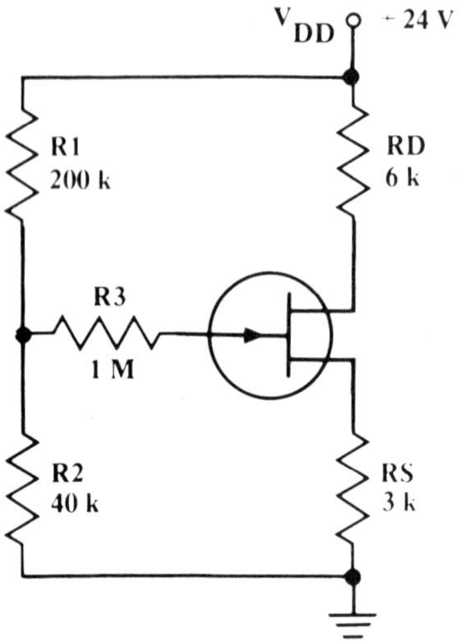

$$V_{R1} = \frac{V_{DD} \times R1}{R1 + R2} = \frac{24\ V \times 200\ k\Omega}{200\ k\Omega + 40\ k\Omega} = 20\ V$$

$$V_{R2} = \frac{V_{DD} \times R2}{R1 + R2} = \frac{24\ V \times 40\ k\Omega}{200\ k\Omega + 40\ k\Omega} = 4\ V$$

$$I_D = I_{DSS}\left(1 - \frac{V_{GS}}{V_p}\right)^2 = 8\ mA\left(1 - \frac{2}{4}\right)^2 = 8 \times 0.5^2 = 2\ mA$$

$$V_{RS} = I_D RS = 2\ mA \times 3\ k\Omega = 6\ V$$

$$V_{RD} = I_D RD = 2\ mA \times 6\ k\Omega = 12\ V$$

$$V_D = V_{DD} - V_{RD} = 24\ V - 12\ V = 12\ V$$

$$V_{DS} = V_D - V_{RS} = 12\ V - 6\ V = 6\ V$$

$$V_{GS} = V_{R2} - V_{RS} = 4\ V - 6\ V = -2\ V$$

$$V_{R3} = I_G R3 = 10\ nA \times 1\ M\Omega = 10^{-8} \times 10^6 = 10^{-2} = 0.01\ V$$

FIGURE 1-39

NOTE: The gate current I_G of 10 nanoamps is estimated to demonstrate the minimal voltage drop to be expected across R3.

Figure 1-40 shows all the DC voltages of the circuit. Remember that all the calculated and measured voltage drops between V_{DD} and ground must equal the applied V_{DD}.

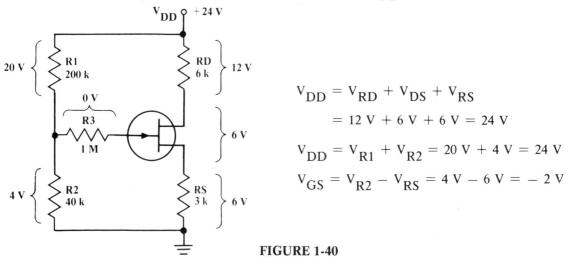

$$V_{DD} = V_{RD} + V_{DS} + V_{RS}$$
$$= 12\ V + 6\ V + 6\ V = 24\ V$$

$$V_{DD} = V_{R1} + V_{R2} = 20\ V + 4\ V = 24\ V$$

$$V_{GS} = V_{R2} - V_{RS} = 4\ V - 6\ V = -2\ V$$

FIGURE 1-40

THE UNIVERSAL BIASED JFET AMPLIFIER CIRCUIT — AC ANALYSIS

Once the DC voltages are known, then the alternating current parameters are found. Voltage gain, input impedance, output impedance, power gain, and maximum peak-to-peak output voltage swing must be determined for each of the common source, common gate, and common drain connections.

THE COMMON SOURCE AMPLIFIER CIRCUIT — ALTERNATING CURRENT ANALYSIS

The connections for the common source amplifier circuit configuration are shown in Figure 1-41. From the information included with the illustration, the alternating current parameters will be solved.

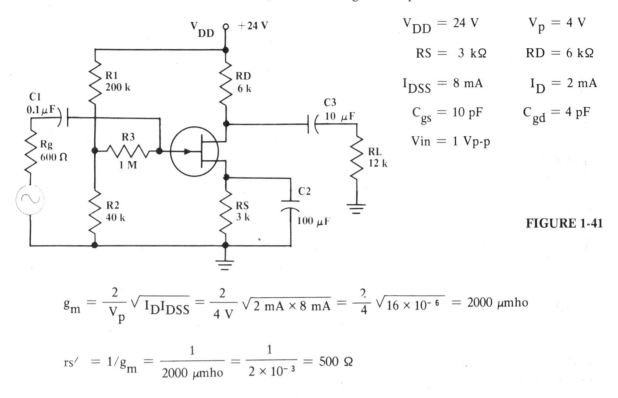

$V_{DD} = 24\ V \qquad V_p = 4\ V$

$RS = 3\ k\Omega \qquad RD = 6\ k\Omega$

$I_{DSS} = 8\ mA \qquad I_D = 2\ mA$

$C_{gs} = 10\ pF \qquad C_{gd} = 4\ pF$

$Vin = 1\ Vp\text{-}p$

FIGURE 1-41

$$g_m = \frac{2}{V_p}\sqrt{I_D I_{DSS}} = \frac{2}{4\ V}\sqrt{2\ mA \times 8\ mA} = \frac{2}{4}\sqrt{16 \times 10^{-6}} = 2000\ \mu mho$$

$$rs' = 1/g_m = \frac{1}{2000\ \mu mho} = \frac{1}{2 \times 10^{-3}} = 500\ \Omega$$

VOLTAGE GAIN:

$$A_v = g_m r_L = g_m(RD \mathbin{/\mkern-5mu/} RL) = 2000\ \mu mho(6\ k\Omega \mathbin{/\mkern-5mu/} 12\ k\Omega) = 2 \times 10^{-3} \times 4 \times 10^3 = 8$$

$$Av = \frac{r_L}{rs'} = \frac{RD \mathbin{/\mkern-5mu/} RL}{rs'} = \frac{6\ k\Omega \mathbin{/\mkern-5mu/} 12\ k\Omega}{500\ \Omega} = \frac{4\ k\Omega}{500\ \Omega} = 8$$

VOLTAGE OUT:

$$Vop\text{-}p = Vinp\text{-}p \times A_v = 1\ Vp\text{-}p \times 8 = 8\ Vp\text{-}p$$

INPUT IMPEDANCE:

$$Zin = R3 + (R1 \mathbin{/\mkern-5mu/} R2) = 1\ M\Omega + (200\ k\Omega \mathbin{/\mkern-5mu/} 40\ k\Omega) = 1\ M\Omega + 33.3\ k\Omega = 1.033\ M\Omega$$

NOTE: R3 is used specifically in this universal circuit to increase the input impedance. Without R3, the input impedance would be 16.6 kΩ. Also, since little gate current exists, little DC voltage is dropped across R3. However, increasing input impedance and increasing R3 to too large a value (much greater than 10 MΩ) would cause a larger voltage drop and provide increased circuit instability. For this reason, extremely large values of R3 are not used.

OUTPUT IMPEDANCE:

$$Zo = RD = 6\ k\Omega$$

POWER GAIN TO THE LOAD RESISTOR:

$$PG = A_v{}^2 \times \frac{Zin}{RL} = 8^2 \times \frac{1033 \text{ M}\Omega}{12 \text{ k}\Omega} = 5509.3$$

FREQUENCY RESPONSE – HIGH CORNER FREQUENCIES:

a.   The first high corner frequency is determined by the input capacitance Cin and the generator resistance Rg in parallel with the input impedance Zin.

$$f_c(\text{high \#1}) = \frac{1}{2\pi(Zin \,/\!/\, Rg)Cin} \approx \frac{0.159}{(1 \text{ M}\Omega \,/\!/\, 600 \text{ }\Omega)46 \text{ pF}} \approx \frac{0.159}{6 \times 10^2 \times 46 \times 10^{-12}}$$

$$= \frac{0.159 \times 10^{-10}}{276} = 5.76 \text{ MHz}$$

NOTE: The input capacitance Cin represents the effect of the Miller effect capacitance and the gate-to-source capacitance. The Miller effect capacitance for the common source circuit is the output capacitance ($Co = C_{gd} = 4$ pF) amplified and reflected into the input by the effective voltage gain of the circuit.

$$C(\text{Miller}) = (1 + A_v)C_{gd} = (1 + 8)4 \text{ pF} = 36 \text{ pF}$$

$$Cin = C(\text{Miller}) + C_{gs} = 36 \text{ pF} + 10 \text{ pF} = 46 \text{ pF}$$

b.   The second high corner frequency is determined by the output and stray capacitances C_{gs} and C_s and the parallel combination of the output resistors RD and RL.

$$f_c(\text{high \#2}) = \frac{1}{2\pi(RD \,/\!/\, RL)(C_{gd} + C_s)} = \frac{0.159}{(6 \text{ k}\Omega \,/\!/\, 12 \text{ k}\Omega)(4 \text{ pF} + 30 \text{ pF})}$$

$$= \frac{0.159}{4 \times 10^3 \times 34 \times 10^{-12}} = \frac{0.159 \times 10^9}{136} \approx 1.17 \text{ MHz}$$

NOTE: Since interaction occurs (less than 10 : 1 removed from each other), the overall high frequency roll-off is solved from:

$$f_c(\text{high}) = \frac{0.159 \times 10^9}{27.6 + 136} = \frac{0.159 \times 10^6}{163.6} \approx 971.8 \text{ kHz}$$

or from: $f_c(\text{high}) = f_c(\text{high \#1}) \,/\!/\, f_c(\text{high \#2}) = 5.76 \text{ MHz} \,/\!/\, 1.17 \text{ MHz} \approx 972 \text{ kHz}$

FREQUENCY RESPONSE — LOW CORNER FREQUENCIES:

a.   The first low corner frequency is determined by the input capacitor C1 and the series resistance of Rg and Zin.

$$f_c(\text{low \#1}) = \frac{1}{2\pi(Rg + Zin)C1} = \frac{0.159}{(600 \text{ }\Omega + 1 \text{ M}\Omega)0.1 \text{ }\mu\text{F}} \approx \frac{0.159}{10^6 \times 10^{-7}} = 1.59 \text{ Hz}$$

b.   The second low corner frequency is determined by the output capacitor C3 and the series resistance of RD and RL.

$$f_c(\text{low \#2}) = \frac{1}{2\pi(R_D + RL)C3} = \frac{0.159}{(6 \text{ k}\Omega + 12 \text{ k}\Omega)10 \text{ }\mu\text{F}} = \frac{0.159}{1.8 \times 10^4 \times 10^{-5}}$$

$$= \frac{0.159 \times 10}{1.8} \approx 0.883 \text{ Hz}$$

c. The third low corner frequency is determined by the source bypass capacitor C2 and the parallel combination of the source resistor R3 and the inverse device transconductance $1/g_m = rs'$. The inverse transconductance $1/g_m$ is the resistance of the device.

$$f_c(\text{low \#3}) = \frac{1}{2\pi(\text{RS} \mathbin{/\mkern-5mu/} 1/g_m)\text{C2}} = \frac{1}{2\pi(\text{RS} \mathbin{/\mkern-5mu/} rs)\text{C2}} = \frac{0.159}{(1 \text{ k}\Omega \mathbin{/\mkern-5mu/} 500 \text{ }\Omega)100 \text{ }\mu\text{F}}$$

$$= \frac{0.159}{333.3 \text{ }\Omega \times 10^{-4}} = \frac{1590}{333.3 \text{ }\Omega} = 4.77 \text{ Hz}$$

NOTE: The interaction of the three low corner freqencies cause the effective addition of all three and an overall roll-off of approximately 7 Hz.

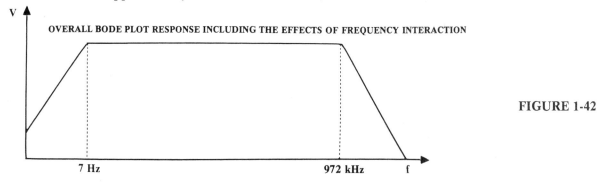

FIGURE 1-42

THE COMMON GATE AMPLIFIER CIRCUIT CONFIGURATION — AC ANALYSIS

The connections for the common gate amplifier circuit configuration are shown in Figure 1-43. From the information include with the illustration, the alternating current parameters will be solved.

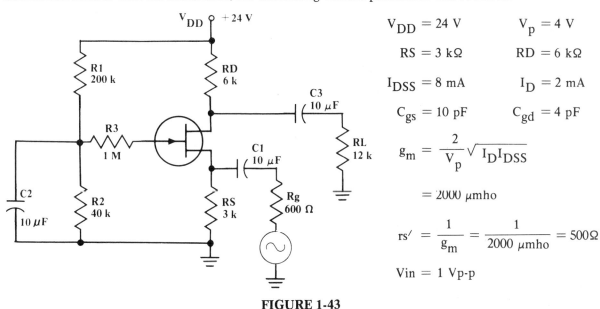

$V_{DD} = 24 \text{ V}$ $V_p = 4 \text{ V}$

$RS = 3 \text{ k}\Omega$ $RD = 6 \text{ k}\Omega$

$I_{DSS} = 8 \text{ mA}$ $I_D = 2 \text{ mA}$

$C_{gs} = 10 \text{ pF}$ $C_{gd} = 4 \text{ pF}$

$$g_m = \frac{2}{V_p}\sqrt{I_D I_{DSS}}$$

$$= 2000 \text{ }\mu\text{mho}$$

$$rs' = \frac{1}{g_m} = \frac{1}{2000 \text{ }\mu\text{mho}} = 500\Omega$$

$Vin = 1 \text{ Vp-p}$

FIGURE 1-43

VOLTAGE GAIN:

$$A_v = g_m r_L = g_m(\text{RD} \mathbin{/\mkern-5mu/} \text{RL}) = 2000 \text{ }\mu\text{mho}(6 \text{ k}\Omega \mathbin{/\mkern-5mu/} 12 \text{ k}\Omega) = 2 \times 10^{-3} \times 4 \times 10^3 = 8$$

$$A_v = \frac{r_L}{rs'} \approx \frac{\text{RD} \mathbin{/\mkern-5mu/} \text{RL}}{rs'} = \frac{6 \text{ k}\Omega \mathbin{/\mkern-5mu/} 12 \text{ k}\Omega}{500 \text{ }\Omega} = \frac{4 \text{ k}\Omega}{500\Omega} = 8$$

VOLTAGE OUT:

$$\text{Vop-p} = \text{Vinp-p} \times A_v = 1 \text{ Vp-p} \times 8 = 8 \text{ Vp-p}$$

INPUT IMPEDANCE:

$$\text{Zin} = \text{RS} \,//\, 1/g_m = \text{RS} \,//\, rs' = 3 \text{ k}\Omega \,//\, 500\Omega \approx 428.6 \ \Omega$$

OUTPUT IMPEDANCE:

$$\text{Zo} = \text{RD} = 6 \text{ k}\Omega$$

POWER GAIN TO THE LOAD:

$$\text{PG} = A_v{}^2 \times \frac{\text{Zin}}{\text{RL}} = 8^2 \times \frac{428.6 \ \Omega}{12 \text{ k}\Omega} = 2.285$$

FREQUENCY RESPONSE — HIGH CORNER FREQUENCIES:

a.   The first high corner frequency is determined by the input capacitance Cin and the generator resistance Rg in parallel with the input impedance.

$$f_c(\text{high \#1}) = \frac{1}{2\pi(\text{Zin} \,//\, \text{Rg})\text{Cin}} = \frac{0.159}{(428.6 \ \Omega \,//\, 600 \ \Omega)10\text{pF}} = \frac{0.159}{2.5 \times 10^2 \times 10^{-11}}$$

$$= \frac{0.159 \times 10^{-9}}{2.5} \approx 63.6 \text{ MHz}$$

b.   The second high corner frequency is determined by the output and stray capacitances and the parallel combination of the output resistors RD and RL.

$$f_c(\text{high \#2}) = \frac{1}{2\pi(\text{RD} \,//\, \text{RL})(C_{gd} + C_s)} = \frac{0.159}{(6 \text{ k}\Omega \,//\, 12 \text{ k}\Omega)(4 \text{ pF} + 30 \text{ pF})} = \frac{0.159}{4 \text{ k}\Omega \times 34 \text{ pF}}$$

$$= \frac{0.159}{4 \times 10^3 \times 34 \times 10^{-12}} = \frac{0.159}{136 \times 10^{-9}} \approx 1.17 \text{ MHz}$$

FREQUENCY RESPONSE — LOW CORNER FREQUENCIES:

a.   The first low corner frequency is determined by the input capacitor C1 and the series resistance of Rg and Zin.

$$f_c(\text{low \#1}) = \frac{1}{2\pi(\text{Rg} + \text{Zin})\text{C1}} = \frac{0.159}{(600 \ \Omega + 423.6 \ \Omega)10 \ \mu\text{F}} = \frac{0.159}{1023 \ \Omega \times 10^{-5}} = \frac{0.159 \times 10^5}{1.023 \times 10^3} = 15.5 \text{ Hz}$$

b.   The second low corner frequency is determined by the output capacitor C2 and the series resistance of RD and RL.

$$f_c(\text{low \#2}) = \frac{1}{2\pi(\text{RD} + \text{RL})\text{C3}} = \frac{0.159}{(6 \text{ k}\Omega + 12 \text{ k}\Omega)10 \ \mu\text{F}} = \frac{0.159}{1.8 \times 10^4 \times 10^{-5}} = \frac{0.159 \times 10^1}{1.8} = 0.883 \text{ Hz}$$

c.   The third low corner frequency is determined by the C2 bypass capacitor and the associated gate resistance.

$$f_c(\text{low \#3}) = \frac{1}{2\pi(\text{R3} \,//\, [\text{R1} \,//\, \text{R2}])\text{C2}} = \frac{0.159}{(1 \text{ M}\Omega \,//\, [200 \text{ k}\Omega \,//\, 40 \text{ k}\Omega])10 \ \mu\text{F}}$$

$$\approx \frac{0.159}{(200 \text{ k}\Omega \,//\, 40 \text{ k}\Omega)10 \ \mu\text{F}} = \frac{0.159}{33.33 \times 10^3 \times 10^{-5}} = \frac{0.159 \times 10^2}{33.33} \approx 0.48 \text{ Hz}$$

NOTE: The R3 resistor at 1 MΩ is too high to be considered in the calculations because, if it is included, it will only change the roll-off frequency from ≈ 0.48 Hz to 0.49 Hz.

d. Obviously the low corner frequency will be f_c(low #1) and this occurs at 15.5 Hz. If a lower value is required, then a larger C1 capacitor must be used. Also, if the other low corner frequencies are included, then the additive value will be approximately 17 Hz.

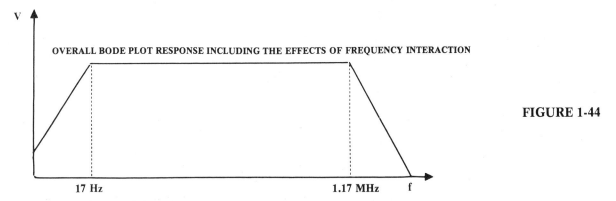

FIGURE 1-44

THE COMMON DRAIN AMPLIFIER CONFIGURATION — AC ANALYSIS

The connections for the common drain amplifier circuit configuration are shown in Figure 1-45. From the information included with the illustration, the alternating current parameters will be solved. The DC analysis for the circuit is similar to the preceding universal biased circuits except for the voltage drop across the gate to source. This is solved from $V_{DD} - V_{RS} = 24 V - 6 V = 18 V$.

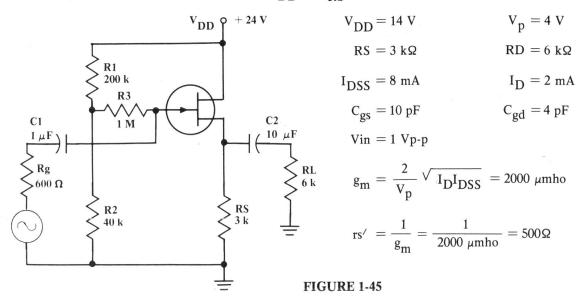

$V_{DD} = 14 V$ $V_p = 4 V$

$RS = 3 k\Omega$ $RD = 6 k\Omega$

$I_{DSS} = 8 mA$ $I_D = 2 mA$

$C_{gs} = 10 pF$ $C_{gd} = 4 pF$

$Vin = 1 Vp\text{-}p$

$$g_m = \frac{2}{V_p} \sqrt{I_D I_{DSS}} = 2000 \ \mu mho$$

$$rs' = \frac{1}{g_m} = \frac{1}{2000 \ \mu mho} = 500 \Omega$$

FIGURE 1-45

VOLTAGE GAIN:

$$A_v = \frac{g_m r_L}{1 + g_m r_L} = \frac{g_m (RS \mathbin{/\mkern-5mu/} RL)}{1 + g_m (RS \mathbin{/\mkern-5mu/} RL)} = \frac{2000 \ \mu mho (3 \ k\Omega \mathbin{/\mkern-5mu/} 6 \ k\Omega)}{1 + 2000 \ \mu mho \ (3 \ k\Omega \mathbin{/\mkern-5mu/} 6 \ k\Omega)}$$

$$= \frac{(2 \times 10^{-3})(2 \times 10^3)}{1 + (2 \times 10^{-3})(2 \times 10^3)} = \frac{4}{1 + 4} = 0.8$$

$$A_v = \frac{r_L}{rs' + r_L} = \frac{RS \mathbin{/\mkern-5mu/} RL}{rs' + (RS \mathbin{/\mkern-5mu/} RL)} = \frac{3 \ k\Omega \mathbin{/\mkern-5mu/} 6 \ k\Omega}{500 \ \Omega + (3 \ k\Omega \mathbin{/\mkern-5mu/} 6 \ k\Omega)} = \frac{2 \ k\Omega}{2.5 \ k\Omega} = 0.8$$

35

VOLTAGE OUT:

$$V\text{op-p} = V\text{inp-p} \times A_v = 1 \text{ Vp-p} \times 0.8 = 0.8 \text{ Vp-p}$$

INPUT IMPEDANCE:

$$Z\text{in} = R3 + (R1 \parallel R2) = 1 \text{ M}\Omega + (100 \text{ k}\Omega \parallel 20 \text{ k}\Omega) = 1016.6 \text{ K}\Omega$$

OUTPUT IMPEDANCE:

$$Z\text{o} = RS \parallel 1/g_m = RS \parallel rs' = 3 \text{ k}\Omega \parallel 500 \ \Omega = 428.6 \ \Omega$$

POWER GAIN TO THE LOAD:

$$PG = A_v{}^2 \times \frac{Z\text{in}}{RL} = 0.8^2 \times \frac{1016.6 \text{ k}\Omega}{6 \text{ k}\Omega} \approx 108.4$$

FREQUENCY RESPONSE — HIGH CORNER FREQUENCIES:

a.   The first high corner frequency is determined by the input capacitance Cin and the generator resistance Rg in parallel with the input impedance Zin.

$$f_c(\text{high \#1}) = \frac{1}{2\pi(Z\text{in} \parallel Rg)C\text{in}} = \frac{0.159}{(1016.6 \text{ M}\Omega \parallel 600\Omega)6 \text{ pF}}$$

$$\approx \frac{0.159}{6 \times 10^2 \times 6 \times 10^{-12}} = \frac{0.159 \times 10^{10}}{36} = 44.16 \text{ MHz}$$

where: $C\text{in} = C_{gd} + (1 - A_v)C_{gs} = 4 \text{ pF} + (1 - 0.8)10 \text{ pF} = 6 \text{ pF}$

b.   The second high corner frequency is determined by the output and stray capacitance and the output impedance in parallel with the load.

$$f_c(\text{high \#2}) = \frac{1}{2\pi(Z\text{o} \parallel RL)(C\text{o} + C_s)} = \frac{0.159}{(428.6 \ \Omega \parallel 6 \text{ k}\Omega)(10 \text{ pF} + 30 \text{ pF})}$$

$$= \frac{0.159}{4 \times 10^2 \times 40 \times 10^{-12}} = \frac{0.159 \times 10^{10}}{160} \approx 9.94 \text{ MHz}$$

NOTE: Slight interaction causes the roll-off to be slightly lower than 9.94 MHz. Therefore, the overall high corner frequency is solved from:

$$f_c(\text{high}) = \frac{0.159 \times 10^{10}}{160 + 36} = \frac{0.159 \times 10^{10}}{196} = 8.11 \text{ MHz, or from:}$$

$$f_c(\text{high}) = f_c(\text{high \#1}) \parallel f_c(\text{high \#2}) = 44.16 \text{ MHz} \parallel 9.94 \text{ MHz} \approx 8.11 \text{ MHz}$$

FREQUENCY RESPONSE — LOW CORNER FREQUENCIES:

a.   The first low corner frequency is determined by the input capacitor C1 and the input impedance in series with the generator impedance Rg.

$$F_c(\text{low \#1}) = \frac{1}{2\pi(Rg + Z\text{in})C1} = \frac{0.159}{(600 \ \Omega + 1.0166 \text{ M}\Omega)1 \ \mu F} = \frac{0.159}{1.0172 \times 10^6 \times 10^{-6}} = 0.156 \text{ Hz}$$

b.   The second low corner frequency is determined by the output capacitor C2 and the series output and load resistances.

$$f_c(\text{low \#2}) = \frac{1}{2\pi(Z_o + R_L)C_2} = \frac{0.159}{(428.6\ \Omega + 6\ k\Omega)10\ \mu F} = \frac{0.159}{6.428 \times 10^3 \times 10^{-5}}$$

$$= \frac{0.159 \times 10^2}{6.428} = 2.47\ Hz \approx 2.5\ Hz$$

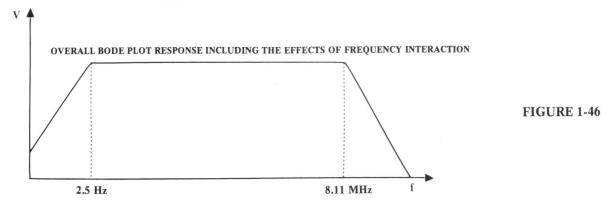

OVERALL BODE PLOT RESPONSE INCLUDING THE EFFECTS OF FREQUENCY INTERACTION

2.5 Hz   8.11 MHz

FIGURE 1-46

THE TWO POWER SUPPLY JFET AMPLIFIER — DIRECT CURRENT ANALYSIS

The DC analysis for the two power supply circuit is only slightly different to that of the self-biased JFET circuit, and the AC analysis is the same. Therefore, only the DC analysis need be done. The DC analysis for the circuit of Figure 1-47 is solved in terms of drain current using the device characteristics of pinch-off voltage V_p and saturation current I_{DSS}.

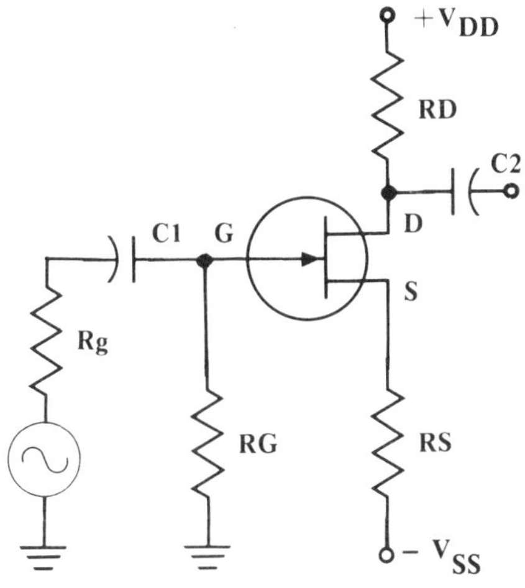

$$I_D = I_{DSS}\left(1 - \frac{V_{GS}}{V_p}\right)^2$$

$$V_{DD} = I_D RD + V_{DS} + I_D RS + (-V_{SS})$$

$$V_{DD} - (-V_{SS}) = I_D RD + V_{DS} + I_D RS$$

$$V_{DD} + V_{SS} = V_{RD} + V_{DS} + V_{SS}$$

where: $V_{RS} = V_{GS} + V_{SS}$ and where: $V_{RG} = 0\ V$

FIGURE 1-47

The direct current analysis begins by solving for the drain current (I_D) in terms of the device's pinch off voltage (V_p), the saturation current (I_{DSS}), and the selected gate-to-source voltage (V_{GS}). Next, the voltages across the source resistor (V_{RS}) and the drain resistor (V_{RD}) are found, and then the voltage drop across the drain to source (V_{DS}) is solved.

Since zero volts exist at the gate terminal, because little or no gate current flows through the gate resistor (RG), and since postive voltage exists at the source terminal, then the reverse biased condition of the gate to source diode junction is satisfied. Also, since the voltage at the source terminal must be positive with respect to the zero volts condition at the gate, then an unusually large voltage is dropped across the source resistor (RS), which in turn forces the V_{DD} voltage to be relatively larger than V_{SS}, if a moderate drain-to-source (V_{DS}) device voltage is to be maintained.

37

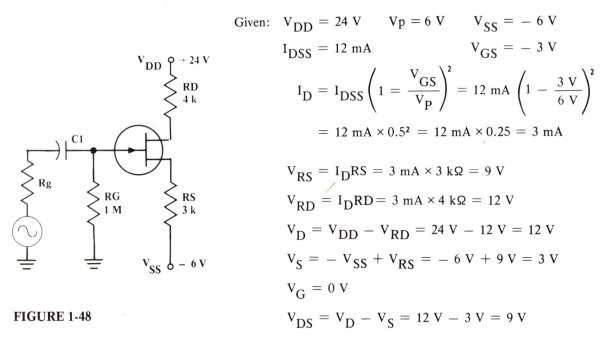

Given: $V_{DD} = 24\ V \qquad V_P = 6\ V \qquad V_{SS} = -6\ V$

$I_{DSS} = 12\ mA \qquad\qquad V_{GS} = -3\ V$

$$I_D = I_{DSS}\left(1 = \frac{V_{GS}}{V_P}\right)^2 = 12\ mA\left(1 - \frac{3\ V}{6\ V}\right)^2$$

$$= 12\ mA \times 0.5^2 = 12\ mA \times 0.25 = 3\ mA$$

$V_{RS} = I_D RS = 3\ mA \times 3\ k\Omega = 9\ V$

$V_{RD} = I_D RD = 3\ mA \times 4\ k\Omega = 12\ V$

$V_D = V_{DD} - V_{RD} = 24\ V - 12\ V = 12\ V$

$V_S = -V_{SS} + V_{RS} = -6\ V + 9\ V = 3\ V$

$V_G = 0\ V$

$V_{DS} = V_D - V_S = 12\ V - 3\ V = 9\ V$

FIGURE 1-48

Figure 1-49 shows the DC distribution of the circuit. It is important to remember that all calculated and measured voltage drops between V_{DD} and $-V_{SS}$ must be equal to the applied voltage.

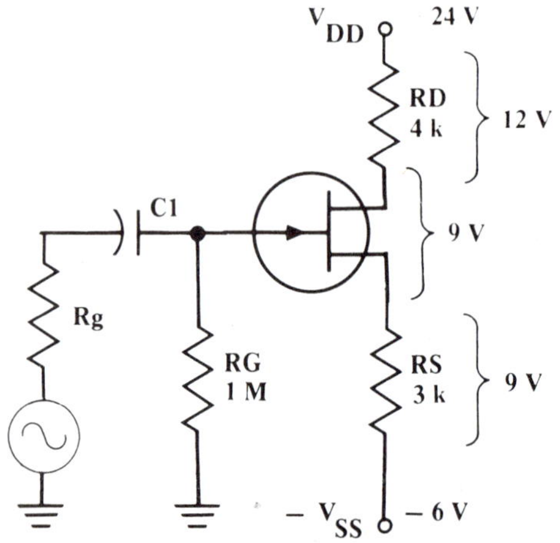

$V_{DD} = V_{RD} + V_{DS} + V_{RS} - V_{SS}$

$24\ V = 12\ V + 9\ V + 9\ V - 6\ V \quad$ or

$V_{DD} + V_{SS} = V_{RD} + V_{DS} + V_{RS}$

$24\ V + 6\ V = 12\ V + 9\ V + 9\ V$

$V_{GS} = V_G - V_S = 0\ V - 3\ V = -3\ V$

FIGURE 1-49

NOTE: The DC analysis is applicable to all of the common source, common gate and common drain two power supply circuits. The AC analysis for the basic circuits is similar to those already solved for the self-biased circuits, so they are not repeated.

METAL OXIDE SEMICONDUCTOR FIELD EFFECT TRANSISTOR (MOSFET)

The MOSFET devices have a much higher input impedance than JFET's and, because they can be either enhancement or depletion types, they have much simpler biasing capabilities. However, in frequencies below 20 MHz the MOSFET is more noisy than the JFET, except in cases where input impedances of more than 100 MΩ are required. Therefore, in all but these rare instances, the preferred device in linear amplifier applications below 10 MHz is the JFET.

At frequencies above 10 MHz, the noise figure for the JFET and MOSFET are comparable and the MOSFET has decided advantages over the JFET because of high input impedance, low feedback capacitance, and low drift conditions with temperature. Additionally, the square law characteristics of the field effect tranistor along with the inherent device characteristics makes the MOSFET ideally suited to high frequency applications. A comparative list of nominal device characteristics follows:

INPUT IMPEDANCE — The input impedance of the discrete MOSFET is nominally greater than 10^{14}, the JFET is greater than 10^{11}, and the bipolar transistor is greater than 10^3.

VOLTAGE GAIN — The voltage gains of both the MOSFET and JFET are low because the transconductances are low. For the bipolar transistor the transconductance and gain is considerably higher. A nominal voltage gain for a MOSFET or JFET is about 10 while that for the bipolar transistor can be 500.

INPUT CURRENT — The input current for the MOSFET is usually greater than 10^{-14}A, for the JFET 10^{-11}A, and for the bipolar transistor 10^{-3}A. However, for the bipolar, the input current will vary considerably with biasing and the beta of the transistor.

SQUARE LAW CHARACTERISTICS — The field effect transistor is a square law device and the bipolar transistor is not. Therefore, the MOSFET and JFET are excellent devices for use in the front end of receivers where, as tuned circuit amplifiers, selectivity is easily achieved. For the non-linear diode input of the bipolar, intermodulation and cross modulation distortion create filtering problems.

TEMPERATURE — MOSFETS are least effected by temperature changes, followed by JFETS, with bipolars last. Both the MOSFET and the JFET can be biased, theoretically, for zero temperature drift but, because the gate to source of the JFET is a reverse biased diode, the leakage current will increase exponentially with temperature increases. The MOSFET, on the other hand, has insulated gate-to-channel properties and gate leakage current between the channel and substrate will increase exponentially with temperature because of these substrate to channel (PN) properties.

OUTPUT IMPEDANCE — The output impedances of the MOSFET, JFET, and bipolar transistor are comparable and all can be optimized with effective biasing. An output impedance of 100 kΩ is nominal.

DEVICE CAPACITANCES — The device capacitances for the MOSFET are normally lower than that for the JFET but, since both resistance and capacitance play a role in frequency response, the clear choice of which device to use is predicated on many factors.

MOSFET HANDLING PRECAUTIONS — The MOSFET, because of its extremely low leakage current and high input impedance, is highly susceptible to electrostatic charges, which can break down the very thin gate-to-channel insulated region. For this reason, precaution is exercised in shipping and installation. In shipping, for instance, the leads of the device are shorted together, while in installation, grounded soldering irons, grounded tools, and even grounded hands (using a grounding strap) are used to guard against the possibility of electrostatic charges destroying the device. This electrostatic charge problem can also be circumvented by the addition of back-to-back diodes between the gate and drain of the device. However, the inclusion of the protective diodes can cause the frequency response of the device to change because they contribute a slight amount of distributive capacitance.

MOSFET TYPES

The discrete MOSFET can be either an enhancement or depletion type device. However, the enhancement device must be operated in the enhancement mode only, while the depletion type can be successfully operated in both the depletion and enhancement modes.

DEPLETION TYPE MOSFETS

The depletion type MOSFET can be connected into all of the JFET biased circuit configurations as well as other circuits. The added flexibility of the depletion type MOSFET over the JFET (the depletion type only) is due to the insulated gate properties of the MOSFET, which allows the gate to source to be either forward biased or reverse biased without effecting the high input impedance of the device. The JFET, on the other hand, must always be reverse biased in maintaining a high input impedance. Therefore, the depletion type MOSFET can be either reverse biased (depletion mode), forward biased (enhancement mode), or operated in a no-bias condition where the gate-to-source voltage is at zero volts.

The circuits of Figure 1-50 demonstrate the similarities between MOSFET and JFET biasing techniques with only Figure 1-50(a) being uniquely different. The circuit of Figure 1-50(a) is connected to provide a zero gate-to-source voltage ($V_{GS} = 0$ V), a circuit configuration not possible with JFET's. The advantage of this configuration is simplistic biasing, with I_{DSS} of the device providing the operating (drain) current for the circuit.

DEPLETION TYPE MOSFET CHARACTERISTICS — The depletion type MOSFET is quite capable of operating in both the depletion and enhancement modes but, in practical circuit operation, it is rarely used beyond the zero gate-to-source biased condition. The reason for this is that the enhancement mode curves are a near mirror image to the depletion mode curves and enhancement mode operation requires a considerably higher operating current. Therefore, the depletion type MOSFET is operated in the enhancement mode in rare situations only. The complete characteristice curves for the depletion type MOSFET are shown in Figure 1-51, where the operating current at $V_{GS} = 0$ volts is the I_{DSS} current. This is

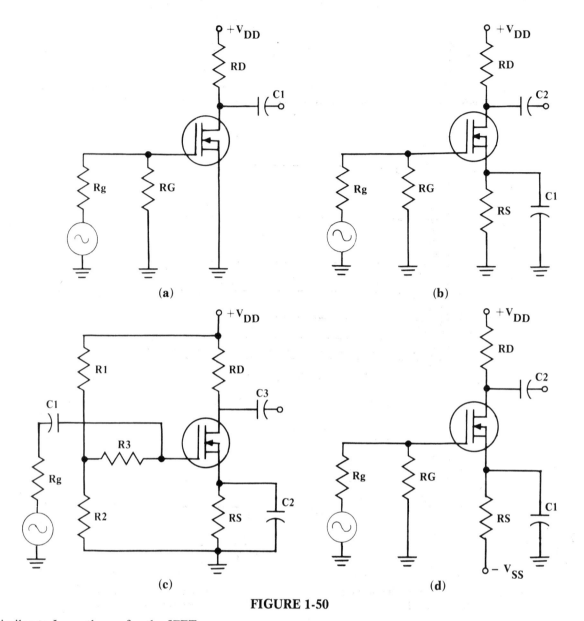

FIGURE 1-50

similar to I_{DSS} theory for the JFET.

LINEAR DC CONSIDERATION — The gate-to-source junction of the depletion type MOSFET can be forward biased, reverse biased, or at zero bias condition in controlling the amount of drain current flow through the device. The formula used to solve for the drain current in the depletion mode is similar to that of JFET's, as shown in Figure 1-10, and is solved from:

$$I_D = I_{DSS}\left(1 - \frac{V_{GS}}{V_p}\right)^2,$$

where IDSS, like that of the JFET, is the drain current at the zero gate-to-source biased condtion. However, in the enhancement mode, the MOSFET drain current is solved from:

$$I_D = I_{DSS}\left(1 + \frac{V_{GS}}{V_p}\right)^2$$

which reflects the forward biased gate-to-source condition of the enhancement mode.

Therefore, solving for the drain current for the depletion type, N-Channel MOSFET having device characteristics of $I_{DSS} = 6$ mA and $V_p = 4$ volts, where the gate-to-source voltage is varied from -4

40

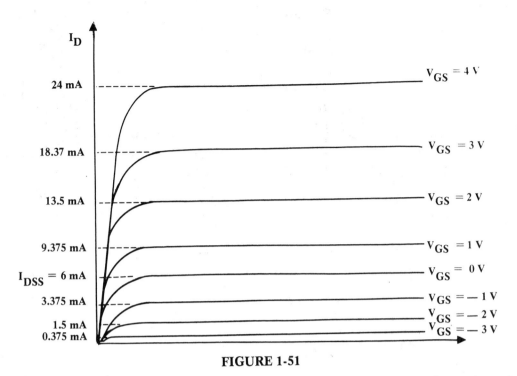

FIGURE 1-51

volts to 0 volts in the depletion mode and 0 volts to 4 volts in the enhancement mode, as shown in Figure 1-52, is as follows:

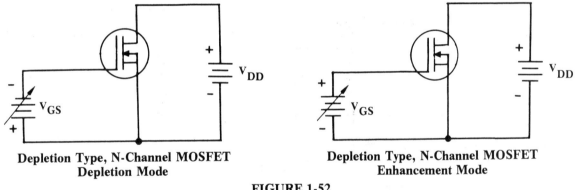

Depletion Type, N-Channel MOSFET
Depletion Mode

Depletion Type, N-Channel MOSFET
Enhancement Mode

FIGURE 1-52

1. DEPLETION MODE:

$V_{GS} = -4$ V $\qquad I_D = I_{DSS}(1 - V_{GS}/V_p)^2 = 6$ mA$(1 - 4$ V$/4$ V$)^2 = 0$ mA

$V_{GS} = -3$ V $\qquad I_D = I_{DSS}(1 - V_{GS}/V_p)^2 = 6$ mA$(1 - 3$ V$/4$ V$)^2 = 0.375$ mA

$V_{GS} = -2$ V $\qquad I_D = I_{DSS}(1 - V_{GS}/V_p)^2 = 6$ mA$(1 - 2$ V$/4$ V$)^2 = 1.5$ mA

$V_{GS} = -1$ V $\qquad I_D = I_{DSS}(1 - V_{GS}/V_p)^2 = 6$ mA$(1 - 1$ V$/4$ V$)^2 = 3.375$ mA

$V_{GS} = 0$ V $\qquad I_D = I_{DSS}(1 - V_{GS}/V_p)^2 = 6$ mA$(1 - 0$ V$/4$ V$)^2 = 6$ mA

2. ENHANCEMENT MODE:

$V_{GS} = 0$ V $\qquad I_D = I_{DSS}(1 + V_{GS}/V_P)^2 = 6$ mA$(1 + 0$ V$/4$ V$)^2 = 6$ mA

NOTE: $V_{GS} = 0$ volts is neither in the depletion or enhancement modes. Therefore, either formula is valid and the reason for showing both is for clarity purposes only.

41

$$V_{GS} = 1 \text{ V} \qquad I_D = I_{DSS}(1 + V_{GS}/V_p)^2 = 6 \text{ mA}(1 + 1 \text{ V}/4 \text{ V})^2) = 9.375 \text{ mA}$$

$$V_{GS} = 2 \text{ V} \qquad I_D = I_{DSS}(1 + V_{GS}/V_p)^2 = 6 \text{ mA}(1 + 2 \text{ V}/4\text{V})^2 = 13.5 \text{ mA}$$

$$V_{GS} = 3 \text{ V} \qquad I_D = I_{DSS}(1 + V_{GS}/V_p)^2 = 6 \text{ mA}(1 + 3 \text{ V}/4 \text{ V})^2 = 18.375 \text{ mA}$$

$$V_{GS} = 4 \text{ V} \qquad I_D = I_{DSS}(1 + V_{GS}/V_p)^2 = 6 \text{ mA}(1 + 4 \text{ V}/4 \text{ V})^2 = 24 \text{ mA}$$

NOTE: The calculated drain currents for the N-Channel depletion type MOSFET device, calculated for both the depletion and enhancement modes, are shown in the characteristic curves of Figure 1-51.

THE DEPLETION TYPE MOSFET — CIRCUIT ANALYSIS

The DC and AC analysis for the self-biased and zero gate-to-source biased MOSFET circuits are included in this section. However, only the common source configurations are presented, because the use of the depletion type MOSFET in a self-biased circuit is similar to JFET analysis and further duplication is not necessary. Also, the zero gate-to-source voltage biasing is limited to the common source circuit only, because of the absence of the source resistor.

THE SELF-BIASED DEPLETION TYPE MOSFET — DC ANALYSIS

The DC analysis of the common source self-biased depletion type MOSFET circuit of Figure 1-53 is solved in terms of the drain current. However, because of the dependence on the device characteristics, the pinch off voltage (V_p) and the saturation current (I_{DSS}) must be known. The analysis is similar to that used for the JFET.

The DC analysis begins by solving for the drain current (I_D) in terms of the device's saturation current (I_{DSS}), the pinch off voltage (V_p), and the selected gate-to-source voltage (V_{GS}). Next the voltage drops across the source resistor (V_{RS}) and drain resistor (V_{RD}) are solved. Since minimum gate current flows (10^{-14} A), little or no voltage is developed across the gate resistor. Therefore, the voltage at the gate is zero volts and, with respect to the positive voltage at the source, provides a depletion-mode, reverse-biased, condition.

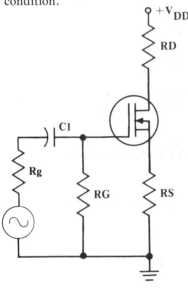

$$I_D = I_{DSS}(1 - V_{GS}/V_p)^2$$

$$V_{DD} = I_D RD + V_{DS} + I_D RS$$

$$V_{DD} = V_{RD} + V_{DS} + V_{RS}$$

$$V_{RS} = - V_{GS} \qquad \text{where:} \quad V_{RG} = 0 \text{ volts}$$

FIGURE 1-53

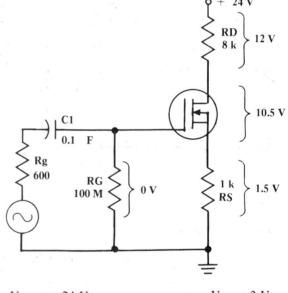

$$V_{DD} = 24 \text{ V} \qquad\qquad V_p = 3 \text{ V}$$

$$RD = 8 \text{ k}\Omega \qquad\qquad RS = 1 \text{ k}\Omega$$

$$RG = 100 \text{ M}\Omega \qquad\qquad I_{DSS} = 6 \text{ mA}$$

$$V_{GS} = 1.5 \text{ V}$$

FIGURE 1-54

$$I_D = I_{DSS}\left(1 - \frac{V_{GS}}{V_p}\right)^2 = 6\text{ mA}\left(1 - \frac{1.5}{3\text{ V}}\right)^2 = 6\text{ mA} \times 0.5^2 = 1.5\text{ mA}$$

$$V_{RS} = I_D RS = 1.5\text{ mA} \times 1\text{ k}\Omega = 1.5\text{ V}$$

$$V_{RD} = I_D RD = 1.5\text{ mA} \times 8\text{ k}\Omega = 12\text{ V}$$

$$V_D = V_{DD} - V_{RD} = 24\text{ V} - 12\text{ V} = 12\text{ V}$$

$$V_S = V_{RS} = 1.5\text{ V}$$

$$V_G = V_{RG} = 0\text{ V}$$

$$V_{GS} = V_G - V_S = 0\text{ V} - 1.5\text{ V} = -1.5\text{ V}$$

$$V_{DS} = V_D - V_S = 12\text{ V} - 1.5\text{ V} = 10.5\text{ V}$$

THE SELF-BIASED DEPLETION TYPE MOSFET — AC ANALYSIS

Once the DC voltage drops are known, then the AC circuit parameters are found. Voltage gain, input impedance, output impedance, power gain, and the voltage swing at the output are determined for each of the common source, common gate, and common drain connections.

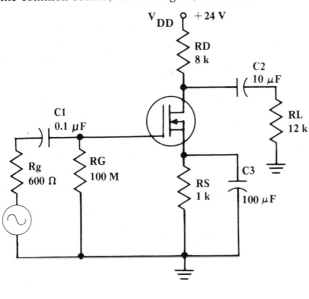

FIGURE 1-55

$V_{DD} = 24$ V	$V_p = 3$ V
$RG = 100$ MΩ	$RD = 8$ kΩ
$RS = 1$ kΩ	$RL = 12$ kΩ
$I_{DSS} = 6$ mA	$I_D = 1.5$ mA
$C_{gs} = 8.5$ pF	$C_{gd} = 1.5$ pF
$V_{in} = 1$ Vp-p	

$$g_m = \frac{2}{V_p}\sqrt{I_D I_{DSS}} = \frac{2}{3}\sqrt{6\text{ mA} \times 1.5\text{ mA}}$$

$$= \frac{2}{3}\sqrt{9 \times 10^{-6}} = 2000\ \mu\text{mho}$$

$$rs' = \frac{1}{g_m} = \frac{1}{2000\ \mu\text{mho}} = \frac{1}{2 \times 10^{-3}} = 500\ \Omega$$

VOLTAGE GAIN:

$$A_v = g_m(RD /\!/ RL) = 2000\ \mu\text{mho}\ (8\text{ k}\Omega /\!/ 12\text{ k}\Omega) = 2 \times 10^{-3} \times 4.8 \times 10^3 = 9.6, \text{ or}$$

$$A_v = \frac{RD /\!/ RL}{rs'} = \frac{8\text{ k}\Omega /\!/ 12\text{ k}\Omega}{500\ \Omega} = \frac{4.8\text{ k}\Omega}{500\ \Omega} = 9.6$$

VOLTAGE OUT:

$$Vop\text{-}p = Vinp\text{-}p \times A_v = 1\text{ Vp-p} \times 9.6 = 9.6\text{ Vp-p}$$

INPUT IMPEDANCE:

$$Zin \approx RG = 100\text{ M}\Omega$$

43

OUTPUT IMPEDANCE:

$$Z_o \approx R_D = 8 \text{ k}\Omega$$

POWER GAIN TO THE LOAD RESISTOR:

$$PG = A_v{}^2 \times \frac{Z_{in}}{R_L} = 9.6^2 \times \frac{100 \text{ M}\Omega}{12 \text{ k}\Omega} = \frac{92.16 \times 10^8}{1.2 \times 10^4} = 7.68 \times 10^4$$

FREQUENCY RESPONSE — HIGH CORNER FREQUENCY:

a.   The first high corner frequency is determined by the capacitance Cin and the generator resistance Rg in parallel with the input impedance Zin ≈ Rin = RG.

$$f_c(\text{high \#1}) = \frac{1}{2\pi(R_{in} /\!/ R_g)C_{in}} = \frac{0.159}{(100 \text{ M}\Omega /\!/ 600 \text{ }\Omega)24.4 \text{ pF}} = \frac{0.159}{(6 \times 10^2)(2.44 \times 10^{-11})}$$

$$= \frac{0.159 \times 10^9}{14.64} = 10.86 \text{ MHz}$$

where: $C_{in} = (1 + A_v)C_{gd} + C_{gs} = (1 + 9.6)1.5 \text{ pF} + 8.5 \text{ pF} = (10.6 \times 1.5 \text{ pF}) + 8.5 \text{ pF} = 24.4 \text{ pF}$

b.   The second high corner frequency is determined by the output and stray capacitance and the parallel combination of the RD and RL resistors.

$$f_c(\text{high \#2}) = \frac{1}{2\pi(R_D /\!/ R_L)(C_{gd} + C_s)} = \frac{0.159}{(8 \text{ k}\Omega /\!/ 12 \text{ k}\Omega)(1.5 \text{ pF} + 30 \text{ pF})}$$

$$= \frac{0.159}{4.8 \times 10^3 \times 31.5 \times 10^{-12}} = \frac{0.159 \times 10^9}{151.2} \approx 1.052 \text{ MHz}$$

NOTE: The corner frequencies are greater than 10 : 1 removed from each other. Therefore, the roll-off frequency is considered to be the lowest high corner frequency or approximately 1.05 MHz. However, greater accuracy can be achieved if the parallel of 10.86 MHz and 1.052 MHz is taken at 959 kHz.

FREQUENCY RESPONSE — LOW CORNER FREQUENCY:

a.   The first low corner frequency is determined by the input capacitor C1 and the series resistances of Rg and Zin.

$$f_c(\text{low \#1}) = \frac{1}{2\pi(R_g + Z_{in})C1} = \frac{0.159}{(600 \text{ }\Omega + 100 \text{ M}\Omega)0.1 \text{ }\mu\text{F}} \approx \frac{0.159}{10^8 \times 10^{-7}} = 0.0159 \text{ Hz}$$

b.   The second low corner frequency is determined by the output coupling capacitor C2 and the series resistance of RD and RL.

$$f_c(\text{low \#2}) = \frac{1}{2\pi(R_D + R_L)C2} = \frac{0.159}{(8 \text{ k}\Omega + 12 \text{ k}\Omega)10 \text{ }\mu\text{F}} = \frac{0.159}{2 \times 10^4 \times 10^{-5}} = 0.8 \text{ Hz}$$

c.   The third low corner frequency is determined by the source bypass capacitor C3 and the parallel combination of the source resistor RS and the inverse device transconductance ($1/g_m = r_s{}'$).

$$f_c(\text{low \#3}) = \frac{1}{2\pi(R_S /\!/ r_s{}')C3} = \frac{0.159}{(1 \text{ k}\Omega /\!/ 500 \text{ }\Omega)100\mu\text{F}} = \frac{0.159}{333.3 \times 10^{-4}} = 4.77 \text{ Hz}$$

NOTE: Since the low corner frequencies are less than 10 : 1 removed from each other, they effectively add. Therefore, the low corner frequency is approximately 5.6 Hz.

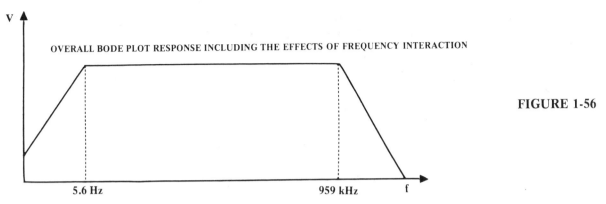

OVERALL BODE PLOT RESPONSE INCLUDING THE EFFECTS OF FREQUENCY INTERACTION

FIGURE 1-56

5.6 Hz   959 kHz   f

THE ZERO VOLT GATE-TO-SOURCE BIASED, DEPLETION TYPE MOSFET DIRECT CURRENT ANALYSIS

The DC analysis of the uniquely biased zero gate-to-source voltage for the depletion type MOSFET is solved in terms of the drain current but, because of the dependence on the device characteristics, the saturation current (I_{DSS}) must be known.

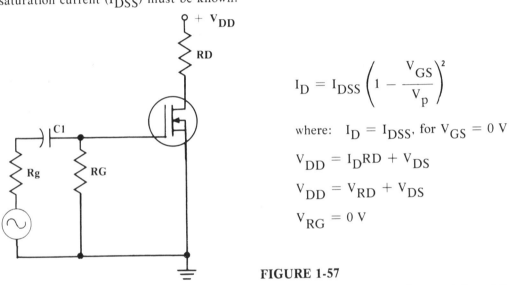

$$I_D = I_{DSS} \left(1 - \frac{V_{GS}}{V_p} \right)^2$$

where: $I_D = I_{DSS}$, for $V_{GS} = 0$ V

$$V_{DD} = I_D RD + V_{DS}$$

$$V_{DD} = V_{RD} + V_{DS}$$

$$V_{RG} = 0 \text{ V}$$

FIGURE 1-57

The DC analysis begins by knowing the drain current from the I_{DSS} characteristics of the device where, at zero gate-to-source voltage, $I_D = I_{DSS}$. Next, the voltage drop across the drain resistor is solved and then the drain-to-source voltage (V_{DS}) from $V_{DD} - V_{RD}$

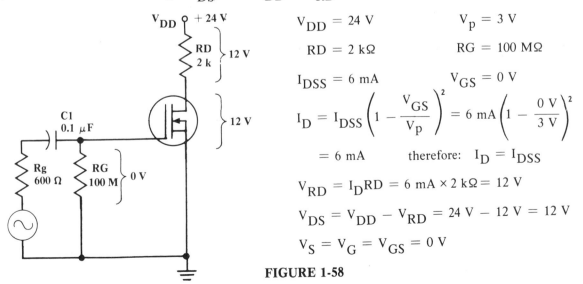

$V_{DD} = 24$ V $V_p = 3$ V

$RD = 2$ kΩ $RG = 100$ MΩ

$I_{DSS} = 6$ mA $V_{GS} = 0$ V

$$I_D = I_{DSS} \left(1 - \frac{V_{GS}}{V_p} \right)^2 = 6 \text{ mA} \left(1 - \frac{0 \text{ V}}{3 \text{ V}} \right)^2$$

$= 6$ mA   therefore: $I_D = I_{DSS}$

$$V_{RD} = I_D RD = 6 \text{ mA} \times 2 \text{ k}\Omega = 12 \text{ V}$$

$$V_{DS} = V_{DD} - V_{RD} = 24 \text{ V} - 12 \text{ V} = 12 \text{ V}$$

$$V_S = V_G = V_{GS} = 0 \text{ V}$$

FIGURE 1-58

45

THE ZERO VOLT GATE-TO-SOURCE BIASED, DEPLETION TYPE MOSFET ALTERNATING CURRENT ANALYSIS

Once the DC voltage drops are known, the alternating current circuit parameters are found. Voltage gain, input impedance, output impedance, and power gain will be determined only for the common source configuration.

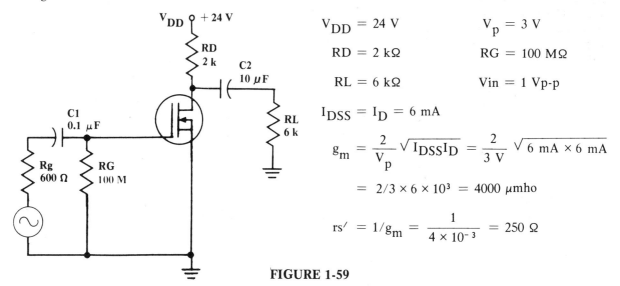

$V_{DD} = 24$ V $V_p = 3$ V

$RD = 2$ kΩ $RG = 100$ MΩ

$RL = 6$ kΩ $Vin = 1$ Vp-p

$I_{DSS} = I_D = 6$ mA

$g_m = \dfrac{2}{V_p}\sqrt{I_{DSS}I_D} = \dfrac{2}{3\,V}\sqrt{6\text{ mA} \times 6\text{ mA}}$

$= 2/3 \times 6 \times 10^3 = 4000\ \mu mho$

$rs' = 1/g_m = \dfrac{1}{4 \times 10^{-3}} = 250\ \Omega$

FIGURE 1-59

VOLTAGE GAIN:

$$A_v = g_m(RD \,/\!/\, RL) = 4000 \text{ mho } (2 \text{ k}\Omega \,/\!/\, 6 \text{ k}\Omega) = 4 \times 10^{-3} \times 1.5 \times 10^3 = 6$$

$$A_v = \frac{RD \,/\!/\, RL}{rs'} = \frac{2\text{ k}\Omega \,/\!/\, 6\text{ k}\Omega}{250\ \Omega} = \frac{1500\ \Omega}{250\ \Omega} = 6$$

VOLTAGE OUT:

$$Vop\text{-}p = Vinp\text{-}p \times A_v = 1\text{ Vp-p} \times 6 = 6\text{ Vp-p}$$

INPUT IMPEDANCE:

$$Zin \approx RG = 100\text{ M}\Omega$$

OUTPUT IMPEDANCE

$$Zo \approx RD = 2\text{ k}\Omega$$

POWER GAIN TO THE LOAD RESISTOR:

$$PG = A_v{}^2 \times \frac{Zin}{RL} = 6^2 \times \frac{100\text{ M}\Omega}{6\text{ k}\Omega} = \frac{36 \times 10^8}{6 \times 10^3} = 6 \times 10^5$$

FREQUENCY RESPONSE:

The frequency response solutions are similar to those for the self-biased common source circuit so they are not repeated.

ENHANCEMENT TYPE MOSFET

The enhancement type MOSFET device is a normally "off" device that requires a certain level of turn-on voltage to provide drain current flow. In this regard, enhancement MOSFET biasing and bipolar transistor biasing are similar. However, while the turn-on, foward-biasing voltage for the bipolar is approximately 0.6 volt, biasing for the enhancement MOSFET can be as low as 0.7 volt or as high as 7 volts. The turn-on, or threshold voltage, for the enhancement type MOSFET is given in manufacturing data sheets as V_T or V_p. Also, for enhancement type devices, the V_p voltage is the opposite polarity to that of depletion type devices. Two techniques used to provide biasing for the discrete enhancement type MOSFET are shown in Figure 1-60.

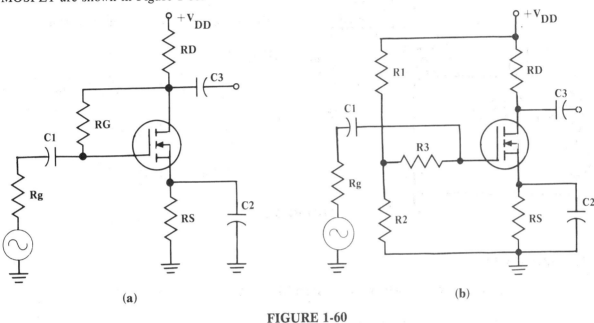

(a)   (b)

FIGURE 1-60

ENHANCEMENT TYPE MOSFET CHARACTERISTICS

The enhancement type MOSFET is operated in the enhancement mode only and it requires a turn-on voltage (V_T) to provide drain current flow. Below the turn-on voltage, no drain current flows, but above the turn-on voltage, the drain current curves have the same square law characteristics as those of any field effect transistor. The characteristics curves for the enhancement type MOSFET are shown in Figure 1-61.

LINEAR DIRECT CURRENT CONSIDERATIONS

The gate-to-source junction of the enhancement type MOSFET must be forward biased to at least a minimum turn-on voltage and beyond to control the amount of drain current flow through the device. One formula for solving for the drain current of the device is $I_D = I_{Don} (1 - V_{GS}/V_T)^2$, where I_{Don} is the saturated value of drain current at the knee of the curve for any of the characteristic drain curves where $V_{DS} = V_{GS} - V_T$. Also, the drain current can be solved from $I_D = K(V_{GS} - V_T)^2$, where K is a function of the device's physical parameters which includes both channel length and width.

Figure 1-61 shows the characteristic drain currents of the enhancement N-Channel type MOSFET device using the $I_D = I_{Don}(1 - V_{GS}/V_T)^2$ formula, where $I_{Don} = 6$ mA, $V_T = 3$ volts, and the gate-to-source voltage is varied from a minimum $V_{GS} = 3$ volts to a maximum of 9 volts. The minimum gate-to-source voltage for threshold condition is 3 volts because $V_{DS} = V_{GS} - V_T$ and $V_{DS} = 3 V - 3 V = 0$ volts. Therefore, the V_{DS} of zero volts demonstrates the limitation, where V_{GS} must be equal to V_T for the threshold voltage condition and greater than V_T for drain current flow. The calculations and circuit representation for the calculated characteristic curves of Figure 1-61 are shown in Figure 1-62.

NOTE: The saturation current for the depletion type MOSFET device is I_{DSS} and it is calculated when $V_{GS} = 0$ volts. For the enhancement type MOSFET device, the saturation curve is I_{Don}, and it is

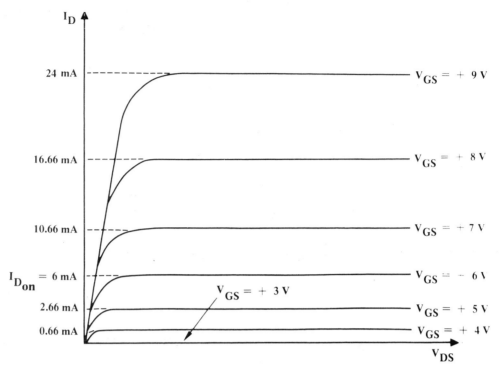

FIGURE 1-61

$V_{GS} = 3$ V $I_D = 6$ mA$(1 - 3/3)^2 = 0$ mA

$V_{GS} = 4$ V $I_D = 6$ mA$(1 - 4/3)^2 = 0.666$ mA

$V_{GS} = 5$ V $I_D = 6$ mA$(1 - 5/3)^2 = 2.666$ mA

$V_{GS} = 6$ V $I_D = 6$ mA$(1 - 6/3)^2 = 6$ mA

$V_{GS} = 7$ V $I_D = 6$ mA$(1 - 7/3)^2 = 10.66$ mA

$V_{GS} = 8$ V $I_D = 6$ mA$(1 - 8/3)^2 = 16.66$ mA

$V_{GS} = 9$ V $I_D = 6$ mA$(1 - 9/3)^2 = 24$ mA

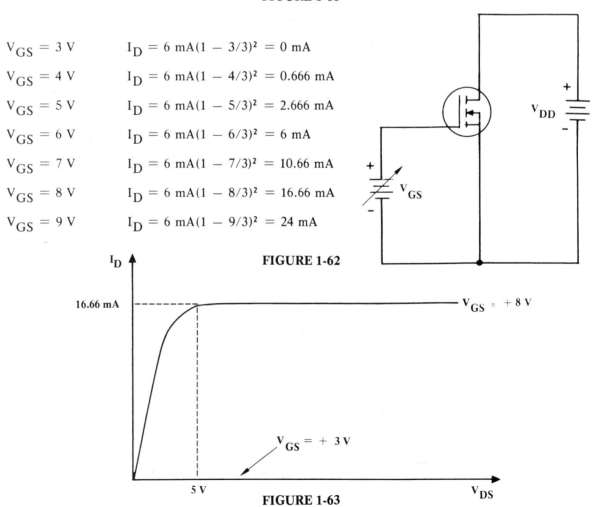

FIGURE 1-62

FIGURE 1-63

48

calculated when $V_{GS} = V_{DS} + V_T$ or, in terms of V_{DS}, as $V_{GS} - V_T$. Therefore, for the enhancement device, like the depletion device, the saturation current is monitored just at the knee of the curve as shown in Figure 1-63.

For instance, with reference to Figure 1-63, where $V_{GS} = 8$ volts and $V_T = 3$ volts, then $V_{DS} = 5$ volts from $V_{DS} = V_{GS} - V_T$. Therefore,

$$I_{Don} = \frac{I_D'}{\left(1 - \frac{V_{GS}}{V_T}\right)^2} = \frac{16.667}{\left(1 - \frac{8\,V}{3\,V}\right)^2} = \frac{16.667}{\left(-\frac{5}{3}\right)^2} = \frac{16.667}{2.777} \approx 6 \text{ mA}$$

This same method of solution can be applied to the other drain characteristic curves of Figure 1-61, and as calculated in Figure 1-62. Therefore, the operating drain current can be found from the device parameters or through interpolation of the characteristice curves of the device.

THE ENHANCEMENT TYPE MOSFET — CIRCUIT ANALYSIS

The direct current and alternating current analysis for the enhancement type MOSFET device, in both the drain feedback and universal circuits, are included in this section. Again, only the common source configurations are analyzed.

THE DRAIN FEEDBACK CIRCUIT USING THE ENHANCEMENT TYPE DEVICE
DIRECT CURRENT ANALYSIS

The direct current biasing for the drain feedback circuit of Figure 1-64 is comparatively simple because the voltage at the drain is the same as the voltage at the gate. The reason little or no voltage is dropped across the gate-to-drain resistor RG is because of the extremely small gate current flow (10^{-14} A) of MOSFET devices. Therefore, even with an RG resistor value of 100 MΩ, the voltage drop across the resistor is minute.

The drain current for the circuit is solved in terms of $I_D = I_{Don}(1 - V_{GS}/V_T)^2$, where both I_{Don} and V_T can be given in data books as I_{DSS} and Vp, respectively, and the only other difference is the polartity change.

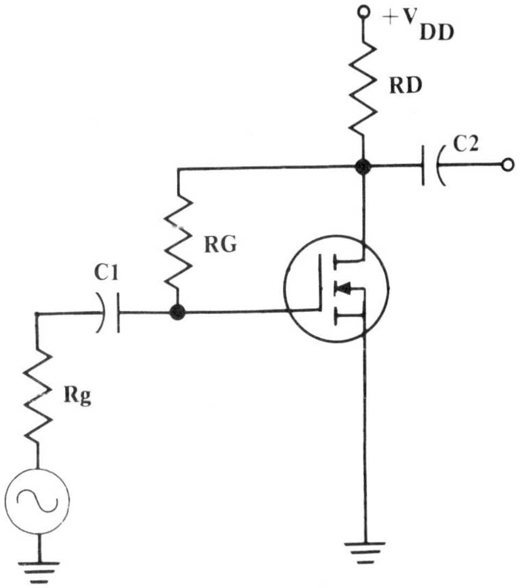

$$I_D = I_{Don}\left(1 - \frac{V_{GS}}{V_T}\right)^2$$

$$V_{DD} = I_D RD + V_{DS}$$

$$V_{DD} = V_{RD} + V_{DS}$$

FIGURE 1-64

The direct current analysis begins by solving for the drain current (I_D) in terms of the device's saturation current (I_{Don}), the device's turn-on voltage (V_T), and the selected gate-to-source voltage (V_{GS}). Next, the voltage drop across the drain resistor V_{RD} is found but, because the voltage at the drain equals the voltage at the gate, the drain resistor value must be selected to reflect the gate-to-source voltage which, in this circuit, is the drain-to-source voltage.

49

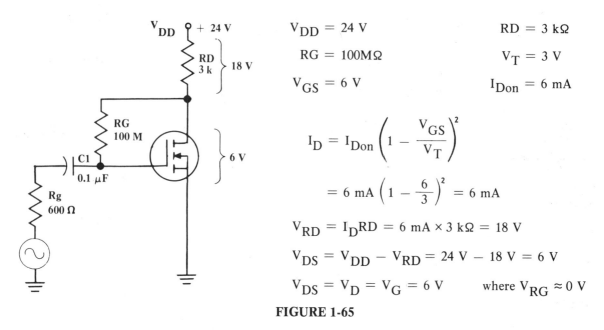

$$V_{DD} = 24 \text{ V} \qquad RD = 3 \text{ k}\Omega$$

$$RG = 100 \text{M}\Omega \qquad V_T = 3 \text{ V}$$

$$V_{GS} = 6 \text{ V} \qquad I_{Don} = 6 \text{ mA}$$

$$I_D = I_{Don}\left(1 - \frac{V_{GS}}{V_T}\right)^2$$

$$= 6 \text{ mA}\left(1 - \frac{6}{3}\right)^2 = 6 \text{ mA}$$

$$V_{RD} = I_D RD = 6 \text{ mA} \times 3 \text{ k}\Omega = 18 \text{ V}$$

$$V_{DS} = V_{DD} - V_{RD} = 24 \text{ V} - 18 \text{ V} = 6 \text{ V}$$

$$V_{DS} = V_D = V_G = 6 \text{ V} \qquad \text{where } V_{RG} \approx 0 \text{ V}$$

FIGURE 1-65

NOTE: For improved circuit stability, a source resistor can be added. However, if the drain current is to be maintained as shown in Figure 1-65, then the drain resistor will have to be changed to accomodate the new voltage drop across the source resistor RS.   This is shown in Figure 1-66.

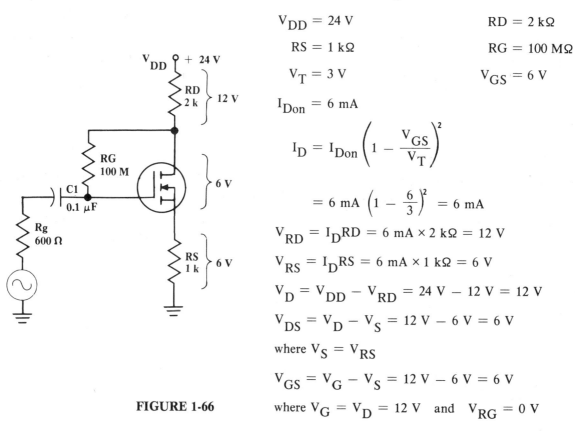

$$V_{DD} = 24 \text{ V} \qquad RD = 2 \text{ k}\Omega$$

$$RS = 1 \text{ k}\Omega \qquad RG = 100 \text{ M}\Omega$$

$$V_T = 3 \text{ V} \qquad V_{GS} = 6 \text{ V}$$

$$I_{Don} = 6 \text{ mA}$$

$$I_D = I_{Don}\left(1 - \frac{V_{GS}}{V_T}\right)^2$$

$$= 6 \text{ mA}\left(1 - \frac{6}{3}\right)^2 = 6 \text{ mA}$$

$$V_{RD} = I_D RD = 6 \text{ mA} \times 2 \text{ k}\Omega = 12 \text{ V}$$

$$V_{RS} = I_D RS = 6 \text{ mA} \times 1 \text{ k}\Omega = 6 \text{ V}$$

$$V_D = V_{DD} - V_{RD} = 24 \text{ V} - 12 \text{ V} = 12 \text{ V}$$

$$V_{DS} = V_D - V_S = 12 \text{ V} - 6 \text{ V} = 6 \text{ V}$$

where $V_S = V_{RS}$

$$V_{GS} = V_G - V_S = 12 \text{ V} - 6 \text{ V} = 6 \text{ V}$$

FIGURE 1-66   where $V_G = V_D = 12 \text{ V}$ and $V_{RG} = 0 \text{ V}$

THE DRAIN FEEDBACK CIRCUIT USING THE ENHANCEMENT TYPE DEVICE
ALTERNATING CURRENT ANALYSIS

Once the direct current voltage drops are known, the alternating current parameters are found. Voltage gain, input impedance, output impedance, peak-to-peak voltage out, and power gain will be determined for this common source configuration only.

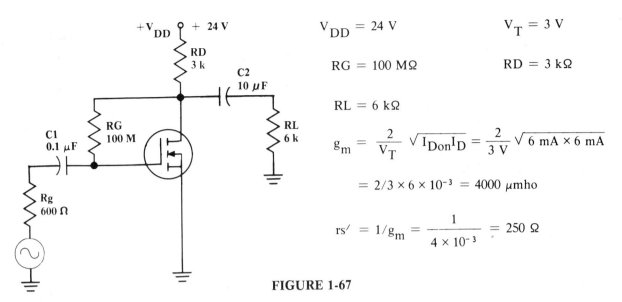

$$V_{DD} = 24 \text{ V} \qquad V_T = 3 \text{ V}$$

$$RG = 100 \text{ M}\Omega \qquad RD = 3 \text{ k}\Omega$$

$$RL = 6 \text{ k}\Omega$$

$$g_m = \frac{2}{V_T} \sqrt{I_{Don}I_D} = \frac{2}{3 \text{ V}} \sqrt{6 \text{ mA} \times 6 \text{ mA}}$$

$$= 2/3 \times 6 \times 10^{-3} = 4000 \ \mu\text{mho}$$

$$rs' = 1/g_m = \frac{1}{4 \times 10^{-3}} = 250 \ \Omega$$

FIGURE 1-67

VOLTAGE GAIN:

$$A_v = g_m(RD \mathbin{/\mkern-5mu/} RL) = 4000 \ \mu\text{mho}(3 \text{ k}\Omega \mathbin{/\mkern-5mu/} 6 \text{ k}\Omega) = 4 \times 10^{-3} \times 2 \times 10^3 = 8$$

$$A_v = \frac{RD \mathbin{/\mkern-5mu/} RL}{rs'} = \frac{3 \text{ k}\Omega \mathbin{/\mkern-5mu/} 6 \text{ k}\Omega}{250 \ \Omega} = \frac{2 \text{ k}\Omega}{250 \ \Omega} = 8$$

VOLTAGE OUT:

$$V_{op\text{-}p} = V_{inp\text{-}p} \times A_v = 1 \text{ Vp-p} \times 8 \text{ Vp-p}$$

POWER GAIN TO THE LOAD RESISTOR:

$$PG = A_v{}^2 \times \frac{Zin}{RL} = 8^2 \times \frac{100 \text{ M}\Omega}{6 \text{ k}\Omega} = \frac{64 \times 10^8}{6 \times 10^3} \approx 10.7 \times 10^5$$

INPUT IMPEDANCE:

$$Zin = RG = 100 \text{ M}\Omega$$

OUTPUT IMPEDANCE:

$$Zo = RD = 3 \text{ k}\Omega$$

FREQUENCY RESPONSE:

The frequency response is similar to that of the self-biased, common source circuit and is not repeated.

UNIVERSAL CIRCUIT USING THE ENHANCEMENT TYPE MOSFET DEVICE DIRECT CURRENT ANALYSIS

The direct current analysis of the universal biased circuit, using the enhancement type MOSFET device, is solved in terms of the drain current and the voltage divider of resistors R1 and R2.

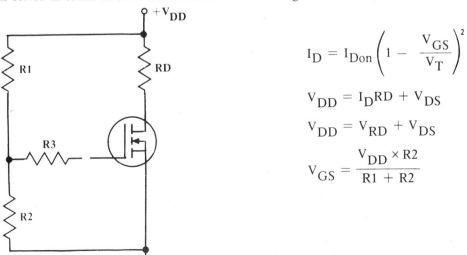

$$I_D = I_{Don}\left(1 - \frac{V_{GS}}{V_T}\right)^2$$

$$V_{DD} = I_D RD + V_{DS}$$

$$V_{DD} = V_{RD} + V_{DS}$$

$$V_{GS} = \frac{V_{DD} \times R2}{R1 + R2}$$

FIGURE 1-68

51

The direct current analysis begins by solving for the drain current I_D in terms of the device's saturation current I_{Don}, the device's turn-on voltage V_T, and the selected gate-to-source voltage V_{GS}. The selected gate-to-source voltage is obtained from the voltage divider circuit of R1 and R2. Next, the voltage drop across the drain resistor V_{RD} and the voltage drop across the drain to source V_{DS} are found. Since little or no current flows through R3, $V_{RS} \approx 0$ volts.

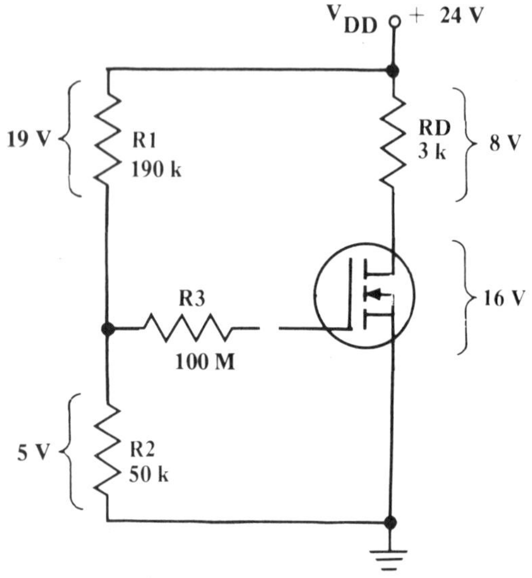

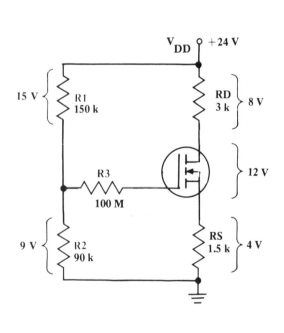

FIGURE 1-69

$$V_{DD} = 24\text{ V} \qquad V_T = 3\text{ V} \qquad RD = 3\text{ k}\Omega$$

$$R1 = 190\text{ k}\Omega \qquad R2 = 50\text{ k}\Omega \qquad R3 = 100\text{ M}\Omega$$

$$I_{Don} = 6\text{ mA}$$

$$I_D = I_{Don}\left(1 - \frac{V_{GS}}{V_T}\right)^2 = 6\text{ mA}\left(1 - \frac{5\text{ V}}{3\text{ V}}\right)^2$$

$$= 6\text{ mA} \times 0.667^2 = 2.66\text{ mA}$$

$$V_{RD} = I_D RD = 2.66\text{ mA} \times 3\text{ k}\Omega \approx 8\text{ V}$$

$$V_D = V_{DD} - V_{RD} = 24\text{ V} - 8\text{ V} = 16\text{ V}$$

$$V_G = V_{R2} = 5\text{ V} \quad\text{and}\quad V_S = 0\text{ V}$$

$$V_{DS} = V_D - V_S = 16\text{ V} - 0\text{ V} = 16\text{ V}$$

$$V_{GS} = V_{R2} = \frac{V_{DD} \times R2}{R1 + R2} = \frac{24\text{ V} \times 50\text{ k}\Omega}{190\text{ k}\Omega + 50\text{ k}\Omega} = 5\text{ V}$$

NOTE: For improved circuit stability, a source resistor can be added as shown in Figure 1-70. However, if the drain current is to be maintained, then the ratio of R1 and R2 must be changed — with regard to the circuit of Figure 1-69.

$$V_{DD} = 24\text{ V} \qquad V_T = 3\text{ V} \qquad RD = 3\text{ k}\Omega$$

$$R1 = 150\text{ k}\Omega \qquad R2 = 90\text{ k}\Omega \qquad R3 = 100\text{ M}\Omega$$

$$R_S = 1.5\text{ k} \qquad\qquad I_{Don} = 6\text{ mA}$$

$$I_D = I_{Don}\left(1 - \frac{V_{GS}}{V_T}\right)^2 = 6\text{ mA}\left(1 - \frac{5\text{ V}}{3\text{ V}}\right)^2$$

$$= 2.66\text{ mA}$$

$$V_{RD} = I_D RD = 2.66\text{ mA} \times 3\text{ k}\Omega \approx 8\text{ V}$$

$$V_{RS} = I_D RS = 2.66\text{ mA} \times 1.5\text{ k}\Omega \approx 4\text{ V}$$

$$V_{R2} = \frac{V_{DD} \times R2}{R1 + R2} = \frac{24\text{ V} \times 90\text{ k}\Omega}{150\text{ k}\Omega + 90\text{ k}\Omega} = 9\text{ V}$$

$$V_D = V_{DD} - V_{RD} = 24\text{ V} - 8\text{ V} = 16\text{ V}$$

$$V_G = V_{R2} = 9\text{ V} \quad\text{and}\quad V_S = V_{RS} = 4\text{ V}$$

$$V_{DS} = V_D - V_S = 16\text{ V} - 4\text{ V} = 12\text{ V}$$

$$V_{GS} = V_G - V_S = 9\text{ V} - 4\text{ V} = 5\text{ V}$$

$$V_{R3} = I_G \times R3 = 10^{-14} \times 10^8 \approx 0\text{ V}$$

FIGURE 1-70

UNIVERSAL CIRCUIT USING THE ENHANCEMENT TYPE MOSFET DEVICE
ALTERNATING CURRENT ANALYSIS

Once the direct current voltage drops are known, the alternating current circuit parameters are found. Voltage gain, peak-to-peak voltage out, input impedance, output impedance, and power gain to the load resistor are solved for the common source circuit configuration only.

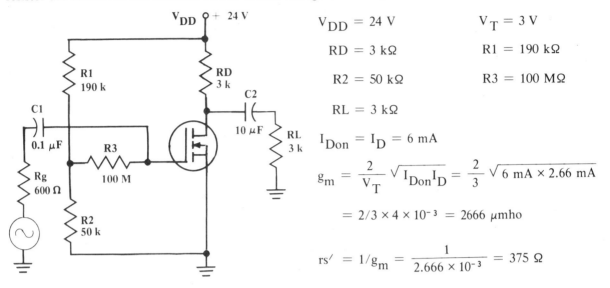

$$V_{DD} = 24 \text{ V} \qquad V_T = 3 \text{ V}$$

$$RD = 3 \text{ k}\Omega \qquad R1 = 190 \text{ k}\Omega$$

$$R2 = 50 \text{ k}\Omega \qquad R3 = 100 \text{ M}\Omega$$

$$RL = 3 \text{ k}\Omega$$

$$I_{Don} = I_D = 6 \text{ mA}$$

$$g_m = \frac{2}{V_T}\sqrt{I_{Don}I_D} = \frac{2}{3}\sqrt{6 \text{ mA} \times 2.66 \text{ mA}}$$

$$= 2/3 \times 4 \times 10^{-3} = 2666 \text{ } \mu\text{mho}$$

$$rs' = 1/g_m = \frac{1}{2.666 \times 10^{-3}} = 375 \text{ } \Omega$$

FIGURE 1-71

VOLTAGE GAIN:

$$A_v = g_m(RD /\!/ RL) = 2666 \text{ } \mu\text{mho} \times (3\text{k}\Omega /\!/ 3 \text{ k}\Omega) = 2.66 \times 10^{-3} \times 1.5 \times 10^3 = 4$$

$$A_v = \frac{RD /\!/ RL}{rs'} = \frac{3 \text{ k}\Omega /\!/ 3 \text{ k}\Omega}{375 \text{ } \Omega} = \frac{1500 \text{ } \Omega}{375 \text{ } \Omega} = 4$$

VOLTAGE OUT:

$$Vop\text{-}p = Vinp\text{-}p \times A_v = 1 \text{ Vp-p} \times 4 = 4 \text{ Vp-p}$$

INPUT IMPEDANCE:

$$Zin = (R1 /\!/ R2) + R3 = (150 \text{ k}\Omega /\!/ 90 \text{ k}\Omega) + 100 \text{ M}\Omega \approx 100 \text{ M}\Omega$$

OUTPUT IMPEDANCE:

$$Zo = RD = 3 \text{ k}\Omega$$

POWER GAIN TO THE LOAD RESISTOR:

$$PG = A_v{}^2 \times \frac{Zin}{RL} = 4^2 \times \frac{100 \text{ M}\Omega}{3 \text{ k}\Omega} = 16 \times \frac{10^8}{3 \times 10^3} \approx 5.33 \times 10^5$$

FREQUENCY RESPONSE:

The frequency response is similar to the self-biased common source circuit and is not repeated.

53

THE FET AS A VOLTAGE CONTROLLED RESISTANCE

In this chapter, the FET has been biased in the operating region where it is best suited for use in amplification functions, However, when biased in the ohmic region, the FET device takes on a whole new set of characteristics, and it can be operated as a voltage controlled resistance.

In the ohmic region, the characteristic curves of the FET are very linear and a constance resistance is provided. This resistance can be varied simply by varying the gate-to-source voltage. The characteristic curves in the ohmic region are shown in Figure 1-72(a) and (b), where a magnified ohmic region illustrates the linearity of the slopes for each of the V_{GS} static curves. Ohmic region applications will be covered in this text as part of the feedback circuit of the Wein bridge oscillator.

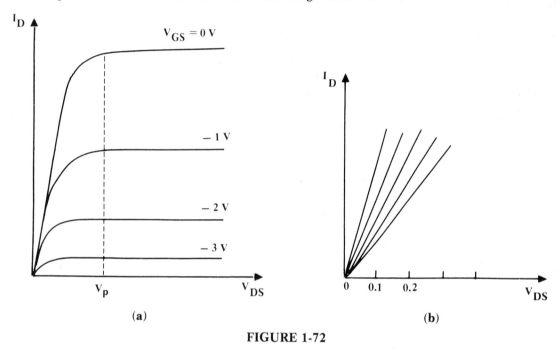

(a)   (b)

FIGURE 1-72

2 | MULTISTAGE DISCRETE HYBRID AND INTEGRATED CIRCUIT AMPLIFIERS

GENERAL DISCUSSION

Multistage linear amplifiers, designed with discrete semiconductor devices, traditionally used bipolar transistors only. However, as field effect transistors became more readily available, they were combined with bipolar transistors to provide simpler circuits (fewer devices) with superior characteristics, compared to circuits designed from all bipolar or all FET devices alone. Additionally, integrated circuits (IC's), ranging from a complete circuit in a single package (op-amps as an example) to an array of discrete devices on a monolithic chip, became available for selection.

In this chapter, two and three stage cascaded, cascode, and differential amplifiers using discrete bipolar and field effect transistor devices, dual gate MOSFET's, and an IC package that utilizes the differential amplifier are presented for analysis.

PART 1 — THE CASCADED AMPLIFIER CIRCUITS

THE COMMON DRAIN, COMMON EMITTER CASCADED AMPLIFIER

The direct coupled amplifier circuit of Figure 2-1, a common drain stage driving a common emitter stage, provides high power gain through both voltage and current gains. The voltage gain is provided by the Q2 stage and the high input impedance by the Q1 stage. The high input impedance can be increased even further by the addition of bootstrapping. Therefore, the circuit is characterized by extremely high input impedance and moderately low output impedance which, even with a modest voltage gain, provides high power gain.

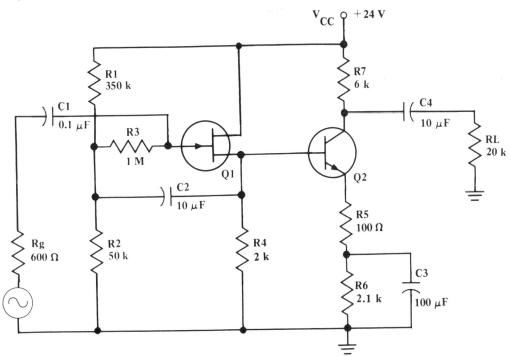

FIGURE 2-1

55

DIRECT CURRENT VOLTAGE ANALYSIS

To begin the direct current analysis of the circuit of Figure 2-1, the voltage drops across the R1 and R2 voltage divider resistors are solved. Then, the voltage drop across R3 is calculated, and it is approximately zero volts because of the extremely low gate current of Q1. Next, the drain current of Q1 is solved in terms of the devices pinch off voltage (V_p) and saturation current (I_{DSS}). Then the voltage drop across R4, the voltage drop across the series R5 and R6 resistors, and the current flow through R_{E2} (R5 + R6), are solved. Next, the voltage drop across R7 is calculated from $(I_{E2})R7$, and then the voltage drops across the devices are found.

1. Solving for the volage drop across R1 from:

$$V_{R1} = \frac{V_{CC} \times R1}{R1 + R2} = \frac{24\,V \times 350\,k\Omega}{350\,k\Omega + 50\,k\Omega} = 21\,V$$

2. Solving for the voltage drop across R2 from:

$$V_{R2} = \frac{V_{CC} \times R2}{R1 + R2} = \frac{24\,V \times 50\,K\Omega}{350\,k\Omega + 50\,k\Omega} = 3\,V$$

3. Solving for the voltage drop across R3 from:

$$V_{R3} = I_{G1} \times R3 = 0.01\,\mu A \times 1\,M\Omega = 10^{-8} \times 10^{-6} = 0.01\,volts$$

NOTE: The gate current of 0.01 μA is a nominal value selected to demonstrate the negligible voltage drop across gate resistor R3.

4. Solving for the drain current of Q1 knowing pinch off voltage and saturation current of the device and selecting V_{GS} at 2 volts, so that the ratio of V_{GS}/V_p can be maintained at 0.5:

$$I_D(Q1) = I_{DSS}\left(1 - \frac{V_{GS}}{V_p}\right)^2 = 10\,mA\left(1 - \frac{2\,V}{4\,V}\right)^2$$

$$= 10\,mA \times 0.5\,V^2 = 10\,mA \times 0.25 = 2.5\,mA$$

NOTE: The V_{GS} value of 2 volts was selected to insure a low current value for Q1. Therefore, knowing the device characteristics of I_{DSS} and V_p, solve for the effective I_D current and then select the R4 resistor value that verifies the needed 5 V voltage drop of V_{R4}. For 2.5 mA, 2 $k\Omega$ is required.

5. Solving for the voltage drop across R4 from:

$$V_{R4} = I_{D1} \times R4 = 2.5\,mA \times 2\,k\Omega = 5\,V$$

NOTE: The 5 volt drop across V_{R4} insures a reverse gate-to-source voltage of 2 volts. Since $V_{R2} = 3$ volts, then V_{R4} must be 5 volts.

6. Solving for the voltage drop across the R5 and R6 series resistance combination:

$$V_{E2} = V_{(R5 + R6)} = V_{S1} - V_{BE2} = 5\,V - 0.6\,V = 4.4\,V$$

7. Solving for the current flow through (R5 + R6):

$$I_{E2} = \frac{V_{E2}}{R_{E2}} = \frac{V_{(R5 + R6)}}{R5 + R6} = \frac{4.4\,V}{100\,\Omega + 2.1\,k\Omega} = 2\,mA$$

8. Solving for the voltage drop across R7:

$$V_{R7} = R7 \times I_{E2} = 6\,k\Omega \times 2\,mA = 12\,V$$

NOTE: The current flow through R7 is $I_{C2} \approx I_{E2}$.

9. Solving for the single point transistor voltages with respect to ground:

$$V_{C2} = V_{CC} - V_{R7} = 24\ V - 12\ V = 12\ V$$

$$V_{E2} = V_{R5} + V_{R6} = 4.4\ V$$

$$V_{B2} = V_{S1} = V_{R4} = 5\ V$$

10. Solving for the voltage drop across the drain-source of Q1:

$$V_{DS1} = V_{D1} - V_{S1} = 24\ V - 5\ V = 19\ V$$

11. Solving for the voltage drop across the collector emitter of Q2:

$$V_{CE2} = V_{C2} - V_{E2} = 12\ V - 4.4\ V = 7.6\ V$$

NOTE: The main drawback to almost any circuit involving field effect transistors, where a constant current source is not used, is that the published device characteristics must be relied upon. This can be circumvented to some degree by "educated guesses", but they are, at best, only "ballpark" derivations. However, if any voltage measurement can be made, that will give the voltage drop across R4, the voltage drops of the circuit can be found precisely. A practical approach to the problem is to replace R4 with a variable resistor and adjust to the exact source voltage required.

ALTERNATING CURRENT CONSIDERATIONS — THE SIGNAL PROCESS

The alternating current peak-to-peak signal from the generator is applied to the gate of Q1 with no effective loss across the 600 ohm internal impedance of the signal generator. The signal is then processed across to the source of Q1, in phase with the input signal and with some signal loss across the gate-to-source junction of the JFET. Next, the signal is processed across the base-emitter junction of Q2, again in phase with the input signal and with only minimal loss across the low base-emitter junction resistance of Q2. The signal is then developed across the series combination of the base-emitter diode resistance of Q2 and the unbypassed resistor R5, and the developed signal current then flows through and is developed across the collector-drain combination resistor R7. At this point, the voltage at the collector is 180° out of phase. (This 180° phase shift is characteristic of all common emitter, or common source, stages.) Therefore, the input signal was processed across the Q1 and Q2 stages, amplified by slightly less than the $(R7\,/\!/\,RL)/(r_{e_2} + R5)$ ratio and obtained an overall 180° phase shift.

ALTERNATING CURRENT ANALYSIS

1. VOLTAGE GAIN: The overall voltage gain of the circuit includes the voltage gain of Q1, the common drain stage, and the voltage gain of Q2, the common emitter stage. Therefore:

$$A_V(\text{overall}) = A_V(Q1) \times A_V(Q2) \approx 0.81 \times 40.8 \approx 33.05, \text{ calculated from:}$$

$$A_V(Q1) = \frac{g_m r_L}{1 + g_m r_L} = \frac{r_L}{rs' + r_L} = \frac{R4\,/\!/\,\beta_2(r_{e_2} + R5)}{rs' + [R4\,/\!/\,\beta_2(r_{e_2} + R5)]}$$

$$= \frac{2\ k\Omega\,/\!/\,100(13\ \Omega + 100\Omega)}{400\ \Omega + [2\ k\Omega\,/\!/\,100(13\ \Omega + 100\ \Omega)]} = \frac{2\ k\Omega\,/\!/\,11.3\ k\Omega}{400\ \Omega + (2\ k\Omega\,/\!/\,11.3\ k\Omega)}$$

$$= \frac{1699}{400 + 1699} \approx 0.81$$

$$\text{where:}\quad rs' = 1/gm = \frac{1}{2500\ \mu mho} = \frac{1}{2.5 \times 10^{-3}} = 400\ \Omega$$

and $\quad g_m = \dfrac{2}{V_p} \sqrt{I_D I_{DSS}} = \dfrac{2}{4 \text{ V}} \sqrt{2.5 \text{ mA} \times 10 \text{ mA}} = 0.5(5 \times 10^{-3}) = 2500 \text{ } \mu\text{mho}$

$$A_v(Q2) = \dfrac{r_L}{r_{e_2} + R5} = \dfrac{R7 \mathbin{/\!/} RL}{r_{e_2} + R5} = \dfrac{6 \text{ k}\Omega \mathbin{/\!/} 20 \text{ k}\Omega}{13 \text{ }\Omega + 100 \text{ }\Omega} = \dfrac{4.62 \text{ k}\Omega}{113 \text{ }\Omega} \approx 40.8$$

2. INPUT IMPEDANCE:

$$Zin \approx [R3 + (R1 \mathbin{/\!/} R2)] \mathbin{/\!/} r_{gs} \approx R3 + (R1 \mathbin{/\!/} R2) = 1 \text{ M}\Omega + (350 \text{ k}\Omega \mathbin{/\!/} 50 \text{ k}\Omega) \approx 1.044 \text{ M}\Omega$$

NOTE: The r_{gs} is normally in the high megohm region, and it can be neglected in calculating the input impedance. However, if r_{gs} is less than 10:1 of the input resistance, $R3 + (R1 \mathbin{/\!/} R2)$, then bootstrapping will improve the effect of r_{gs} and, hence, Zin.

3. OUTPUT IMPEDANCE:

$$Zo = R7 = 5 \text{ k}\Omega$$

4. POWER GAIN TO THE LOAD:

$$PG = A_v{}^2 \times \dfrac{Zin}{RL} = 33.05^2 \times \dfrac{1.044 \text{ M}\Omega}{20 \text{ k}\Omega} \approx 1092.3 \times 52.2 = 57018.06$$

PG in dB: $10 \log 57018.06 = 47.56 \text{ dB}$

5. MAXIMUM PEAK-TO-PEAK OUTPUT VOLTAGE SWING: The maximum peak-to-peak output voltage swing of the Q2 stage with a 20 kΩ load resistor is $2 V_{CE2}$ or $2I_C r_L$, whichever is smaller. Since $2V_{CE2} = 2 \times 7.6 \text{ V} = 15.2 \text{ Vp-p}$ and $2 I_C r_L = 2[2 \text{ mA}(6 \text{ k}\Omega \mathbin{/\!/} 20 \text{ k}\Omega)] = 2(2 \text{ MA} \times 4.62 \text{ k}\Omega) \approx 18.48 \text{ Vp-p}$, then the smaller of the two, 15.2 Vp-p, is the output voltage swing. Note also that the maximum peak-to-peak signal voltage is limited by the Q2 stage, since the signal at the source of Q1 is too small to be effected by the biasing of Q1.

EXPERIMENTAL OBJECTIVES

Investigate the common drain, common emitter cascaded amplifier circuit configuration through analysis and measurements.

LIST OF MATERIALS AND EQUIPMENT

1. Transistors (one each): 2N3904 and 2N3819 (or equivalents)
2. Capacitors (15 volt — one each): 0.1 μF, 10 μF, and 100 μF
3. Resistors (value in ohms — one each except where indicated):
   120 Ω 220 Ω 3.3 kΩ 6.8 kΩ 10 kΩ 47 kΩ 330 kΩ 470 kΩ 10 kΩ (variable)
4. Power Supply (24 volt)
5. VTVM
6. Oscilloscope
7. Signal Generator

EXPERIMENTAL PROCEDURE

1. Construct the common drain, common emitter direct coupled amplifier circuit of Figure 2-2.
2. Calculate the direct current voltage drops around the circuit. Include all transistors and resistors. Calculate From:

(a) $V_{R1} = \dfrac{V_{CC} \times R1}{R1 + R2}$

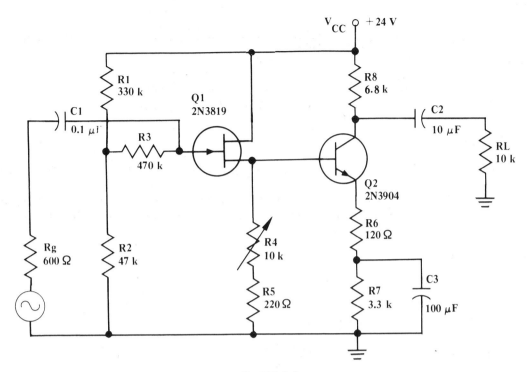

FIGURE 2-2

(b) $VR2 = \dfrac{V_{CC} \times R2}{R1 + R2}$

(c)   Vary resistor R4 and monitor the voltage at the source of Q1 (across the series R4 and R5 resistors), to be approximately 5 volts, taken with respect to ground. Measure the I_D of Q1 by connecting an ammeter in series with the drain, or by measuring the series R4 and R5 resistance, and calculate I_D from:

$$I_D = V(R4 + R5)/(R4 + R5)$$

(d)   The drain current I_D of Q1 can also be solved from the classical $I_D = I_{DSS} (1 - V_{GS}/Vp)^2$ formula, if V_{GS}, I_{DSS}, and Vp are known.   These parameters can be found by using the techniques of Appendix A, where the device is removed from the circuit, or by using similar techniques, but where the device remains in the circuit. If the latter method is used, then use the following steps to measure I_D, I_{DSS}, and V_{GS}. Then calculate $V_{\mathbf{p}}$.
  1.   Connect a milliammeter in series with the drain and monitor I_D.
  2.   Measure the V_{GS}.
  3.   Connect the gate to ground, and also connect the source to ground by connecting a short directly across the series R4 and R5 source resistors. Monitor I_{DSS} on the milliammeter.

  4.   Calculate Vp from: $Vp = \dfrac{V_{GS}}{1 - \sqrt{I_D/I_{DSS}}}$

(e)   Calculate $V_{E2} = V(R6 + R7)$ from: $V_{S1} - V_{BE2}$

(f)   Calculate V_{R8} from: $V_{R8} = I_{E2} \times R8$ where: $I_{E2} = V(R6 + R7)/(R6 + R7)$.

(g)   Solve for the single point transistor voltages, taken with respect to ground, from:

  1. $V_{D1} = V_{CC}$

  2. $V_{E2} = V(R6 + R7) = V_{S1} - V_{BE2}$

59

3. $V_{C2} = V_{CC} - V_{RB}$

(h) Solve for the voltage drop across the drain-source of Q1 from:

$$V_{DS1} = V_{D1} - V_{S1}$$

(i) Solve for the voltage drop across the collector-emitter of Q2 from:

$$V_{CE2} = V_{C2} - V_{E2}$$

3. Measure the DC voltage drops across all the circuit resistors, capacitors, and transistors.
4. Insert the calculates and measured values, as indicated, into Table 2-1.

TABLE 2-1	V_{R1}	V_{R2}	V_{S1}	V_{E2}	V_{GS}	I_{D1}	V_p	I_{DSS}	V_{R8}	V_{DS1}	V_{CE2}
CALCULATED			/////		/////			/////			
MEASURED							/////				

5. Voltage Gain:
   (a) Solve for the (loaded) voltage gain from:

$A_v(overall) = A_v(Q1) \times A_v(Q2)$, calculated from:

$$A_v(Q1) = \frac{r_L}{rs' + r_L}$$

where: $rs' = 1/g_m$ where $g_m = \frac{2}{V_p} \sqrt{I_D I_{DSS}}$

where: $r_L = ([R4 + R5] \mathbin{/\mkern-5mu/} \beta[r_{e_2} + R6])$ where: $\beta = 100$

$$A_v(Q2) = \frac{R8 \mathbin{/\mkern-5mu/} RL}{r_{e_2} + R6}$$

   (b) Measure the loaded voltage gain using an input signal of 100 mVp-p at a frequency of 1 kHz.

6. INPUT IMPEDANCE:

   (a) Calculate the input impedance from: $Zin \approx R3 + (R1 \mathbin{/\mkern-5mu/} R2)$

   (b) Measure the input impedance using the series resistance technique. Use 100 mVp-p at 1 kHz.

NOTE: Care must be used in taking proper measurements. For instance, the high series resistance used can introduce noise to the output signal and 1 MΩ input scope impedance will result in erroneous readings because of loading.

7. POWER GAIN: Calculate the power gain under loaded conditions from:

   $PG = A_v{}^2 \times Zin/RL$ where: $RL = 10$ kΩ

8. OUTPUT VOLTAGE SWING:

   (a) Calculate the output voltage swing for 100 mVp-p of input signal. Calculate from:

   $Vop\text{-}p = A_v \times Vinp\text{-}p$

(b)   Measure the output voltage swing using 100 mVp-p of input signal at 1 kHz.

9.   MAXIMUM PEAK-TO-PEAK OUTPUT VOLTAGE SWING:

(a)   Calculate the maximum peak-to-peak output voltage swing from $2V_{CE2}$ or $2I_C r_L$, whichever is smaller.

NOTE: The pinch-off voltage line of Q1 is not a factor in limiting the peak-to-peak voltage swing, because of the no voltage gain of the Q1 stage.

(b)   Measure the maximum peak-to-peak voltage swing using a 1 kHz input signal voltage.

10.   Insert the calculated and measured values, as indicated, into table 2-2.

TABLE 2-2	VOLTAGE GAIN	Vop-p for 100 mVp-p Input	MAXp-p Vout	Zin	POWER OUT	POWER GAIN in dB
CALCULATED						
MEASURED					/////////	/////////

11.   Plot a frequency response and monitor the upper and lower 3 dB points.

THE COMMON SOURCE, COMMON EMITTER CASCADED AMPLIFIER

The direct coupled amplifier circuit of Figure 2-3, a common source stage driving a common emitter stage driving a common collector stage, provides an amplifier having a high input impedance, low output impedance, and high power gain capabilities. The voltage gain is provided by both Q1 and Q2, while Q3 provides buffering in maintaining the high voltage gain. Q2 and Q3 also provide current gain. Therefore, the circuit is characterized by high input impedance, moderately low output impedance, and excellent power gains.

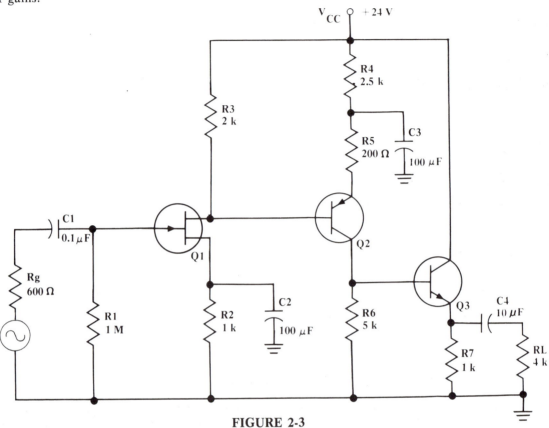

FIGURE 2-3

DIRECT CURRENT VOLTAGE ANALYSIS

To begin the direct current voltage analysis of the circuit of Figure 2-3, calculate the drain current I_D from $I_{D(Q1)} = I_{DSS}(1 - V_{GS}/V_p)^2$, where V_p and I_{DSS} are known and the ratio of V_{GS} to V_p is selected at 0.5, where V_p equals 6 volts and V_{GS} is approximately 3 volts. Next, solve for the voltage drop across R2 from $I_{D1} \times R2$ and the voltage drop across R3 from $I_{D1} \times R3$. The voltage drop across the series R4 and R5 resistors is solved from $V_{R3} - V_{BE2}$ and the emitter current of Q2 from $I_{E2} = V_{(R4 + R5)} / (R4 + R5)$. Then, the voltage drop across R6 is found, approximately, from $I_{E2} \times R6$ and the voltage drop across R7 from $V_{R6} - V_{BE3}$.

1. $I_{D1} = I_{DSS} \left(1 - \dfrac{V_{GS}}{V_p}\right)^2 = 12 \text{ mA} \left(1 - \dfrac{3 \text{ V}}{6 \text{ V}}\right)^2 = 12 \text{ mA} \times 0.5^2 = 3 \text{ mA}$

2. $V_{R2} = I_{D1} \times R2 = 3 \text{ mA} \times 1 \text{ k}\Omega = 3 \text{ V}$

NOTE: The value of R2 was chosen to establish a V_{GS} of 3 volts.

3. $V_{R3} = I_{D1} \times R3 = 2 \text{ k}\Omega \times 3 \text{ mA} = 6 \text{ V}$

4. $V_{(R4 + R5)} = V_{R3} - V_{BE2} = 6 \text{ V} - 0.6 \text{ V} = 5.4 \text{ V}$

62

5. $I_{E2} = I_{(R4 + R5)} = \dfrac{V_{(R4 + R5)}}{R4 + R5} = \dfrac{5.4\text{ V}}{2500\ \Omega + 200\ \Omega} = 2\text{ mA}$

6. $V_{R6} \approx I_{E2} \times R6 = 2\text{ mA} \times 5\text{ k}\Omega = 10\text{ V}$

7. $V_{R7} = V_{R6} - V_{BE3} = 10\text{ V} - 0.6\text{ V} = 9.4\text{ V}$

8. Solving for the single point circuit voltages with respect to ground:

   (a) $V_{D1} = V_{CC} - V_{R3} = 24\text{ V} - 6\text{ V} = 18\text{ V}$

   (b) $V_{S1} = V_{R2} = 3\text{ V}$

   (c) $V_{E2} = V_{CC} - V_{(R4 + R5)} = 24\text{ V} - 5.4\text{ V} = 18.6\text{ V}$

   (d) $V_{C2} = V_{R6} \approx 10\text{ V}$

   (e) $V_{E3} = V_{R7} = 9.4\text{ V}$

   (f) $V_{C3} = V_{CC} = 24\text{ V}$

9. Solving for the voltage drop across the drain-source of Q1:

$$V_{DS1} = V_{D1} - V_{S1} = 18\text{ V} - 3\text{ V} = 15\text{ V}$$

10. Solving for the voltage drop across the collector-emitter of Q2:

$$V_{CE2} = V_{E2} - V_{C2} = 18.6\text{ V} - 10\text{ V} = 8.6\text{ V}$$

11. Solving for the voltage drop across the collector-emitter of Q3:

$$V_{CE3} = V_{C3} - V_{E3} = 24\text{ V} - 9.4\text{ V} = 14.6\text{ V}$$

ALTERNATING CURRENT CONSIDERATIONS — THE SIGNAL PROCESS

The alternating current peak-to-peak signal from the generator is applied to the gate of Q1 with no effective loss across the 600 ohm internal impedance of the signal generator. The signal is then processed to the drain of Q1 (because of the transconductance of the field effect transistor), developed across resistor R3, and processed to the emitter of Q2 with some loss across the base-emitter junction of Q2.

The signal voltage at the emitter of Q2 is developed across the series base-emitter resistance of Q2 and the unbypassed R5 emitter resistance. The signal current is then processed to the collector of Q2, where it is developed across the R6 collector resistance, and it is then processed to the parallel combination of the emitter resistance of Q3 and the load resistance. There is a slight signal (voltage) loss across the base-emitter resistance of Q3.

ALTERNATING CURRENT ANALYSIS

1. VOLTAGE GAIN: The overall loaded voltage gain is found from:

$$A_v(Q1) = \dfrac{R3 \mathbin{/\mkern-4mu/} \beta_2 (r_{e_2} + R5)}{rs'} = \dfrac{2\text{ k}\Omega \mathbin{/\mkern-4mu/} 100(13\ \Omega + 200\ \Omega)}{500\ \Omega} \approx 3.65$$

where: $r_{e_2} = \dfrac{26\text{ mV}}{I_E} = \dfrac{26\text{ mV}}{2\text{ mA}} = 13\ \Omega$ and $rs' = 1/g_m = \dfrac{1}{2 \times 10^{-3}} = 500\ \Omega$

and $g_m = \dfrac{2}{V_p} \sqrt{I_D I_{DSS}} = \dfrac{2}{6\text{ V}} \sqrt{3\text{ mA} \times 12\text{ mA}} = 1/3\text{ V} \times 6\text{ mA} = 2000\ \mu\text{mho}$

63

$$A_v(Q2) = \frac{R6 \mathbin{/\!/} \beta_2[r_{e_3} + (R7 \mathbin{/\!/} RL)]}{r_{e_2} + R5} = \frac{5\ k\Omega \mathbin{/\!/} 100[3\ \Omega + (1\ k\Omega \mathbin{/\!/} 4\ k\Omega)]}{13\ \Omega + 200\ \Omega}$$

$$= \frac{5\ k\Omega \mathbin{/\!/} 80.3\ k\Omega}{213\ \Omega} \approx 22.1$$

where: $r_{e_3} = \dfrac{26\ mV}{I_{E3}} = \dfrac{26\ mV}{9.4\ mA} = 2.76\ \Omega \approx 3\ \Omega$ and $I_{E3} = \dfrac{V_{E3}}{R7} = \dfrac{9.4\ V}{1\ k\Omega} = 9.4\ mA$

$$A_v(Q3) = \frac{R7 \mathbin{/\!/} RL}{r_{e_3} + R7 \mathbin{/\!/} RL} = \frac{1\ k\Omega \mathbin{/\!/} 4\ k\Omega}{3\ \Omega + (1\ k\Omega \mathbin{/\!/} 4\ k\Omega)} = \frac{800\ \Omega}{803\ \Omega} \approx 0.996$$

$$A_v(\text{overall}) = A_v(Q1) \times A_v(Q2) \times A_v(Q3) = 3.65 \times 22.1 \times 0.996 \approx 80.4$$

2. INPUT IMPEDANCE: The input impedance is found from:

$$Zin = R1 \mathbin{/\!/} R_{DS1} \approx R1 = 1\ M\Omega$$

3. POWER GAIN TO THE LOAD:

$$PG = A_{v^2} \times Zin/RL = 80.4^2 \times 1\ M\Omega/4\ k\Omega \approx 1.62 \times 10^6 \approx 62\ dB$$

4. MAXIMUM PEAK-TO-PEAK VOLTAGE SWING: Because of the signal voltage gain, the maximum peak-to-peak output voltage swing considerations are for Q2 and Q3 only.

(a) $Vp\text{-}p(max)Q2 = 2V_{CE2}$ or $2I_{C2}r_L$.

$$2V_{CE2} = 2 \times 8.6\ V = 17.2\ Vp\text{-}p$$

$$2I_{C2}r_L = 2I_{C2}(R6 \mathbin{/\!/} \beta_3[r_{e_3} + (R7 \mathbin{/\!/} RL)]) = 2 \times 2\ mA(5\ k\Omega \mathbin{/\!/} 100[3\ \Omega + (1\ k\Omega \mathbin{/\!/} 4\ k\Omega)])$$

$$\approx 2(2\ mA \times 4.7\ k) \approx 18.8\ Vp\text{-}p$$

(b) $Vp\text{-}p(max)Q3 = 2V_{CE3}$ or $2I_{C3}r_L$

$$2V_{CE3} = 2 \times 14.6\ V = 29.2\ Vp\text{-}p$$

$$2I_{C3}r_L = 2I_{C3}(R7 \mathbin{/\!/} RL) = 2(9.4\ mA \times 800\ \Omega) = 15.04\ Vp\text{-}p \approx 15\ Vp\text{-}p$$

Since the maximum peak-to-peak output voltage swing is limited to the smaller of $2V_{CE}$ or $2I_C r_L$, 15 Vp-p is the maximum peak-to-peak voltage swing.

EXPERIMENTAL OBJECTIVES

Investigate the common source, common emitter, common collector, N-channel, PNP-NPN, amplifier circuit configuration through analysis and measurement.

LIST OF MATERIALS AND EQUIPMENT

1. Transistors (one each or equivalents): 2N3819 (or 2N5951), 2N3906, and 2N3904
2. Capacitors (15 volt): 0.1 μF (one), 10 μF (one), and 100 μF (two)
3. Resistors (value in ohms — one each except where indicated):
   220 Ω 2.2 kΩ (three)   4.7 kΩ (two)   100 kΩ 10 kΩ potentiometer
4. Power Supply (24 volt)
5. VTVM
6. Oscilloscope
7. Signal Generator

EXPERIMENTAL PROCEDURE

1. Construct the common source, common emitter, common collector amplifier circuit as shown in Figure 2-4.

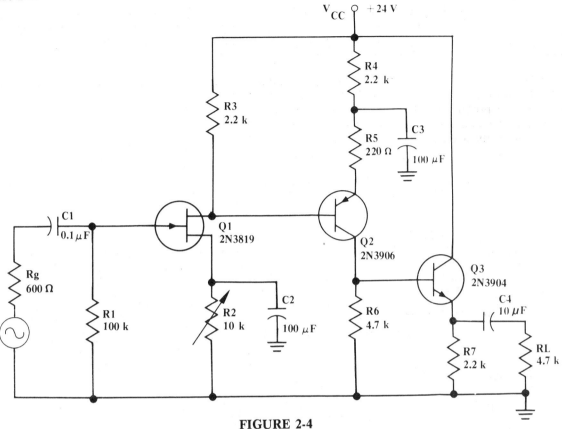

FIGURE 2-4

2. Calculate the DC voltage drops around the circuit. Include all resistors, capacitors, and transistors.
    (a) Adust resistor R2 and monitor approximately 20 volts at the drain terminal of Q1, taken with respect to ground. Measure the resistance of R2 and calculate I_D from $I_D = V_{R3}/R3$.

NOTE: I_D can also be solved from $I_D = I_{DSS}(1 - V_{GS}/V_p)^2$, if the parameters of the device and associated circuitry are known precisely. A nominal value for R2 was indicated in the solved example.

    (b) Calculate V_{R2} from: $\quad V_{R2} = I_{D1} \times R2$

    (c) Calculate V_{R3} from: $\quad V_{R3} = I_{D1} \times R3 \quad$ (verify)

    (d) Calculate $V_{(R4 + R5)}$ from: $\quad V_{(R4 + R5)} = V_{R3} - V_{BE2}$

    (e) Calculate I_{E2} from: $\quad I_{E2} = V_{(R4 + R5)}/(R4 + R5)$

    (f) Calculate V_{R4} from: $\quad V_{R4} = I_{E2} \times R4$

    (g) Calculate V_{R5} from: $\quad V_{R5} = I_{E2} \times R5$

    (h) Calculate V_{R6} from: $\quad V_{R6} \approx I_{E2} \times R6$

    (i) Calculate V_{R7} from: $\quad V_{R7} = V_{R6} - V_{BE3}$

    (j) Calculate the single point voltages with respect to ground from:

$$V_{G1} = V_{R1} \approx 0 \text{ V}$$

$$V_{S1} = V_{R2}$$

$$V_{D1} = V_{CC} - V_{R3}$$

$$V_{E2} = V_{CC} - V_{(R4 + R5)}, \quad \text{or} \quad V_{E2} = V_{D1} + V_{BE2}$$

$$V_{C2} = V_{R6}$$

$$V_{E3} = V_{R7}$$

$$V_{C3} = V_{CC}$$

(k)   Calculate the voltage drop across the drain-source of Q1 from:

$$V_{DS1} = V_{D1} - V_{S1}$$

(i)   Calculate the voltage drop across the collector-emitter of Q2 from:

$$V_{CE2} = V_{E2} - V_{C2}$$

(m)   Calculate the voltage drop across the collector-emitter of Q3 from:

$$V_{CE3} = V_{C3} - V_{E3}$$

3.   Measure the voltage drops around the circuit. Include all of the resistors, capacitors, and transistors.
4.   Insert the calculated and measured values, as indicated, into Table 2-3.

TABLE 2-3	V_{R1}	V_{R2}	V_{R3}	V_{R4}	V_{R5}	V_{R6}	V_{R7}	V_{DS1}	V_{CE2}	V_{CE3}	I_{D1}	V_{GS1}
CALCULATED												
MEASURED												

5.   VOLTAGE GAIN:
   (a)   Calculate the loaded voltage gain from:

$$A_v(\text{overall}) = A_v(Q1) \times A_v(Q2) \times A_v(Q3) \quad \text{where}$$

$$A_v(Q1) = \frac{R3 \mathbin{/\!/} \beta_2(r_{e_2} + R5)}{rs_1'} \quad \text{where:} \quad rs_1' = 1/g_m \quad \text{and} \quad g_m = 2/V_p \sqrt{I_D I_{DSS}}$$

$$A_v(Q2) = \frac{R6 \mathbin{/\!/} \beta_3[r_{e_3} + (R7 \mathbin{/\!/} RL)]}{R5 + r_{e_2}} \quad \text{where:} \quad r_{e_2} = 26 \text{ mV}/I_{E2}$$

$$A_v(Q3) = \frac{R7 \mathbin{/\!/} RL}{r_{e_3} + (R7 \mathbin{/\!/} RL)} \quad \text{where:} \quad r_{e_3} = 26 \text{ mV}/I_{E3}$$

   (b)   Measure the loaded voltage gain using an input voltage of 50 mVp-p at a frequency of 1 kHz.

NOTE: Resistor R1 was deliberately set at 100 kΩ to minimize the possibility of oscillatory conditions. The combination of high input impedance and improper circuit layout (no shielding against positive feedback from the collector of Q2 to the gate of Q1) with moderate in-phase voltage gain, is all that is necessary for unwanted oscillations to occur. In addition, connecting a nominal 100 pF capacitor between the base and emitter of the common collector Q3 stage will help minimize oscillatory conditions due to corner frequency interaction.

6. INPUT IMPEDANCE:
    (a) Calculate the input impedance from:

$$Zin \approx R1$$

    (b) Measure the input impedance using the series resistor technique.

NOTE: A scope impedance of approximately 1 MΩ will effect only slightly the measured Zin.

7. PEAK-TO-PEAK OUTPUT VOLTAGE SWING:

    (a) Calculate the output voltage swing for 50 mVp-p of input from:

$$Vop\text{-}p = A_v \times Vinp\text{-}p$$

    (b) Measure the output voltage swing for 50 mVp-p of input signal at 2 kHz.

8. MAXIMUM PEAK-TO-PEAK VOLTAGE SWING:
    (a) Calculate the maximum peak-to-peak voltage swing from $2I_{C2}r_L$, $2I_{C3}r_L$, $2V_{CE2}$, or $2V_{CE3}$, whichever is smaller.
    (b) Measure the maximum loaded peak-to-peak voltage swing at the load where RL = 4.7 kΩ.

9. Calculate the power gain to the load from:

$$PG = A_v{}^2 \times Zin / RL$$

10. Insert both the calculated and measured values, as indicated, into Table 2-4.

TABLE 2-4	A_v(overall)	A_v			Zin	Vop-p for 50 mVp-p input	Vop-p maximum	POWER
		Q1	Q2	Q3				
CALCULATED								
MEASURED								///////

11. Plot the frequency response and monitor the upper and lower frequency 3 dB points.

PART II — THE CASCODE AMPLIFIER CIRCUITS

THE COMMON SOURCE, COMMON BASE CASCODE AMPLIFIER CIRCUIT

The direct coupled amplifier of Figure 2-5 has a common source stage driving a common base input stage to provide high input impedance, moderate voltage gains, and high frequency response capabilities. The voltage gain is provided by Q2, the bipolar stage, and the high impedance is provided by Q1, the FET stage. This cascode circuit is also designed so that the Miller effect, a cause of minimized high frequency response with gain for both the common source or common emitter stages, is not a factor. This is because the gain of less than one, for the Q1 common source stage, cannot reflect into the input as amplified gate-to-source capacitance. Therefore, frequency response rolloff, because of Miller effect, is negated.

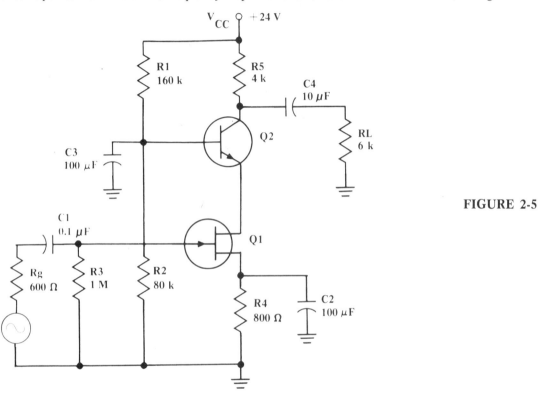

FIGURE 2-5

DIRECT CURRENT VOLTAGE ANALYSIS

To begin the direct current analysis of the circuit of Figure 2-5, consider R1 and R2 as a simple voltage divider and solve for the voltage drop across each. Next, solve for the drain current of Q1 in terms of the device's pinch-off voltage V_p and saturation current I_{DSS}. Then, the voltage drop across R5 and the V_{CE2} and V_{DS1} can be found.

1. $V_{R1} = \dfrac{V_{CC} \times R1}{R1 + R2} = \dfrac{24\ V \times 160\ k\Omega}{160\ k\Omega + 80\ k\Omega} = 16\ V$

2. $V_{R2} = \dfrac{V_{CC} \times R2}{R1 + R2} = \dfrac{24\ V \times 80\ k\Omega}{160\ k\Omega + 80\ k\Omega} = 8\ V$

3. $I_D(Q1) = I_{DSS}(1 - V_{GS}/V_p)^2 = 10\ mA(1 - 2\ V/4\ V)^2 = 10\ mA \times 0.5^2 = 2.5\ mA$

NOTE: The V_{GS} value of 2 volts for a V_{GS}/V_p ratio of 0.5 was selected with known V_p and I_{DSS} device parameters. The value of R4 was then selected to maintain the drain current I_D for this self-baised Q1 stage.

4. $V_{R3} = I_{G1} \times R3 = 0.01\ \mu A \times 1\ M\Omega = 10^{-8} \times 10^6 = 0.01\ V \approx 0\ V$

68

NOTE: The gate current of 0.01 μA or 10^{-8} amps was estimated to demonstrate that little or no voltage is dropped across R3.

5. $V_{R4} = I_{D1} \times R4 = 2.5 \text{ mA} \times 800 \, \Omega = 2 \text{ V}$

NOTE: Since $V_{G1} \approx 0$ V and $V_{R4} = 2$ V, the V_{GS} of -2 V is satisfied.

6. $V_{R5} = I_{D1} \times R5 = 2.5 \text{ mA} \times 4 \text{ k}\Omega = 10 \text{ V}$

7. Solving for the single point circuit voltages with respect to ground:

$$V_{C2} = V_{CC} - V_{R5} = 24 \text{ V} - 10 \text{ V} = 14 \text{ V}$$

$$V_{E2} = V_{D1} = V_{R2} - V_{BE1} = 8 \text{ V} - 0.6 \text{ V} = 7.4 \text{ V}$$

$$V_{S1} = V_{R4} = 2 \text{ V}$$

8. Solving for the voltage drop across the drain-source of Q1:

$$V_{DS1} = V_{D1} - V_{S1} = 7.4 \text{ V} - 2 \text{ V} = 5.4 \text{ V}$$

9. Solving for the voltage drop across the collector-emitter of Q2:

$$V_{CE2} = V_{C2} - V_{E2} = 14 \text{ V} - 7.4 \text{ V} = 6.6 \text{ V}$$

ALTERNATING CURRENT CONSIDERATIONS — THE SIGNAL PROCESS

The alternating current peak-to-peak signal from the generator is applied to the gate of Q1 with no effective loss across the 600 ohm internal impedance of the signal generator. It is then processed to the drain of Q1, because of the transconductances of the field effect transistor, but it has a signal loss as well as a 180° phase shift. The signal is then developed across the base-emitter resistance of Q1 and the signal current is developed across the collector resistor of Q2. Essentially, the voltage gain is a product of $A_V(Q1) \times A_V(Q2) = A_V(\text{overall})$, where Av(overall) also equals either $g_m r_L$ or r_L / rs'.

ALTERNATING CURRENT ANALYSIS

1. VOLTAGE GAIN: The overall voltage gain can be calculated from:

(a) $A_V(\text{overall}) = A_V(Q2) \times A_V(Q1) = 240 \times 0.025 = 6$ where:

$$A_V(Q1) = g_m r_L = g_m r_{e_2} = 2.5 \times 10^{-3} \times 10 = 2.5 \times 10^{-2} = 0.025$$

$$A_V(Q1) = \frac{r_L}{rs'} = \frac{r_{e_2}}{rs'} = \frac{10 \, \Omega}{400 \, \Omega} = 0.025$$

where: $rs' = \dfrac{1}{g_m} = \dfrac{1}{2.5 \times 10^{-3}} = 400\Omega$ and $r_{e_2} = \dfrac{26 \text{ mV}}{I_{E2}} = \dfrac{26 \text{ mV}}{2.5 \text{ mA}} \approx 10 \, \Omega$

$$g_m = \frac{2}{V_p} \sqrt{I_D I_{DSS}} = \frac{2}{4 \text{ V}} \sqrt{2.5 \text{ mA} \times 10 \text{ mA}} = 0.5 \sqrt{25 \times 10^{-6}}$$

$$= 0.5 \times 5 \times 10^{-3} = 2.5 \text{ mmho} = 2500 \, \mu\text{mho}$$

$$A_V(Q2) = \frac{r_L}{r_{e_2}} = \frac{R5 /\!/ RL}{r_{e_2}} = \frac{4 \text{ k}\Omega /\!/ 6 \text{ k}\Omega}{10 \, \Omega} = \frac{2.4 \text{ k}\Omega}{10 \, \Omega} = 240$$

(b) $A_V(\text{overall}) = g_m r_L = g_m (R5 /\!/ RL) = 2.5 \times 10^{-3} \times 2.4 \times 10^3 = 6$

(c) $A_v(\text{overall}) = \dfrac{r_L}{rs'} = \dfrac{R5 \mathbin{/\!/} RL}{rs'} = \dfrac{2400\ \Omega}{400\Omega} = 6$

NOTE: The three methods of solving for A_v were shown to prove that all three techniques are valid.

2. INPUT IMPEDANCE:

$$Zin = R3 \mathbin{/\!/} r_{ds} \approx R3 = 1\ M\Omega$$

3. OUTPUT IMPEDANCE:

$$Zo = R5 = 4\ k\Omega$$

4. POWER GAIN TO THE LOAD:

$$PG = A_v{}^2 \times Zin/RL = 6^2 \times 1\ M\Omega / 6\ k\Omega = 36 \times \dfrac{10^6}{6 \times 10^3} = 6000$$

5. MAXIMUM PEAK-TO-PEAK VOLTAGE SWING: The $Vp\text{-}p(\text{max}) = 2V_{CE2}$ or $2I_C r_L$, whichever is smaller.

$$2V_{CE2} = 2 \times 6.6\ V = 13.2\ Vp\text{-}p$$

$$2I_C r_L = 2 \times 2.5\ mA \times (4k\Omega \mathbin{/\!/} 6\ k\Omega) = 2 \times 2.5\ mA \times 2.4\ k\Omega = 12\ Vp\text{-}p$$

Therefore, the maximum peak-to-peak voltage swing is 12 Vp-p.

EXPERIMENTAL OBJECTIVES

Investigate the common source, common base cascode amplifier configuration through analysis and measurements.

LIST OF MATERIALS AND EQUIPMENT

1. Transistors (one each): 2N3904 and 2N3819 (or 2N5951) or equivalents
2. Capacitors (15 volts): 0.1 μF 10 μF (two)   100 μF
3. Resistors (Value in ohms — one each except where indicated):
     4.7 kΩ 10 kΩ 47 kΩ 100 kΩ 1 MΩ 10 kΩ (potentiometer)
4. Power Supply (24 volts)
5. VTVM
6. Oscilloscope
7. Signal Generator

EXPERIMENTAL PROCEDURE

1. Construct the common source, common base direct couple cascoded amplifier circuit of Figure 2-6.
2. Calculate the DC voltage drops around the circuit. Include all transistors and resistors.

(a) Calculate V_{R1} from $\dfrac{V_{CC} \times R1}{R1 + R2}$

(b) Calculate V_{R2} from $\dfrac{V_{CC} \times R2}{R1 + R2}$

(c) Adjust resistor R4 and monitor 16 volts at the collector terminal of Q2, taken with respect to ground. Measure the resistance of R4 and calculate I_D from: $I_D = V_{R5}/R5$.

NOTE: A nominal value for R4 is indicated in the solved example.

(d) Calculate V_{R4} from $I_D R4$

(e) Calculate V_{R5} from $I_D R5$ (verify)

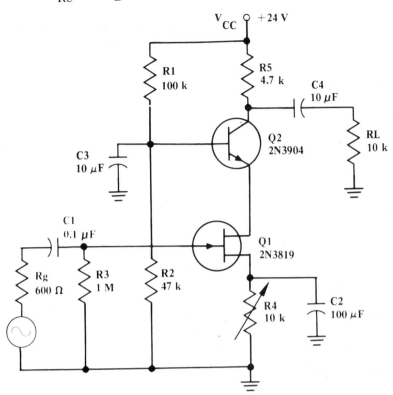

FIGURE 2-6

(f) Calculate the single point circuit voltages, taken with respect to ground, from:

$$V_{B2} = V_{R2}$$

$$V_{C2} = V_{CC} - V_{R5}$$

$$V_{E2} = V_{D1} = V_{R2} - V_{BE2}$$

$$V_{S1} = V_{R4}$$

$$V_{G1} = V_{R3} \approx 0 \text{ V}$$

(g) Calculate the voltage drop across the collector-emitter of Q2 from: $V_{CE2} = V_{C2} - V_{E2}$

(h) Calculate the voltage drop across the drain-source of Q1 from: $V_{DS1} = V_{D1} - V_{S1}$

3. Measure the DC voltage drops around the circuit.   Include all transistors and resistors.
4. Insert the calculated and measured values, as indicated, into Table 2-5.

TABLE 2-5	V_{R1}	V_{R2}	V_{R3}	V_{R4}	I_{D1}	V_{R5}	V_{CE2}	V_{DS1}
CALCULATED								
MEASURED								

71

5. VOLTAGE GAIN:
   (a) Solve for the (loaded) voltage gain from:

$$A_v = g_m r_L \quad \text{where:} \quad r_L = R5 \mathbin{/\mkern-5mu/} RL \quad \text{or} \quad A_v = r_L / rs' \quad \text{where:} \quad rs' = 1/g_m$$

NOTE: $g_m = \dfrac{2}{V_p} \sqrt{I_D I_{DSS}}$ where: I_{DSS} and V_p of the device were previously found

   (b) Measure the loaded voltage gain using an input signal of 100 mVp-p at a frequency of 1 kHz.
6. INPUT IMPEDANCE:
   (a) Calculate the input impedance from $Zin \approx R3 \mathbin{/\mkern-5mu/} Z\text{(equipment)}$
   (b) Measure the input impedance using the series resistance technique. Use 100 mVp-p at 1 kHz.

NOTE: The high series resistance ($\approx$ 1 MΩ) can introduce noise to the output signal. Also, an oscilloscope impedance of 1 MΩ will effect the measured Zin greatly.

7. POWER GAIN: Calculate the power gain under loaded conditions from:

$$PG = A_v{}^2 \times Zin / RL \quad \text{where:} \quad RL = 10 \text{ k}\Omega$$

8. OUTPUT VOLTAGE SWING:
   (a) Calculate the output voltage swing for 100 mVp-p of input signal. Calculate from:

$$Vop\text{-}p = A_v \times Vinp\text{-}p$$

   (b) Measure the output voltage swing using 100 mVp-p of input signal at 1 kHz.
9. MAXIMUM PEAK-TO-PEAK OUTPUT VOLTAGE SWING:
   (a) Calculate the maximum peak-to-peak output voltage swing from $2V_{CE2}$ or $2I_C r_L$, whichever is smaller.

NOTE: $r_L = R5 \mathbin{/\mkern-5mu/} RL$

   (b) Measure the maximum peak-to-peak output voltage swing using a 1 kHz input signal voltage.
10. Insert the calculated and measured values into Table 2-6.

TABLE 2-6	VOLTAGE GAIN	Vop-p for input of 100 mVp-p	MAXp-p Vout	INPUT IMPEDANCE	POWER GAIN
CALCULATED					
MEASURED					/////////

11. Plot a frequency response and monitor the upper and lower 3 dB points.

THE DUAL GATE MOSFET IN THE CASCODE AMPLIFIER CONFIGURATION

The dual gate MOSFET device has all of the parameter advantages of the single gate field effect transistor, and it has the additional advantages of improved AGC capabilities, lower feedback capacitance, better cross modulation characteristics, and higher overall voltage gain. The dual gate MOSFET device is, essentially, two devices or units cascoded in one package as shown in Figure 2-7. Therefore, the device can be connected two ways: (1) where the input signal is applied to gate #1 and, (2) where the input signal is applied to gate #2. In the gate #1 connection, shown in Figure 2-7(b), unit #2 acts as the load for unit #1, and in the gate #2 connection, shown in Figure 2-7(c), unit #1 acts as the source resistance for unit #2.

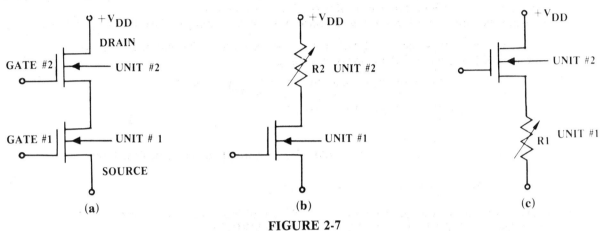

FIGURE 2-7

Of the two connections, the cascode configuration with the signal being applied to gate #1 is the most desirable because it retains all of the advantages of the cascode connection: low feedback capacitance, high frequency response capabilities without neutralization, and overall higher voltage gains. The draw-backs are less AGC capabilities and the dynamic range and cross modulation performances are slightly degraded — with respect to the signal applied to the gate #2 connection.

DIRECT CURRENT VOLTAGE ANALYSIS

For the cascode connection of Figure 2-8(a), the direct current analysis begins by solving for the voltage drop across the voltage divider resistors R1 and R2. Next, the drain current for the device is solved in terms of $I_D = I_{DSS}(1 - V_{GS}/V_p)^2$, where the parameters given in manufacturer data sheets are for V_p test conditions — gate #2 is at 4 volts, gate #1 is at 0 volts, and I_{DSS} is given in terms of gate #1. Next, the voltage across R4 is found from $I_{D1} \times R4$ and the voltage drop across R5 is found from $I_{D1} \times R5$. Then the voltage drop across the drain to source is solved.

1. Solving for the voltage drop across R1 from:

$$V_{R1} = \frac{V_{DD} \times R1}{R1 + R2} = \frac{24 \text{ V} \times 200 \text{ k}\Omega}{200 \text{ k}\Omega + 40 \text{ k}\Omega} = 20 \text{ V}$$

2. Solving for the voltage drop across R2 from:

$$V_{R2} = \frac{V_{DD} \times R2}{R1 + R2} = \frac{24 \text{ V} \times 40 \text{ k}\Omega}{200 \text{ k}\Omega + 40 \text{ k}\Omega} = 4 \text{ V}$$

NOTE: The four (4) volts across R2 establish the manufacturer's test conditions of 4 volts at gate #2. Also, on some devices, as shown in Figure 2-8(b), diode pellets (back-to-back diodes) are fabricated in the same monolithic chip as the MOSFET to protect the gates against electrostatic charges. The protective circuit is used on the CA40673 device, but not on the CA40600 device, as an example.

3. Solving for the device drain current from:

$$I_D = I_{DSS}\left(1 - \frac{V_{GS}}{V_p}\right)^2 = 8 \text{ mA}\left(1 - \frac{0.5}{1 \text{ V}}\right)^2 = 2 \text{ mA}$$

NOTE: The VGS and Vp are given for the gate #1 to source (unit #1) and the IDSS is solved in terms of Unit #1.

73

4. Solving for the voltage drop across R3 from:

$$V_{R3} = I_{G1} \times R3 \approx 0 \text{ mA} \times 5 \text{ M}\Omega = 0 \text{ V}$$

5. Solving for the voltage drop across R4 from:

$$V_{R4} = I_D \times R4 = 2 \text{ mA} \times 250 \text{ }\Omega = 0.5 \text{ V}$$

NOTE: The 250 ohm value was chosen to establish a V_{GS} of 0.5 volts across the source to gate #1 of the device.

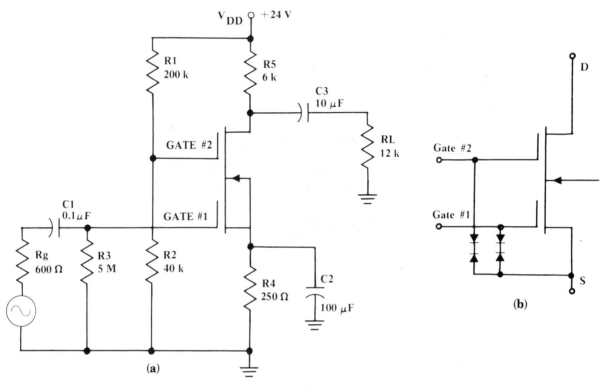

FIGURE 2-8

6. Solving for the voltage drop across R5 from:

$$V_{R5} = I_D \times R5 = 2 \text{ mA} \times 6 \text{ k}\Omega = 12 \text{ V}$$

7. Solving for the single point voltages with respect to ground:

$$V_{D1} = V_{DD} - V_{R5} = 24 \text{ V} - 12 \text{ V} = 12 \text{ V}$$

$$V_{S1} = V_{R4} = 0.5 \text{ V}$$

$$V_{G1} = V_{R3} = 0 \text{ V}$$

$$V_{G2} = V_{R2} = 4 \text{ V}$$

8. Solving for the voltage drop across the drain source of the device from:

$$V_{DS} = V_{D1} - V_{S1} = 12 \text{ V} - 0.5 \text{ V} = 11.5 \text{ V}$$

ALTERNATING CURRENT CONSIDERATIONS — THE SIGNAL PROCESS

The alternating current signal from the generator is applied to gate #1, with no loss across the signal generator, and processed across the gate #1 to the source — in phase with the signal. The signal current is then developed across the drain resistance and amplified by the ratio of (RD $/\!/$ RL) to rs$'$, and 180° out of phase with the input signal. (The amplified gain can also be solved from g_m(RD $/\!/$ RL).)

ALTERNATING CURRENT ANALYSIS

1. VOLTAGE GAIN: Voltage gain can be solved in two ways as follows:

$$A_v = g_m(RD \,/\!/\, RL) = 8000 \;\mu\text{mho} \times (6\;k\Omega \,/\!/\, 12\;k\Omega) = 8 \times 10^{-3} \times 4 \times 10^3 = 32$$

where: $g_m = \dfrac{2}{V_p} \sqrt{I_D I_{DSS}} = \dfrac{2}{1\;V} \sqrt{2\;mA \times 8\;mA} = 2 \times 4 \times 10^{-3} = 8000\;\mu\text{mho}$

$$A_v = \frac{RD \,/\!/\, RL}{rs'} = \frac{6\;k\Omega \,/\!/\, 12\;k\Omega}{125\;\Omega} = \frac{4\;k\Omega}{125\;\Omega} = 32$$

where: $rs' = 1/g_m = \dfrac{1}{8000\;\mu\text{mho}} = \dfrac{1}{8 \times 10^{-3}} = \dfrac{1000\;\Omega}{8} = 125\;\Omega$

NOTE: Transconductance is given in the manufacturer data sheets but it is more accurately calculated from $g_m = 2/V_p \sqrt{I_D I_{DSS}}$ at the operating parameter conditions.

2. INPUT IMPEDANCE: The input impedance is the parallel combination of R3 and the input impedance of the device. However, the input impedance of the device is usually greater than 100 MΩ and can be neglected. Therefore:

$$Z_{in} \approx R3 = 5\;M\Omega$$

3. OUTPUT IMPEDANCE:

$$Z_o = R5 = 6\;k\Omega$$

4. POWER GAIN TO THE LOAD:

$$PG = A_v{}^2 \times \frac{Z_{in}}{RL} = 32^2 \times \frac{5\;M\Omega}{12\;k\Omega} = \frac{1024 \times 5 \times 10^3}{12} = 426{,}666$$

5. MAXIMUM PEAK-TO-PEAK VOLTAGE SWING: The maximum peak-to-peak voltage swing is solved from $2(V_{DS} - V_p + V_{GS})$ or $2I_D r_L$, whichever is smaller. Since $2V_{DS} = 2(11.5\;V - 1\;V + 0.5\;V) = 22$ Vp-p and $2I_D r_L = 2(2\;mA \times 4\;k\Omega) = 16$ Vp-p, the maximum peak-to-peak output swing is 16 Vp-p.

NOTE: The pinch-off voltage line will limit the peak-to-peak voltage swing to less than $2V_{DS}$. A Vp of 1 volt for the dual gate MOSFET is nominal and, if the operating gate-to-source voltage of 0.5 V is included, then the pinch-off line at the operating point is about 0.5 V from $V_p - V_{GS} = 1\;V - 0.5\;V$.

EXPERIMENTAL OBJECTIVES

Investigate the dual gate MOSFET in the cascode configuration.

LIST OF MATERIALS AND EQUIPMENT

1. Dual Gate MOSFET: RCA 40673 or equivalent
2. Capacitors (15 volts — one each):   0.1 μF   10 μF   100 μF
3. Resistors (one each except where indicated):
   5.6 kΩ 10 kΩ 100 kΩ 470 kΩ 1 MΩ 10 kΩ Potentiometer
4. Power Supply (24 volt)
5. VTVM
6. Oscilloscope
7. Signal Generator

EXPERIMENTAL PROCEDURE

1. Construct the dual gate MOSFET cascode amplifier circuit of Figure 2-9.

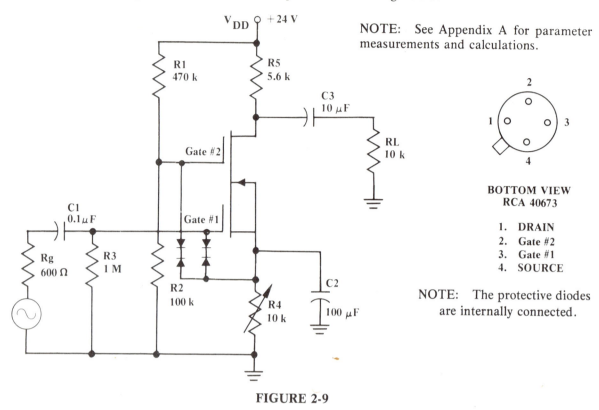

NOTE: See Appendix A for parameter measurements and calculations.

BOTTOM VIEW
RCA 40673

1. DRAIN
2. Gate #2
3. Gate #1
4. SOURCE

NOTE: The protective diodes are internally connected.

FIGURE 2-9

2. Calculate the DC voltage drops around the circuit. Include all the transistors and resistors. Calculate from:

(a) $V_{R1} = \dfrac{V_{DD} \times R1}{R1 + R2}$

(b) $V_{R2} = \dfrac{V_{DD} \times R2}{R1 + R2}$

(c) Vary resistor R4 and monitor 12 V at the drain terminal, taken with respect to ground. Measure the resistance of R4 and calculate I_D from: $I_D = V_{R5}/R5$.

NOTE: A nominal value for R4 in the solved example indicates that a 1 kΩ potentiometer could be used.

(d) Calculate V_{R3} from: $V_{R3} = I_{G1} \times R3$

NOTE: I_{G1} is an extremely low current value which yields approximately zero volts across resistor R3.

(e) Calculate V_{R4} from: $V_{R4} = I_{D1} \times R4$

(f) Calculate V_{R5} from: $V_{R5} = I_{D1} \times R5$ **(verify)**

(g) Solve for the single point transistor voltages taken with respect to ground:

$$V_{D1} = V_{DD} - V_{R5}$$

$$V_{S1} = V_{R4}$$

$$V_{G1} = V_{R3}$$

$$V_{G2} = V_{R2}$$

(h) Solve for the voltage drop across the drain-source of the device from: $V_{DS1} = V_{D1} - V_{S1}$

3. Measure the DC voltage drops across all the circuit transistors, resistors, and capacitors.
4. Insert the calculated and measured values, as indicated, into Table 2-7.

TABLE 2-7	V_{R1}	V_{R2}	V_{R3}	V_{R4}	V_{R5}	V_{D1}	V_{S1}	I_{D1}	V_{G2}	V_{DS}	V_{GS1}
CALCULATED											
MEASURED											

5. VOLTAGE GAIN:
   (a) Solve for the loaded voltage gain from:

$$A_v = g_m(R5 \, /\!/ \, RL) \quad \text{where } g_m = \frac{2}{V_p}\sqrt{I_D I_{DSS}} \quad \text{or from:}$$

$$A_v = \frac{R5 \, /\!/ \, RL}{rs'} \quad \text{where:} \quad rs' = 1/g_m \quad \text{where a nominal } g_m \text{ for the 40673 is 10,000 } \mu mho.*$$

   (b) Measure the loaded voltage gain using an input signal of 100 mVp-p at 1 kHz.
6. INPUT IMPEDANCE:
   (a) Calculate the input impedance from: $Zin \approx R3$
   (b) Measure the input impedance using the series resistance technique.   Use 100 mVp-p at 1 kHz.

NOTE: The high series resistance can introduce noise to the output signal (also, consider scope Z).

7. POWER GAIN: Calculate the power gain to the load from:

$$PG = A_v{}^2 \times Zin / RL \quad \text{where:} \quad A_v \text{ is the loaded voltage gain.}$$

8. OUTPUT VOLTAGE SWING:
   (a) Calculate the output voltage swing for 100 mVp-p of input signal voltage.   Calculate from:

$$Vop\text{-}p = A_v \times Vinp\text{-}p$$

   (b) Measure the output voltage swing using 100 mVp-p of input signal voltage at 1 kHz.
9. MAXIMUM PEAK-TO-PEAK OUTPUT VOLTAGE SWING:
   (a) Calculate the approximate maximum peak-to-peak output voltage swing from $2(V_{DS} - Vp + V_{GS})$ or $2I_D r_L$, whichever is smaller, where $r_L = R5 \, /\!/ \, RL$.
   (b) Measure the maximum peak-to-peak output voltage swing using a 1 kHz input signal voltage.
10. Insert the calculated and measured values, as indicated, into Table 2-8.

TABLE 2-8	VOLTAGE GAIN	POWER GAIN	INPUT IMPEDANCE	Vop-p for 100 mVp-p Input	Vóp-p Max.
CALCULATED					
MEASURED		/////////			

11. Plot a frequency response and monitor the upper and lower 3 dB points.

* **NOTE:** The g_m parameter can also be calculated if I_D, I_{DSS}, and Vp are known. Also, adjusting variable resistor R4 can provide optimum gain.

PART III — THE DIFFERENTIAL AMPLIFIER CIRCUITS

FET DIFFERENTIAL AMPLIFIER — WITH A CONSTANT CURRENT SOURCE

The use of FET devices in the differential amplifier provides high input impedances not normally obtained with bipolar devices. Too, as shown in Figure 2-10, the use of the Q3 bipolar constant current source provides both a controlled current and, because of the high output impedance, a high CMRR. Additionally, a high common mode rejection ratio (CMRR) requires matched devices and, for FET's, the parameters of transconductance (g_m) and gate-to-source voltage (V_{GS}) must be matched. (For bipolar devices, the parameters of beta (β) and base-emitter-diode voltage must be matched.)

An excellent device for constructing the circuit of Figure 2-10 is the dual JFET, which has two devices in one package. It is monolithically constructed to provide excellent parameter matching. The alternative is to use matched discrete devices. However, close matching of discrete devices is difficult to achieve.

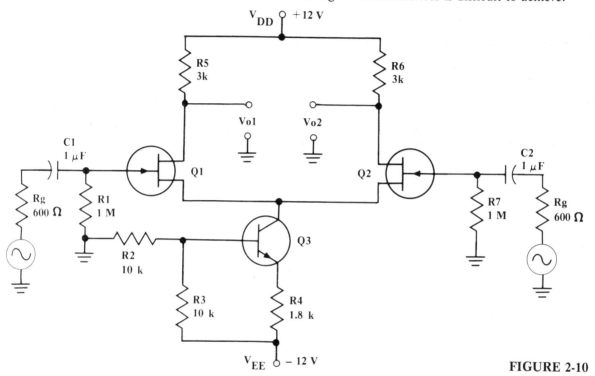

FIGURE 2-10

DIRECT CURRENT VOLTAGE ANALYSIS

To begin the direct current analysis of the circuit of Figure 2-10, solve for the voltage drops across the voltage divider resistors R2 and R3. Then, solve for the voltage drop across R4 from $V_{R3} - V_{BE3}$ and the emitter current flow of Q3 from V_{R4} / I_{R4}. Next, solve for the source current flow of Q1 and Q2 from $I_{E3}/2$ and solve for the voltage drops across R5 and R6 from $I_{D1} \times R5$ and $I_{D2} \times R6$. Since the voltage at the gates of both Q1 and Q2 is at zero volts, the gate-to-source voltage (V_{GS}) can be solved using the device characteristics and the drain current. Then, the voltage drops across all three devices are solved.

1. Solving for the voltage drop across R2:

$$V_{R2} = \frac{V_{EE} \times R2}{R2 + R3} = \frac{12\ V \times 10\ k\Omega}{10\ k\Omega + 10\ k\Omega} = 6\ V$$

2. Solving for the voltage drop across R3:

$$V_{R3} = \frac{V_{EE} \times R3}{R2 + R3} = \frac{12\ V \times 10\ k\Omega}{10\ k\Omega + 10\ k\Omega} = 6\ V$$

3. Solving for the voltage drop across R4:

$$V_{R4} = V_{R3} - V_{BE3} = 6\ V - 0.6\ V = 5.4\ V$$

78

4. Solving for the current flow through R4:

$$I_{E3} = \frac{V_{R4}}{R4} = \frac{5.4\text{ V}}{1800\ \Omega} = 3\text{ mA}$$

5. Solving for the drain current of Q1 and Q2:

$$I_{D1} = I_{D2} = I_{E3}/2 = 3\text{ mA}/2 = 1.5\text{ mA}$$

NOTE: For matched transistors, the current flow of I_{E3} should split equally through Q1 and Q2.

6. Solving for the voltage drop across R5:

$$V_{R5} = I_{D1} \times R5 = 1.5\text{ mA} \times 3\text{ k}\Omega = 4.5\text{ V}$$

7. Solving for the voltage drop across R6:

$$V_{R6} = I_{D2} \times R6 = 1.5\text{ mA} \times 3\text{ k}\Omega = 4.5\text{ V}$$

NOTE: Since the constant current of Q3 splits equally through matched Q1 and Q2 FET's, then the voltage drop across R5 and R6 is equal for equal resistor values.

8. Solving for the gate-to-source voltage with known V_p, I_{DSS}, and I_D:

$$V_{GS} = V_p(1 - \sqrt{I_D/I_{DSS}}) = 4\text{ V }(1 - \sqrt{1.5\text{ mA}/6\text{ mA}}) = 4\text{ V }(1 - \sqrt{0.25})$$

$$= 4\text{ V}(1 - 0.5) = 4\text{ V} \times 0.5 = 2\text{ V}$$

NOTE: Since $I_D = I_{DSS}\left(1 - \dfrac{V_{GS}}{V_p}\right)^2$, then $\dfrac{I_D}{I_{DSS}} = \left(1 - \dfrac{V_{GS}}{V_p}\right)^2$

$$\sqrt{\frac{I_D}{I_{DSS}}} = 1 - \frac{V_{GS}}{V_p}, \quad \text{and} \quad \frac{V_{GS}}{V_p} = 1 - \sqrt{\frac{I_D}{I_{DSS}}}$$

therefore, $\quad V_{GS} = V_p - V_p\sqrt{\dfrac{I_D}{I_{DSS}}}$, and $\quad V_{GS} = V_p\left(1 - \sqrt{\dfrac{I_D}{I_{DSS}}}\right)$

9. Solving for the single point voltages with respect to ground.

$$V_{D1} = V_{DD} - V_{R5} = 12\text{ V} - 4.5\text{ V} = 7.5\text{ V}$$

$$V_{D2} = V_{DD} - V_{R6} = 12\text{ V} - 4.5\text{ V} = 7.5\text{ V}$$

$$V_{S1} = V_{S2} = V_{G2} + V_{GS2} = V_{G1} + V_{GS1} = 0\text{ V} + 2\text{ V} = 2\text{ V}$$

NOTE: $V_{GS1} = V_{GS2}$ is a condition of matched FET devices. The plus 2 volts at the source is necessary to reverse bias the gate-to-source junctions.

$$V_{E3} = V_{EE} - V_{R4} = -12\text{ V} - (-5.4\text{ V}) = -6.6\text{ V}$$

10. Solving for the voltage drop across the drain-source of Q1:

$$V_{DS1} = V_{D1} - V_{S1} = 7.5\text{ V} - 2\text{ V} = 5.5\text{ V}$$

11. Solving for the voltage drop across the drain-source of Q2:

$$V_{DS2} = V_{D2} - V_{S2} = 7.5\text{ V} - 2\text{ V} = 5.5\text{ V}$$

12. Solving for the voltage drop across the collector-emitter of Q3:

$$V_{CE3} = V_{C3} - V_{E3} = 2\,V - (-6.6\,V) = 8.6\,V$$

ALTERNATING CURRENT CONSIDERATIONS — THE SIGNAL PROCESS

The alternating current peak-to-peak signal applied to the high input impedance gate of Q1 is processed to the drain resistance of Q1 with a 180° phase shift and is amplified by the effective drain resistance to the effective source resistances of both Q1 and Q2 — the effective source resistance being the inverse ratio of the Q1 and Q2 transconductances or $1/g_{m_1} + 1/g_{m_2}$. The alternating current peak-to-peak signal applied to the high input impedance gate of Q2 is processed to the drain of Q2 with a 180° phase shift and is also amplified by the ratio of the effective drain resistance of Q2 to the series source resistances of Q1 and Q2. The difference of the output signals of Q1 and Q2 is then calculated and the closer these output signals are, the closer each "half" of the circuit is to each other, and the higher the common mode rejection ratio becomes.

ALTERNATING CURRENT ANALYSIS

1. VOLTAGE GAIN:

(a) $\quad A_v(Q1) = \dfrac{g_{m_1} R_{L_1}}{1 + g_{m_1}/g_{m_2}} = \dfrac{g_{m_1} R5}{1 + g_{m_1}/g_{m_2}} = \dfrac{2.5 \times 10^{-3} \times 3 \times 10^3}{1 + (2.5 \times 10^{-3}/2.5 \times 10^{-3})} = \dfrac{7.5}{1+1} = 3.75$

where: $\quad g_m = 2/V_p \sqrt{I_{DSS} I_D} = 2/4\,V \sqrt{10\,mA \times 2.5\,mA} = 0.5\sqrt{25 \times 10^{-6}}$

$$= 0.5 \times 5 \times 10^{-3} = 2.5 \times 10^{-3} = 2.5\,mmho = 2500\,\mu mho$$

$$A_v(Q1) = \dfrac{R_{L1}}{r_{s_1}' + r_{s_2}'} = \dfrac{R5}{r_{s_1}' + r_{s_2}'} = \dfrac{3\,k\Omega}{400\,\Omega + 400\,\Omega} = 3.75$$

where: $\quad r_{s_1}' = 1/g_{m_1} = \dfrac{1}{2.5 \times 10^{-3}} = 400\,\Omega \quad$ and $\quad r_{s_2}' = 1/g_{m_2} = \dfrac{1}{2.5 \times 10^{-3}} = 400\,\Omega$

NOTE: Since Q1 and Q2 are matched, the transconductance of each is the same. Too, the high output impedance of the constant current source is not used in the calculations. Therefore:

$$A_v(Q1) = \dfrac{g_{m_1} R_{L1}}{1 + g_{m_1}(R_o(Q3)\,/\!/\,1/g_{m_2})} = \dfrac{g_m R_{L1}}{1 + g_{m_1}/g_{m_2}} \quad \text{and equals} \quad \dfrac{g_m R_{L1}}{2} \quad \text{when } g_{m_1} = g_{m_2}, \text{ a}$$

condition of matched devices.

(b) $\quad A_v(Q2) = \dfrac{g_{m_2} R_{L2}}{1 + g_{m_2}/g_{m_1}} = \dfrac{g_{m_1} R6}{2} = \dfrac{2.5 \times 10^{-3} \times 3 \times 10^3}{2} = 3.75$

$$A_v(Q2) = \dfrac{R_{L2}}{r_{s_1}' + r_{s_2}'} = \dfrac{3\,k\Omega}{400\,\Omega + 400\,\Omega} = \dfrac{3\,k\Omega}{800\,\Omega} = 3.75$$

(c) The difference voltage for equal, in-phase, input signals should be zero with the CMRR approaching infinity.

2. COMMON MODE REJECTON RATIO: The common mode rejection ratio is a measurement of how close to exact the Q1 stage is to the Q2 stage. For instance, if 1 Vp-p is applied to both the Q1 and Q2 inputs, and the measured output voltage of Q1 = 3.75 volts and the measured output voltage of Q2 =

3.74 volts, then the CMRR $= \dfrac{V_o}{V_{DIFF}} \approx \dfrac{3.75}{0.01} = 375$ or by $\text{CMRR} = \dfrac{Vo}{V_{DIFF}} \approx \dfrac{3.74}{3.75 - 3.74} = 374.$

If the CMRR is higher then, obviously, the V_{DIFF} is smaller.

3. INPUT IMPEDANCE:
    (a) The input impedance of Q1 is solved from:

    $$Zin(Q1) \approx R_{G1} = 1 \text{ M}\Omega$$

    (b) The input impedance of Q2 is solved from:

    $$Zin(Q2) \approx R_{G2} = 1 \text{ M}\Omega$$

4. OUTPUT IMPEDANCE:
    (a) The output impedance of Q1 is solved from:

    $$Zo_1 = R5 = 3 \text{ k}\Omega$$

    (b) The output impedance of Q2 is solved from:

    $$Zo_2 = R6 = 3 \text{ k}\Omega$$

5. MAXIMUM PEAK-TO-PEAK OUTPUT VOLTAGE SWING: The maximum peak-to-peak output voltage swing is determined by $2(V_{DS2} - V_p + V_{GS})$ or $2 I_D r_L$, whichever is smaller.

$$2(V_{DS1} - V_p + V_{GS}) = 2(5.5 \text{ V} - 4 \text{ V} + 2 \text{ V}) = 7 \text{ Vp-p}$$

$$2I_D r_L = 2(1.5 \text{ mA} \times 3 \text{ k}\Omega) = 9 \text{ Vp-p}$$

Therefore, the maximum peak-to-peak output voltage is ≈ 7 Vp-p, since the pinch-off voltage line can be approximated at a V_{GS} of 2 V to be $V_p - V_{GS}$, and $(V_{DS1} - [V_p - V_{GS}]) = (V_{DS1} - V_p + V_{DS})$.

EXPERIMENTAL OBJECTIVES

Investigate the FET differential amplifier, with a bipolar constant current source, through analysis and measurement.

LIST OF MATERIALS

1. Transistors: J406 (one), 2N3904 (one), or equivalents.
2. Capacitors (15 volt): 0.1 μF (two)
3. Resistors: 2.7 kΩ (one) 4.7 kΩ (two) 10 kΩ (two) 1 MΩ (two)

NOTE: The Siliconix J406 is a low cost monolithic dual JFET, which can be substituted for by another dual N channel device or by two well-matched 2N3319 or 2N5951 discrete devices. However, exact matching of discretes is tedious and not always successful.

EXPERIMENTAL PROCEDURE

1. Construct the differential amplifier circuit of Figure 2-11.
2. Calculate the DC voltage drops around the circuit. Include all resistors and transistors.

    (a) Calculate $V_{R2} = \dfrac{V_{EE} \times R2}{R2 + R3}$ (b) Calculate $V_{R3} = \dfrac{V_{EE} \times R3}{R2 + R3}$

    (c) Calculate $V_{R4} = V_{R3} - V_{BE3}$

    (d) Calculate $I_{E3} = V_{R4}/R4$

    (e) Calculate $I_{S1} = I_{S2} \approx I_{E3}/2$

    (f) Calculate $V_{R5} = I_{S1} \times R5$

(g) Calculate $V_{R6} = I_{S2} \times R6$

(h) Calculate $V_{R1} = V_{R7} \approx 0$ V

(i) Calculate $V_{GS1} \approx V_{GS2} = V_p\left(1 - \sqrt{I_D/I_{DSS}}\right)$

NOTE: $V_{GS1} = V_{GS2}$ only if the devices are matched. Otherwise each device will have separate parameters and separate calculations. If discrete devices are used, the device with the largest V_{GS} will clamp the source voltage, causing a different I_D current to flow in the opposite device and an unbalanced circuit to occur. Monolithic dual FETS minimize this unbalanced condition.

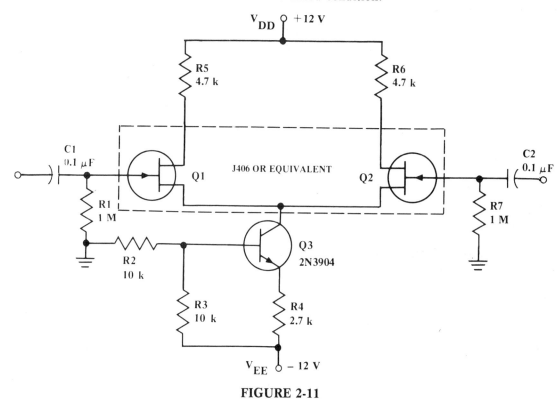

FIGURE 2-11

(j) Solving for the single point voltages with respect to ground:

$$V_{D1} = V_{DD} - V_{R5}$$

$$V_{D2} = V_{DD} - V_{R6}$$

$$V_{C3} = V_{S1} = V_{S2} = V_{R1} - V_{GS1} \quad \text{or} \quad V_{C3} = V_{R7} - V_{GS2}$$

$$V_{E3} = V_{EE} - V_{R4}$$

(k) Solving for the voltage drops across the Q1, Q2, and Q3 transistors:

$$V_{DS1} = V_{D1} - V_{S1}$$

$$V_{DS2} = V_{D2} - V_{S2}$$

$$V_{CE3} = V_{C3} - V_{E3}$$

3. Measure the DC voltage drops around the circuit. Include all the resistors and transistors.
4. Insert the calculated and measured values, as indicated, into Table 2-9.

82

TABLE 2-9	V_{R1}	V_{R2}	V_{R3}	V_{R4}	V_{R5}	V_{R6}	V_{R7}	V_{DS1}	V_{DS2}	V_{CE3}
CALCULATED										
MEASURED										

5. Construct the circuit of Figure 2-12, where the gate of Q2 is terminated in 600 Ω to simulate the loading of another signal generator.

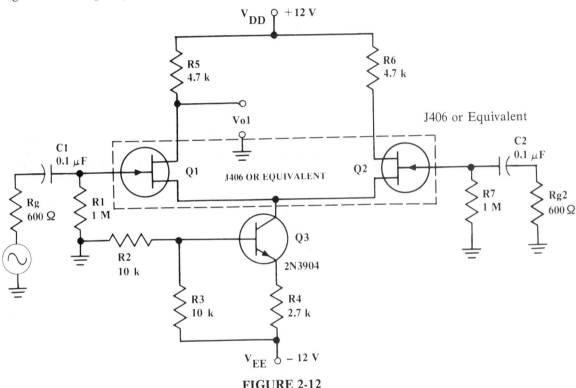

FIGURE 2-12

(b) Calculate the no load voltage gain from:

$$A_v(Q1) = \frac{R5}{r_{s_1}' + r_{s_2}'} \quad \text{where:} \quad r_{s_1}' = 1/g_m \quad \text{and} \quad r_{s_2}' = 1/g_m$$

(c) Measure the voltage gain of Q1 by taking the signal voltage off of the drain of Q1 with respect to ground. The input signal level is 100 mVp-p at 1 kHz.

(d) Construct the circuit of Figure 2-13, where the gate of Q1 is now terminated in 600 Ω to simulate generator loading.

(e) Calculate the no load voltage gain from:

$$A_v(Q2) = \frac{R6}{r_{s_1}' + r_{s_2}'} \quad \text{where:} \quad r_{s_1}' = 1/g_{m_1} \quad \text{and} \quad r_{s_2}' = 1/g_{m_2}$$

(f) Measure the voltage gain of Q2.

6. COMMON MODE REJECTION RATIO (CMMR): Use the measured voltage gain of Q1 and the measured voltage gain of Q2 to calculate the CMRR. Calculate from:

$$CMRR = \frac{Vin \times A_v}{V(\text{difference})} = \frac{Vo}{V(\text{difference})} \quad \text{and convert to dB}$$

83

NOTE: Since the voltage out for both stages, with the same 100 mVp-p input voltage is approximately equal, the voltage out (Vo) for either stage can be used. The voltage difference is the calculated difference of the measured voltages of V_{o_1} and V_{o_2}.

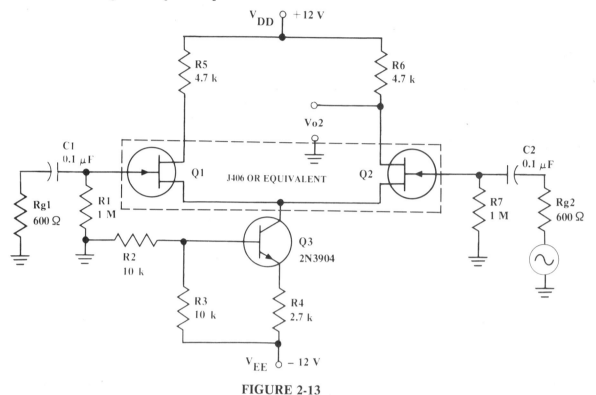

FIGURE 2-13

7. INPUT IMPEDANCE:
   (a) Calculate the input impedance of the Q1 stage from: Zin ≈ R1
   (b) Calculate the input impedance of the Q2 stage from: Zin ≈ R7
   (c) Measure the input impedance of the Q1 and Q2 stages using the known series resistor technique.
8. MAXIMUM PEAK-TO-PEAK OUTPUT VOLTAGE SWING:
   (a) Calculate the maximum peak-to-peak output voltage swing for both the Q1 and Q2 stages. (Since these stages are identical, the peak-to-peak output voltage swing is the same for both.)
   (b) Measure the maximum peak-to-peak output voltage swing for both the Q1 and Q2 stages. Use an input signal at 1 kHz.
9. Calculate and measure the Vop-p for 100 mVp-.p of input signal for both the Q1 and Q2 FET's. Use an input signal at 1 kHz.
10. Insert the calculated and measured values, as indicated, into Table 2-10.

TABLE 2-10	VOLTAGE GAIN Q1	Q2	Vop-p for 100 mVp-p Input Q1	Q2	CMRR in dB	V(max)p-p Output Q1	Q2	INPUT IMPEDANCE Q1	Q2
CALCULATED									
MEASURED					/////				

84

THE INTEGRATED CIRCUIT DIFFERENTIAL AMPLIFIER

Integrated circuits are replacing discrete devices in new circuit design because they occupy less space than the discrete devices and the associated circuitry that they replace, and device parameters can be more closely controlled in monolithic construction. Therefore, in a differential amplifier, for instance, the VBE and beta's will be closely matched and the CMRR will be high. The CA3028 is an example where a "differential pair" and a constant current source with associated circuitry are provided in one package. The circuitry packaged in the CA3028 is shown in Figure 2-14(b). Q1, Q2, Q3, R1, R2, and R3 are all included in the monolithic package.

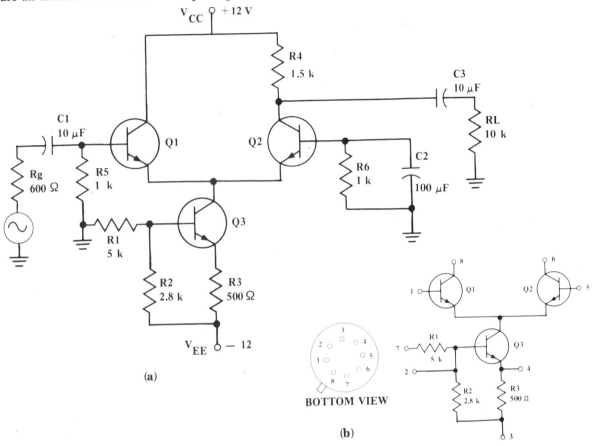

FIGURE 2-14

DIRECT CURRENT VOLTAGE ANALYSIS

To begin the direct current analysis of the differential amplifier circuit of Figure 2-14(a), find the voltage drops across the R1 and R2 voltage dividers. Then, solve for the voltage drop across R3 from $V_{R2} - V_{BE3}$, the emitter current flow of I_{E3} from $V_{R3}/R3$, and the emitter current flow of Q1 and Q2 from $I_{E1} = I_{E2} = I_{E3}/2$. Next, the voltage drop across R4 is found from $I_{E2} \times R4$, and then V_{CE1}, V_{CE2}, and V_{CE3} are solved.

1. Solving for the voltage drop across R1 from:

$$VR1 = \frac{V_{EE} \times R1}{R1 + R2} = \frac{12\,V \times 5\,k\Omega}{5\,k\Omega + 2.8\,k\Omega} = 7.692 \approx 7.7\,V$$

2. Solving for the voltage drop across R2 from:

$$VR2 = \frac{V_{EE} \times R2}{R1 + R2} = \frac{12\,V \times 2.8\,k\Omega}{5\,k\Omega + 2.8\,k\Omega} = 4.307 \approx 4.3\,V$$

3. Solving for the voltage drop across R3 from:

$$V_{R3} = V_{R2} - V_{BE2} = 4.3\ V - 0.6\ V = 3.7\ V$$

4. Solving for the emitter current of Q3 from:

$$I_{E3} = V_{R3}/R3 = 3.7\ V/500\ \Omega = 7.4\ mA$$

5. Solving for the emitter current of Q1 and Q2 from:

$$I_{E1} = I_{E2} = I_{E3}/2 = 7.4\ mA/2 = 3.7\ mA$$

6. Solving for the voltage drop across R4 from:

$$V_{R4} = I_{E2} \times R4 = 3.7\ mA \times 1.5\ k\Omega = 5.55\ V$$

7. Solving for the single point voltages with respect to ground:

$$V_{C1} = V_{CC} = 12\ V$$

$$V_{C2} = V_{CC} - V_{R4} = 12\ V - 5.55\ V = 6.45\ V$$

$$V_{E1} = V_{E2} = V_{C3} \approx -0.6\ V$$

NOTE: There is negligible voltage drop across R5 and R6 and, therefore, the voltage at the emitters of Q1 and Q2 is simply V_{BE1} or V_{BE2} or 0.6 V below ground. Hence, -0.6 volts.

$$V_{E3} = V_{EE} - V_{R3} = -12V - (-3.7\ V) = -8.3\ V$$

8. Solving for the voltage drop across the collector-emitters of Q1, Q2, and Q3 from:

$$V_{CE1} = V_{C1} - V_{E1} = 12\ V - (-0.6\ V) = 12.6\ V$$

$$V_{CE2} = V_{C2} - V_{E2} = 6.45\ V - (-0.6\ V) = 7.05\ V$$

$$V_{CE3} = V_{C3} - V_{E3} = -0.6\ V - (-8.3\ V) = 7.7\ V$$

ALTERNATING CURRENT CONSIDERATIONS — THE SIGNAL PROCESS

As a single-ended amplifier, the signal from the generator is processed to the base of Q1 with some loss across the signal generator. It is then processed, in phase, across the base emitter junction of Q1 and developed across the base emitter junctions of Q1 and Q2. The signal current then flows to the collector of Q2, still in phase, but amplified by the ratios of $(R4 /\!/ RL)/(r_{e_1} + r_{e_2})$. Therefore, the signal is processed by a common colllector stage driving a common base stage, no phase shift occurs, and the amplification is provided by the Q2 common base stage. Also, instead of showing a loss of one half by the Q1 stage and an individual gain by Q2, the total emitter resistance of Q1 and Q2 is lumped, for convenience, since the same result occurs in gain.

ALTERNATING CURRENT ANALYSIS

1. VOLTAGE GAIN: The overall voltage gain of the circuit includes the voltage loss of Q1 and the voltage gain of Q2. However, for convenience, the emitter resistance of Q1 and Q2 can be combined. (Note that a worse case of 52 mV/I_E is used for the monolithic device, instead of 26 mV/I_E, which is normally used for discrete transistors.) Two methods of solution are as follows:

(a) $A_V(Q1) = \dfrac{r_{e_2}}{r_{e_1} + r_{e_2}} = \dfrac{7\ \Omega}{14\ \Omega + 14\ \Omega} = 0.5$ where: $r_{e_1} = r_{e_2} = 52\ mV/3.7\ mA \approx 14\ \Omega$

$A_V(Q2) = \dfrac{r_L}{r_{e_2}} = \dfrac{R4 /\!/ RL}{r_{e_2}} = \dfrac{1.5\ k\Omega /\!/ 10\ k\Omega}{14\ \Omega} \approx \dfrac{1304\ \Omega}{14\ \Omega} \approx 93.14$

$$A_v(\text{overall}) = A_v(Q1) \times A_v(Q2) = 0.5 \times 93.14 = 46.57$$

(b) $\quad A_v(\text{overall}) = \dfrac{r_L}{r_{e_1} + r_{e_2}} = \dfrac{R4 \mathbin{/\mkern-5mu/} RL}{r_{e_1} + r_{e_2}} = \dfrac{1.5\ \text{k}\Omega \mathbin{/\mkern-5mu/} 10\ \text{k}\Omega}{14\ \Omega + 14\ \Omega} = \dfrac{1304\ \Omega}{28\ \Omega} \approx 46,57$

2. INPUT IMPEDANCE: Calculate the input impedance from:

$$Z_{in} = R1 \mathbin{/\mkern-5mu/} \beta_1(r_{e_1} + r_{e_2}) = 1\ \text{k}\Omega \mathbin{/\mkern-5mu/} 100(14\ \Omega + 14\Omega) = 1\ \text{k}\Omega \mathbin{/\mkern-5mu/} 100 \times 28\ \Omega$$

$$= 1\ \text{k}\Omega \mathbin{/\mkern-5mu/} 2.8\ \text{k}\Omega \approx 737\ \Omega$$

3. POWER GAIN TO THE LOAD: Calculate the power gain to the load from:

$$PG = A_v^2 \times Z_{in} / RL = 46.57^2 \times 584\ \Omega / 10\ \text{k}\Omega \approx 159.84$$

4. MAXIMUM PEAK-TO-PEAK OUTPUT VOLTAGE SWING: The maximum peak-to-peak output voltage swing is solved from either $2V_{CE2}$ or $2I_C r_L$, whichever is smaller.

$$2V_{CE2} = 2 \times 7.05 = 14.1\ \text{Vp-p}$$

$$2I_C r_L = 2(3.7\ \text{mA} \times 1.304\Omega) \approx 9.65\ \text{Vp-p} \quad \text{where: } r_L = R4 \mathbin{/\mkern-5mu/} RL = 1.5\ \text{k}\Omega \mathbin{/\mkern-5mu/} 10\ \text{k}\Omega \approx 1304\ \Omega$$

Since $2I_C r_L$ is smaller, 9.65 Vp-p is the maximum peak-to-peak output voltage swing.

EXPERIMENTAL OBJECTIVE

Investigate the single-ended differential amplifier integrated circuit package.

LIST OF MATERIALS AND EQUIPMENT

1. Integrated Circuit: CA3028 or equivalent
2. Capacitors: 10 μF (two)   100 μF (one)
3. Resistors: 1 kΩ (two)   1.8 kΩ (one)   10 kΩ (one)
4. Power Supply: 12 volt (two)
5. VTVM
6. Oscilloscope
7. Signal Generator

EXPERIMENTAL PROCEDURE

1. Construct the single-ended common collector stage driving a common base stage using the CA3028 IC package, as shown in Figure 2-15.
2. Calculate the DC voltage drops around the circuit. Include all of the resistors, capacitors, and transistors. Calculate from:

(a) $\quad V_{R1} = \dfrac{V_{EE} \times R1}{R1 + R2}$

(b) $\quad V_{R2} = \dfrac{V_{EE} \times R2}{R1 + R2}$

(c) $\quad V_{R3} = V_{R2} - V_{BE3}$

(d) $\quad I_{E3} = I_{R3} = V_{R3}/R3$

(e) $\quad I_{E1} = I_{E2} = I_{E3}/2$

(f) $\quad V_{R4} = I_{E2} \times R4$

NOTE: R1, R2 and R3 are nominal resistance values for the monolithic and actual resistance values can differ by as much as 20%.

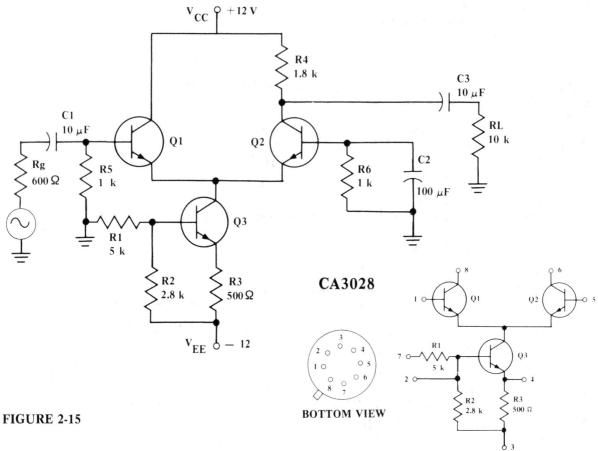

FIGURE 2-15

CA3028

BOTTOM VIEW

(g) Calculate the single point voltages taken with respect to ground:

$$V_{C1} = V_{CC}$$

$$V_{C2} = V_{CC} - V_{R4}$$

$$V_{E1} = V_{E2} = V_{C3} = -V_{BE1} \quad \text{or} \quad -V_{BE2}$$

$$V_{E3} = -V_{EE} - (-V_{R3})$$

$$V_{B1} = R5(I_{E1}/\beta[Q1]) \quad \text{where:} \quad \beta = 100$$

$$V_{B2} = R6(I_{E2}/\beta[Q2]) \quad \text{where:} \quad \beta = 100$$

(h) Solving for the voltage drops across Q1, Q2, and Q3 from:

$$V_{CE1} = V_{CC} = V_{E1} \quad \text{where:} \quad V_{E1} = V_{B1} - V_{BE1}$$

$$V_{CE2} = V_{C2} = V_{E2} \quad \text{where:} \quad V_{E2} = V_{B2} - V_{BE2}$$

$$V_{CE3} = V_{C3} = V_{E2} \quad \text{where:} \quad V_{C3} = V_{E1} = V_{E2}$$

NOTE: The emitters of Q1 and Q2 and the collector of Q3 are not externally connected in the CA3028. Therefore, they must be measured and calculated at the bases of Q1 and Q2, where $V_{BE1} = V_{BE2} \approx$ 0.6 volts.

3. Measure the DC voltage drops across all the circuit resistors, capacitors, and transistors.
4. Insert the calculated and measured values, as indicated, into Table 2-11.

TABLE 2-11	V_{R1}	V_{R2}	V_{R3}	V_{R4}	V_{R5}	V_{R6}	V_{CE1}	V_{CE2}	V_{CE3}
CALCULATED									
MEASURED									

88

5. VOLTAGE GAIN:
   (a) Solve for the voltage gain from:

$$Av = \frac{r_L}{r_{e_1} + r_{e_2}} \quad \text{where:} \quad r_L = R4 \,/\!/\, RL \quad \text{and} \quad r_{e_1} = r_{e_2} = 40 \text{ mV}/I_{E1}$$

   (b) Measure the voltage gain using an input signal of 100 mVp-p at 1 kHz.*
6. INPUT IMPEDANCE:
   (a) Calculate the input impedance from: $\quad Zin = R5 \,/\!/\, \beta_1(r_{e_1} + r_{e_2})$
   (b) Measure the input impedance, using the series resistor technique, at 1 kHz and 100 mVp-p.
7. POWER GAIN TO THE LOAD: Calculate the power gain to the load from:

$$PG = A_v^2 \times Zin/RL$$

8. OUTPUT VOLTAGE SWING:
   (a) Calculate the output voltage swing for 100 mVp-p of input signal voltage a 1 kHz. Calculate from:

$$Vop\text{-}p = A_v \times Vinp\text{-}p$$

   (b) Measure the output voltage swing.
9. MAXIMUM PEAK-TO-PEAK OUTPUT VOLTAGE SWING:
   (a) Calculate the maximum peak-to-peak output voltage swing from $2V_{CE2}$ or $2I_C r_L$, whichever is smaller.
   (b) Measure the maximum peak-to-peak output voltage swing using a 1 kHz input voltage signal.
10. Insert the calculated and measured values into Table 2-13.

TABLE 2-13	VOLTAGE GAIN	INPUT IMPEDANCE	POWER GAIN	Vop-p for 100 mVp-p Input	Vop-p MAX
CALCULATED					
MEASURED			/////////		

11. Plot a frequency response and monitor the upper and lower 3 dB points.

* **NOTE:**   A worse case condition of 52 mA/I_E was used in the solved example , but typical laboratory measured values were found to be about 40 mV/I_E.

89

3 | DIRECT CURRENT POWER SUPPLIES

GENERAL DISCUSSION

Most of today's electronic equipment is powered by DC voltages that are obtained from batteries, DC to DC converters, or AC to DC converters. Both batteries and DC to DC converters imply portability, and the degree of portability is determined by the amount of power consumed. AC to DC converters, on the other hand, are not normally portable since they convert AC line power from the power company to DC power. However, this latter technique for obtaining power for electronic equipment is efficient, convenient, and the method most commonly used.

The AC to DC converter consists of a "bulk" supply and, in most instances, regulating circuitry. This is where the "bulk" supply provides the initial conversion of AC to DC voltage, and where the regulating circuitry then provides an output voltage or current that will remain constant under varying degrees of loading. The regulating circuitry also minimizes the ripple content, that usually exists at the output of the "bulk" supply, from getting to the load.

SECTION I: "BULK" POWER SUPPLIES — A REVIEW

PART 1A — UNREGULATED AC TO DC CONVERTERS

The bulk supply unregulated AC to DC converter consists of a transformer, rectifiers or a bridge rectifier, load resistance, and filtering capacitors. The transformer fills the needs of stepping up or stepping down the line voltage and providing isolation. The diodes, or rectifiers, provide the AC to DC conversion, and the capacitors provide the necessary filtering so that most of the AC voltage can be converted to DC voltage.

The means by which the transformer steps up or steps down the line voltage is simply a ratio of turns on the primary to those of the secondary. For instance, a transformer with 240 turns on the primary and 120 turns on the secondary will step down the line voltage of 120 volts RMS across the primary to 60 volts RMS across the secondary. This is shown in Figure 3-1.

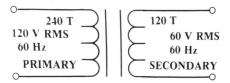

240 T
120 V RMS
60 Hz
PRIMARY

120 T
60 V RMS
60 Hz
SECONDARY

FIGURE 3-1

However, to insure adequate flux cutting, necessary in providing power transformation between primary and secondary transformer windings, iron core material must be used. This is because power transformers operating at line frequencies of 60 Hz can only produce a magnetic field, necessary in the transfer of electrical energy, if close coupling and adequate iron core material are used. Therefore, in power transformers, the secondary is wound directly on top of the primary and the E's and I's of the core material are interleaved to minimize eddy current losses.

An example of a power transformer capable of providing 60 V RMS at 4 amps with a high degree of efficiency is one with 240 turns on the primary, with 120 turns on the secondary, and with three square inches of core material. The bobbin used to wind the wire on the transformer should be 1½ by 2 inches.

Transformer isolation is used to prevent the possibilities of lethal voltages existing between chassis ground and earth ground. The isolation is obtained by physically separating the primary and secondary windings from each other with layers of insulation. High potential voltage tests are usually performed on transformers to insure this isolation.

Autotransformers that use taps in stepping down the line voltage, and AC to DC converters that do not use transformers, have no isolation. They must rely on keyed power plugs to guard against a "hot" chassis condition. This is because the chassis is usually used as the common ground for the circuit, and when power companies provide 120 volts AC, one wire is "hot" at 120 volts AC and the other is at ground condition. Therefore, an inadvertantly reversed plug in the power outlet can cause the chassis to be at 120 volts AC with respect to the power company and earth ground. Some portable TV's, for instance, are designed without transformers to cut down on weight, cost, and size; and these TV's are good examples of "hot" chassis possibilities.

NOTE: If work must be done on the portable (possible) hot chassis TV, where the chassis is exposed, then a simple VOM reading should be made, prior to connecting any grounded test equipment (scope) leads to the chassis. That is, if severe and spectacular ground loops are to be avoided.

One grounded test lead is common practice, with most quality test equipment, where the three prong plug for maximum safety is used. (This is where the power company provides two of the three wires, 120 volts AC and ground, and the user provides an additional safety ground, green in color and connected to a water pipe or ground rod.) The safety ground is then connected to the chassis, references the power company ground, and if at any time the chassis attempts to go "hot" (inadvertently of course), the line fuse will simply blow, until the problem is corrected. Both isolation transformers or keyed plugs can circumvent this problem, but isolation transformers provide the ultimate safety provision.

For rectification purposes, the diode (rectifier) is a device that is an effective switch — a "short" when forward biased and an "open" when reverse biased. See Figure 3-2. Therefore, when a sinusoidal

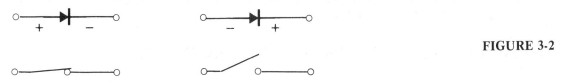

FIGURE 3-2

peak-to-peak wave is applied to the diode, only the positive-going pulses are processed to the load. This is because the positive-going wave turns the diode "on" and the negative-going wave turns the diode "off", as shown in Figure 3-3. However, if the diode were to be reversed, then only the negative-going pulses would be processed to the load.

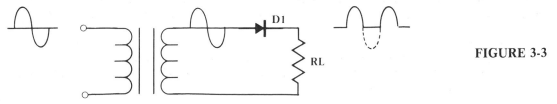

FIGURE 3-3

The reason for processing only one half of the peak-to-peak positive-going and negative-going wave is because the average of the peak-to-peak wave is zero. However, for the positive only or negative only pulse, a zero reference is established and the average voltage is solved by dividing Vpeak by pi (π).

The power company line voltage is normally stated in RMS voltage and can be read off of a standard voltmeter at approximately 120 volts RMS. However, when an oscilloscope is used, the monitored voltage will be in peak-to-peak voltage. Therefore, the relationship between peak-to-peak, peak, and RMS voltage is shown in Figure 3-4, where Vpeak = $\sqrt{2}$ RMS and Vp-p = 2 Vp or, with regard to peak-to-peak voltage, Vp = Vp-p/2 and RMS = $1/\sqrt{2}$ Vp or 0.707 Vp. Hence, 120 volts RMS equals 169.7 Vp and 339.4 Vp-p.

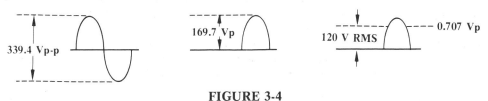

FIGURE 3-4

91

PART 2A — UNFILTERED AC TO DC "BULK" SUPPLIES

THE HALFWAVE RECTIFIER — CIRCUIT ANALYSIS

The halfwave rectifier circuit, shown in Figure 3-5, uses the diode to process only the positive-going pulse to the load. However, at the load, the pulse contains both DC and AC components, where the DC voltage is the average of the pulse and the AC ripple content is the amplitude of the pulse. therefore, the AC ripple content at the load is the remaining AC voltage that has not been converted to DC voltage.

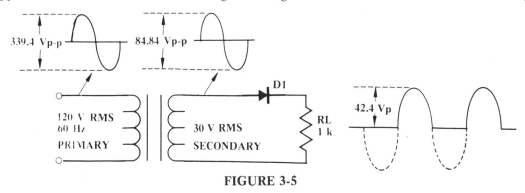

FIGURE 3-5

1. a) Solving for the peak-to-peak voltage across the primary for 120 V RMS of line voltage:

   $$Vp\text{-}p = 2\sqrt{2}\ V\ RMS = (2\sqrt{2})(120\ V) = 339.4\ Vp\text{-}p$$

   b) Solving for the V RMS across the secondary for a 4 : 1 stepdown transformer:

   $$Vsec = Vpri/N = 120\ V\ RMS/4 = 30\ V\ RMS$$

   c) Solving for the Vp-p across the secondary:

   $$Vp\text{-}p(sec) = 2\sqrt{2}\ V\ RMS(sec) = (2\sqrt{2})(30\ V\ RMS) = 84.84\ Vp\text{-}p$$

2. a) Solving for the peak voltage at the load:

   $$Vp = Vp\text{-}p/2 = 84.84\ Vp\text{-}p/2 = 42.42\ Vp$$

NOTE: The peak-to-peak voltage at the secondary is "chopped" in half by the diode and only the peak voltage reaches the load.

   b) Solving for the peak current through the load and, hence, through the series diode:

   $$Ip = Vp/RL = 42.42\ Vp/1\ k\Omega = 42.42\ mA\ peak$$

3. a) Solving for the DC content at the load:

   $$VDC = Vp/\pi = 42.42\ Vp/3.1415 = 13.5\ VDC$$

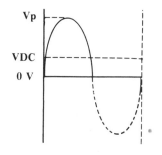 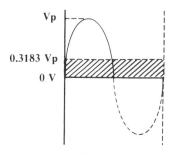 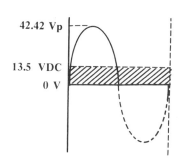

FIGURE 3-6

92

NOTE: The DC voltage at the load is the average voltage for the positive-going pulse when taken over one cycle. Therefore, the VDC is approximately 0.3183 of peak voltage, as shown in Figure 3-6, where $0.3183 = 1/\pi$.

b) The $VDC = Vp/\pi$ formula is derived by using the classical approach of integrating the area under the positive-going pulse, averaged over one complete (2π) cycle.

$$VDC = Vav = \frac{1}{T}\int V\, dt = \frac{1}{2\pi}\int_{0}^{\pi} Vp \sin \omega t\, dt = \frac{Vp}{2\pi}\left[\cos \omega t\right]_{0}^{\pi}$$

$$= \frac{Vp}{2\pi}[-(\cos \pi - \cos 0)] = \frac{Vp}{2\pi}(-\cos \pi + \cos 0) = \frac{Vp}{2\pi}[-(-1)+1] = \frac{Vp}{\pi}$$

c) Another technique used in solving for the DC voltage content at the load, for the halfwave rectifier circuit, is to multiply the RMS voltage value by 0.45. This is where the ratio of the VDC to RMS equal 0.45 from $VDC/V\,RMS = 0.3183/0.707$.

$$IDC = VDC/RL = 13.5\ VDC/1\ k\Omega = 13.5\ mA$$

5. a) Solving for the total RMS voltage at the output load from Vp/2:

$$VRMS = Vp/2 = 42.42\ V/2 = 21.21\ V$$

b) The total RMS at the load can also be solved from:

$$V\,RMS = \sqrt{VDC^2 + VAC(RMS)^2} = \sqrt{13.5^2 + 16.36^2}$$

$$= \sqrt{182.25 + 267.65} = \sqrt{449.9} = 21.21\ V$$

NOTE: The total RMS at the load is comprised of the DC voltage and the RMS value of the AC ripple voltage. Therefore, the V RMS can be solved from the calculus resulting in $V\,RMS = Vp/2$, or it can be solved from the algebraic summation of VDC and VAC(RMS).

6. Solving for the RMS value of the AC ripple voltage at the load:

. $$VAC(RMS) = 0.3856\ Vp = 0.3856 \times 42.42 = 16.36\ V \quad \text{also,}$$

$$VAC(RMS) \approx Vp/2.6 = 42.42/2.6 = 16.32\ V$$

NOTE: The RMS value of the AC ripple voltage is solved, knowing that the total RMS is the algebraic summation of the DC content and the RMS value of the AC ripple voltage at the load. Therefore, manipulating the equations and solving:

$$V\,RMS = \sqrt{VDC^2 + VAC(RMS)^2}, \text{ therefore, } VRMS^2 = VDC^2 + VAC(RMS)^2$$

$$VAC(RMS = \sqrt{V\,RMS^2 - VDC^2} = \sqrt{(Vp/2)^2 - (Vp/\pi)^2} = Vp\sqrt{(\tfrac{1}{2})^2 - (1/\pi)^2}$$

$$= Vp\sqrt{0.5^2 - 0.3183^2} = Vp\sqrt{0.25 - 0.10132} = 0.3856\ Vp$$

Therefore, $VAC(RMS) = 0.3856\ Vp$ or, by using a more convenient approximate number, $VAC(RMS) \approx Vp/2.6$. (The slight difference between 16.36 and the approximate value of 16.32 is well within the standard tolerances of the components being used.)

7. a) Solving for the peak inverse voltage of the diode:

$$PIV = Vp = 42.42\ V$$

NOTE: The PIV is a parameter of the diode, which denotes the maximum reverse voltage that can be applied across the diode before device breakdown occurs. The least expensive diodes have PIV of about 100 volts and diodes rated at 200 PIV to 400 PIV can be obtained for a few pennies more.

b) The peak inverse voltage (PIV) for the diode of the halfwave rectifier circuit of Figure 3-7 takes place when the applied negative-going wave turns diode D1 off causing zero volts to exist at the cathode at the same time as the applied negative-going wave reaches its maximum of −42.42 volts at the anode. At this point in time, 42.42 volts is applied across the diode in a reverse direction, and the PIV parameter of the diode must be higher than 42.42 volts to avoid having the device destroyed. For the halfwave circuit, the PIV occurs 60 times per second.

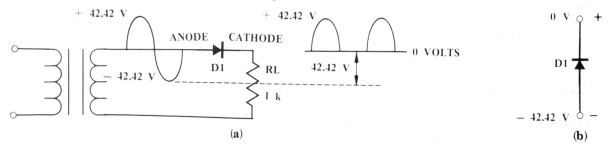

FIGURE 3-7

8. a) Percent ripple is a comparison of the DC voltage content to the RMS value of the AC content at the load. For the ideal power supply, the percent ripple is zero, but for the halfwave rectifier circuit it is 121%. Therefore, solving for the % ripple for the unfiltered halfwave rectifier circuit:

$$\% \text{ ripple} = \frac{\text{VAC(RMS}}{\text{VDC}} \times 100 = \frac{16.36 \text{ V}}{13.5 \text{ V}} \times 100 = 121\%$$

b) Percent ripple can also be solved by using the classical form factor approach:

$$F = \sqrt{\left(\frac{V \text{ RMS}}{VDC}\right)^2 - 1} = \sqrt{\left(\frac{Vp}{2} \Big/ \frac{Vp}{\pi}\right)^2 - 1} = \sqrt{\left(\frac{Vp}{2} \times \frac{\pi}{Vp}\right)^2 - 1}$$

$$= \sqrt{\left(\frac{\pi}{2}\right)^2 - 1} = \sqrt{\left(\frac{3.14}{2}\right)^2 - 1} = \sqrt{1.57^2 - 1} = \sqrt{2.465 - 1} = \sqrt{1.467} = 1.21$$

and, in terms of percent, $1.21 \times 100 = 121 \%$

EXPERIMENTAL OBJECTIVES

The student will investigate the halfwave, unfiltered rectifier circuit.

LIST OF MATERIALS

1. Transformer: 120 V:24 V RMS or equivalent (not exceding 40 volts)
2. Diode: 1N4001 or Equivalent
3. Resistor: 1 kΩ
4. Oscilloscope (Dual if Available)
5. Voltmeter

NOTE: 24 volts is a standard secondary voltage, but any transformer secondary voltage that does not exceed 40 volts can be used successfully.

EXPERIMENTAL PROCEDURE

1. Connect the circuit as shown in Figure 3-8.

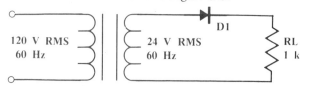

FIGURE 3-8

2. Across the secondary of the transformer:
   a. Measure the V RMS with a voltmeter.
   b) Measure the Vp-p with an oscilloscope.

94

c. Convert the Vp-p scope reading to V RMS from:

$$V\ RMS(sec) = \frac{Vp\text{-}p(sec)}{2\sqrt{2}} \approx \frac{Vp\text{-}p(sec)}{2.828}$$

NOTE: Use the measured values in all of the calculations. Also, calibrate equipment and good measuring techniques will insure proper calculated to measured results.

3. a) Calculate Vp at the load from Vp-p(sec)/2.
   b) **Measure the peak voltage Vp at the load.**
   c) Calculate the peak current (Ip) from: Ip = Vp/RL.
4. **a)** Calculate the DC voltage content at the load from: VDC = Vp/π.
   b) Measure the DC voltage across the load.

NOTE: If the oscilloscope is used to measure the DC voltage, the effect of the AC ripple voltage riding on the DC voltage will be noted.

   c) Calculate the DC current from: IDC = VDC/RL.
   d) Measure the DC current. Use an ammeter or, if the load resistance is accurately known, use the calculations.
5. a) Calculate the RMS value of the AC ripple voltage at the load from: VAC(RMS) = 0.3856 Vp or Vp/2.6.
   b) Measure the RMS value of the AC ripple voltage at the load using a standard RMS voltmeter.
6. a) Calulate the total RMS voltage at the load from:

$$VRMS = Vp/2 \quad \text{or} \quad V\ RMS = \sqrt{VDC^2 + VAC(RMS)^2}$$

NOTE: The total RMS voltage at the load can only be measured accurately if a true RMS meter is used. The standard voltmeter will measure only RMS value of the AC content, while some models of the true RMS voltmeter measure the heating effect of all voltages, both AC and DC, at the load.

7. a) Calculate the peak inverse voltage PIV from: PIV = Vp.
   b) Measure the PIV of D1 with a dual trace oscilloscope. Monitor the output load with one oscilloscope probe and monitor the anode of D1 with the other oscilloscope probe.
8. Calculate the percent ripple at the load from:

$$\%\ ripple = \frac{VAC(RMS)}{VDC} \times 100$$

9. Insert the calculated and measured values into Table 3-1.

TABLE 3-1	Vp-p Sec	V RMS Sec	Vp Load	Ip Load	VDC Load	IDC Load	VAC(RMS) Load	PIV	V RMS Load	%Ripple
CALCULATED	/////									
MEASURED				////					/////////////	

FULL WAVE RECTIFIER CIRCUITS

The advantages of full wave over halfwave rectification are that twice the number of pulses (frequency) exist at the output load, the DC content is doubled, the AC content in decreased, and the percent ripple is lowered. Two methods used to provide full wave rectification are the full wave bridge and the full wave center-tapped rectifier.

THE FULL WAVE CENTER-TAPPED RECTIFIER — CIRCUIT ANALYSIS

The full wave center-tapped rectifier circuit of Figure 3-9 uses two diodes and a "grounded" center-tapped transformer to produce a full-wave condition at the load. The effect of this center-tapped connection is to produce two back-to-back half wave rectifier circuits 180° out of phase with each other. Therefore, the amplitude of the waveshape remains the same, but double the amount of pulses exist at the load. Hence, the average, or DC voltage, is doubled.

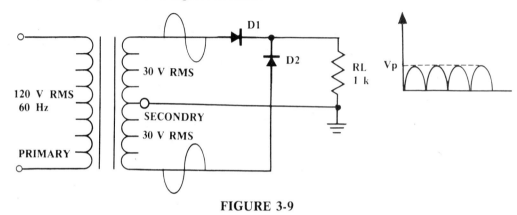

FIGURE 3-9

1. a)   Solving for the p-p voltage at the anode of D1 with respect to the center-tap ground:

   $$Vp\text{-}p = 2\sqrt{2} \text{ V RMS} \approx 2.828 \times 30 \text{ V RMS} = 84.85 \text{ Vp-p}$$

   b)   Solving for the p-p voltage at the anode of D2 with respect to the center-tap ground:

   $$Vp\text{-}p = 2\sqrt{2} \text{ V RMS} \approx 2.828 \times 30 \text{ V RMS} = 84.85 \text{ Vp-p}$$

NOTE: As shown in Figure 3-10, a 180° phase shift difference exists between the p-p voltage at the anode of D1, with respect to the anode of D2.

2. a)   Solving for the peak voltage at the load resulting from D1 and the upper portion (30 V RMS) of the center-tapped transformer:

   $$Vp(D1) = Vp\text{-}p/2 = 84.85 \text{ Vp-p}/2 \approx 42.42 \text{ Vp}$$

   b)   Solving for the peak voltage at the load resulting from D2 and the lower portion (30 V RMS) of the center-tapped transformer:

   $$Vp(D2) = Vp\text{-}p/2 = 84.85 \text{ Vp-p}/2 \approx 42.42 \text{ Vp}$$

NOTE: The peak voltage to the load that is alternately processed by D1 and then by D2 is of the same magnitude as the half wave rectifier circuit, but there are twice as many pulses. Therefore, the frequency increase of the full wave. over the halfwave, doubles — from 60 Hz to 120 Hz.

3. a)   Solving for the peak current through the load resistor:

   $$Ip = Vp/RL = 42.42 \text{ Vp}/1 \text{ k}\Omega = 42.42 \text{ mA}$$

   b)   Solving for the peak current through each diode:

96

$$Ip(D1) = Vp/RL = 42.42\ Vp/1\ k\Omega = 42.42\ mA$$

$$Ip(D2) = Vp/RL = 42.42\ Vp/1\ k\Omega = 42.42\ mA$$

NOTE: This is similar to the halfwave rectifier circuit where the peak current that flows through the load flows through the diode. For the full wave center-tapped rectifier, $Ip(D1) = Ip(D2)$.

4. a) Solving for the DC voltage content of the peak voltage at the load:

$$VDC = 2 \times Vpeak/\pi \approx 2 \times 42.42\ Vp/3.14 = 2 \times 13.5\ V = 27\ V$$

NOTE: The DC voltage is the average of the pulsating positive-going peak voltage and, since twice the number of pulses exist that existed with respect to the half wave, twice as much DC content also exists. The VDC is 0.6366 of the peak voltage for the full wave unregulated rectifier circuit. Therefore, VDC = 0.6366 × 42.42 Vp = 27 VDC, as shown in Figure 3-10.

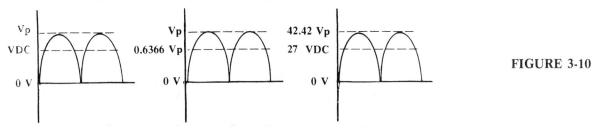

FIGURE 3-10

b) Another technique used in solving for the DC volage content at the load for a full wave rectifier circuit is to multiply the RMS voltage value by 0.9. This is the ratio of VDC to RMS where VDC/RMS = 0.6366/0.707 = 0.9.

$$VDC = 0.9\ V\ RMS = 0.9 \times 30\ V\ RMS = 27\ VDC$$

5. a) Solving for the DC current at the load:

$$IDC = VDC/RL = 27\ V/1\ k\Omega = 27\ mA$$

b) Solving for the DC current through each diode:

$$IDC(D1) = IDC(load)/2 = 27\ mA/2 = 13.5\ mA$$

$$IDC(D2) = IDC(load)/2 = 27\ mA/2 = 13.5\ mA$$

NOTE: Since the full wave center-tapped rectifier circuit is effectively two half wave rectifier circuits 180° removed, each circuit delivers to the load the equivalent of what a halfwave circuit does. The effect is that each diode delivers one half of the DC load current.

6. Solving for the RMS value of the AC ripple voltage at the load:

$$VAC(RMS) = 0.3077\ Vp = 0.3077 \times 42.42\ V = 13.05\ V \quad \text{also,}$$

$$VAC(RMS) = Vp/3.25 = 42.42\ V/3.25 = 13.05\ V$$

NOTE: Rearranging the total V RMS formula and solving for the VAC(RMS) in terms of the full wave VDC and VAC(RMS):

$$V\ RMS = \sqrt{VDC^2 + VAC(RMS)^2} \quad \text{therefore,} \quad V\ RMS^2 = VDC^2 + VAC(RMS)^2$$

$$VAC(RMS) = \sqrt{V\ RMS^2 - VDC^2} = \sqrt{(Vp/\sqrt{2})^2 - (2Vp/\pi)^2}$$

$$= Vp\sqrt{(1/\sqrt{2})^2 - (2/\pi)^2} = Vp\sqrt{0.7071^2 - 0.6366^2} = Vp\sqrt{0.5 - 0.4053} = 0.3077\ Vp$$

Therefore, VAC(RMS) = 0.3077 Vp or, by using a more convenient number, VAC(RMS) = Vp/3.25.

7. a) Solvng for the total RMS at the output load from $V_p/\sqrt{2}$.

$$V\ RMS = V_p\ /\sqrt{2} \approx 42.42\ V\ /\ 1.414 = 29.995\ V \approx 30\ V$$

b) The total RMS at the load can also be solved from:

$$V\ RMS = \sqrt{VDC^2 + VAC(RMS)^2} = \sqrt{27\ V^2 + 13.05\ V^2}$$

$$= \sqrt{729 + 170.3} = \sqrt{899.3} = 29.98\ V \approx 30\ V$$

NOTE: Since both pulses are processed to the load during full wave rectification, the applied V RMS equals the V RMS at the load.

8. Solving for the peak inverse voltage (PIV) of diodes D1 and D2:

$$PIV(D1) = 2\ V_p = 2 \times 42.42\ V \approx 84.84\ V$$

$$PIV(D2) = 2\ V_p = 2 \times 42.42\ V \approx 84.84\ V$$

NOTE: For the full wave center-tapped rectifier circuit, the postive-going wave, processed by Diode D1, coincides with the negative-going wave at the anode of diode D2. Therefore, at that point in time, a peak 42.42 V exists at the output load and a −42.42 V peak exists at the anode of D2, for 84.84 V of reverse biased voltage across D2. Likewise, a positive-going 42.42 V processed by D2 conincides with the negative-going −42.42 V at the anode of D1, causing a PIV for D1 of 84.84 V. This 180° phase shift difference, the output voltage, and the effective PIV are shown in Figure 3-11.

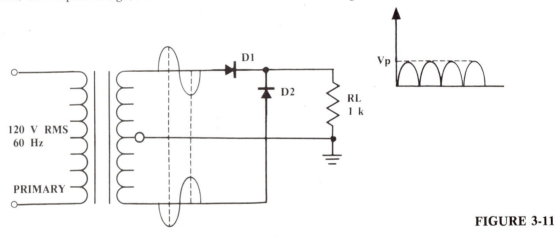

FIGURE 3-11

9. a) Percent ripple is a comparison of the DC voltage content to the RMS value of the AC ripple voltage at the load. For the full-wave rectifier circuit, it is 48.3%. Therefore, solving for the percent ripple for the unfiltered full-wave rectifier circuit:

$$\%\ Ripple = VAC(RMS)/VDC \times 100 = 13.5/27\ V \times 100 = 48.3\%.$$

b) Percent ripple can also be solved by using the classical form factor approach where:

$$F = \sqrt{\left(\frac{V\ RMS}{VDC}\right)^2 - 1} = \sqrt{\left(\frac{V_p}{\sqrt{2}}\Big/\frac{2\ V_p}{\pi}\right)^2 - 1} = \sqrt{\left(\frac{V_p}{\sqrt{2}} \times \frac{\pi}{2\ V_p}\right)^2 - 1}$$

$$= \sqrt{\left(\frac{\pi}{2\sqrt{2}}\right)^2 - 1} = \sqrt{\left(\frac{3.14}{2.828}\right)^2 - 1} = \sqrt{1.11^2 - 1} \approx \sqrt{1.234 - 1} \approx 0.483$$

and, in terms of percent, $0.483 \times 100 = 48.3\%$. Also,

$$F = \sqrt{\left(\frac{V\,RMS}{VDC}\right)^2 - 1} = \sqrt{\left(\frac{0.707}{0.6366}\right)^2 - 1} = \sqrt{1.11^2 - 1} = 0.483$$

and, in terms of percent, $0.483 \times 100 = 48.3\%$

EXPERIMENTAL OBJECTIVES

The student will investigate the full wave center-tapped unfiltered rectifier circuit.

LIST OF MATERIALS

1. Transformer: 120 V : 24 V CT or equivalent (not to exceed 50 V CT)
2. Diode: 1N4001 or Equivalent
3. Resistor: 1 kΩ
4. Oscilloscope (Dual if Available)
5. Voltmeter

EXPERIMENTAL PROCEDURE

1. Connect the circuit as shown in Figure 3-12.

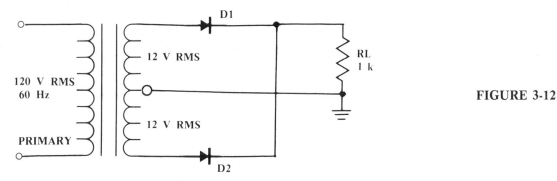

FIGURE 3-12

2. Across the full secondary of the transformer:
   a) Measure the V RMS with a voltmeter.
   b) Measure the Vp-p with an oscilloscope.
3. Across the upper half of the transformer secondary (center-tap ground and the anode of D1):
   a) Measure the V RMS with a voltmeter.
   b) Measure the Vp-p with a scope.
4. Across the lower half of the transformer secondary (center-tap ground and the anode of D2):
   a) Measure the V RMS with a voltmeter.
   b) Measure the Vp-p with an oscilloscope.

NOTE: The center-tapped transformer should provide equal peak-to-peak voltages to be applied to the anodes of diodes D1 and D2. However, if the center-tap is slightly off, then the peak pulses at the load will be, alternately, of slightly different amplitudes.

5. a) Calculate Vp at the load from: Vp = Vp-p/2 (where the center is close to exact)
   b) Measure the Vp at the load.
6. Calculate Ip from: Ip = Vp/RL
7. a) Calculate VDC at the load from: VDC = 2 × Vp/π
   b) Measure the VDC at the load.
8. a) Calculate the DC current flow through the load resistor from: IDC = VDC/RL
   b) Measure the DC current flow through the load resistor. Measure with an ammeter or by monitoring the voltage drop across the known load resistor.
9. a) Calculate the DC current flow through diodes D1 and D2 from: ID1 = ID2 = IRL/2.

NOTE: The total DC current through the load is shared by D1 and D2.

b) Measure the current flow through each diode. Measure by inserting a known 10 ohm resistor in series with each diode and monitoring the voltage drop across each or use ammeters. The circuit using the series resistor technique is shown in Figure 3-13. Another method is to use an ammeter.

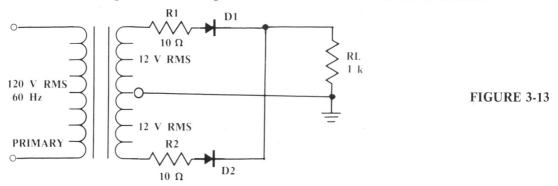

FIGURE 3-13

10. a) Calculate the RMS value of the AC ripple voltage at the load from: VAC(RMS) = 0.3077 Vp or Vp/3.25.

   b) Measure the RMS value of the AC ripple voltage, at the load, using a standard RMS voltmeter.

11. Calculate the total RMS voltage at the load from:

$$V\ RMS = Vp/\sqrt{2} \quad or \quad V\ RMS = \sqrt{VDC^2 + VAC(RMS)^2}.$$

12. a) Calculate the PIV of Diodes D1 and D2 from: PIV(D1) = PIV(D2) = 2 Vp.

   b) Measure the PIV with a dual scope. Monitor the output with one scope probe and, alternately, monitor the anode of D1, and then D2, with the other scope probe.

13. Calculate the percent ripple from:

$$\%\ ripple = \frac{VAC(RMS)}{VDC} \times 100$$

14. Insert the calculated and measured values, as indicated, into Table 3-2.

| TABLE 3-2 | Vp-p | | | V RMS | | | Vp | IP |
	Full Sec	Upper Sec	Lower Sec	Full Sec	Upper Sec	Lower Sec	Load	Load
CALCULATED	/////	/////	/////	/////	/////	/////	/////	
MEASURED								/////

	VDC Load	Load	IDC D1	D2	VAC (RMS)	V RMS Total	PIV D1-D2	% Ripple
CALCULATED								
MEASURED						/////		/////

100

THE FULL WAVE BRIDGE RECTIFIER

The full wave bridge rectifier circuit, such as that shown in Figure 3-14, uses four diodes and a non-center tapped transformer secondary to provide full wave rectification to the load. Therefore, it has the same output as that provided by the full wave center-tapped rectifier but it does not require a center-tapped transformer and the PIV rating of the diodes need be only half that of those used in the latter circuit. The only minor drawback is that two extra diodes are required.

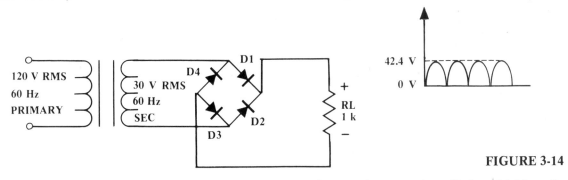

FIGURE 3-14

The full wave operation is best illustrated by applying DC input voltages to show diode switiching. See Figure 3-15(a). The positive battery voltage turns on diodes D1 and D3 and turns off diodes D2 and D4 to provide a series circuit of diodes D1, D3, and the load resistor RL, as shown in Figure 3-15(b). Then a

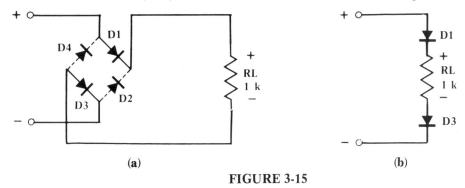

(a)   (b)

FIGURE 3-15

negative battery voltage turns on diodes D2 and D4 and turns off diodes D1 and D3. See Figure 3-16(a). This provides a series circuit of diodes D2, D4, and load resitor RL, as shown in Figure 3-16(b).

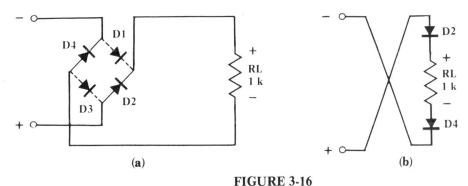

(a)   (b)

FIGURE 3-16

In circuit operation, series diodes D1 and D3, and series diodes D2 and D4, alternately switch. Therefore, no matter what polarity exists at the input to the bridge, the direction of the voltage drop across RL will remain the same. As shown in Figure 3-14, the full wave bridge rectifies the positive-going and negative-going excursions of the sine wave which provides the load with positive-going pulses only.

1. Solving for the peak-to-peak voltage at the secondary:

$$V_{p\text{-}p} = V\ RMS \times 2\sqrt{2} = 30\ V \times 2.828 = 84.84\ V_{p\text{-}p}$$

2. Solving for the peak voltage at the load:

$$Vpeak = Vp\text{-}p/2 = 84.84\ Vp\text{-}p/2 = 42.42\ Vpeak$$

NOTE: The frequency of the full wave waveshape, at the load, is 120 Hz.

3. Solving for the peak current through the load resistor:

$$Ip = Vpeak/RL = 42.42\ V/1\ k\Omega = 42.42\ mA$$

4. Solving for the peak current through the diode series D1 and D3 and the diode series D2 and D4:

$$Ip(D1, D3) = Vpeak/RL = 42.42\ V/1\ k\Omega = 42.42\ mA$$

$$Ip(D2, D4) = Vpeak/RL = 42.42\ V/1\ k\Omega = 42.42\ mA$$

NOTE: With reference to the circuit of Figure 3-17, peak current is a function of peak voltage and the load resistor. However, peak voltage amplitude remains the same, regardless of whether a halfwave or full wave waveshape exists at the load. Therefore, peak current flows through each diode string every other pulse (alternately), and that peak current is 42.42 mA.

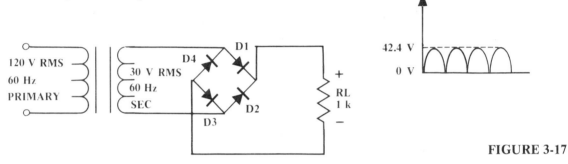

FIGURE 3-17

5. Solving for the DC voltage content at the load:

$$VDC = 2 \times Vp/\pi = 2 \times 42.42\ V/3.14 = 2 \times 13.5\ V = 27\ VDC$$

Another technique used to solve for the DC content at the load, for the full wave rectifier circuit, is to multiply the RMS voltage value by 0.9. Therefore, $VDC = 0.9 \times V\ RMS = 0.9 \times 30\ V\ RMS = 27\ V\ DC$. (The 0.9 constant was derived in the previous discussion on the center-tapped full wave rectifier circuit.)

6. Solving for the DC current at the load:

$$IDC = VDC/RL = 27\ V/1\ k\Omega = 27\ mA$$

7 Solving for the DC current through each diode of the series diodes D1, D3 and D2, D4:

$$IDC(D1, D2) = IDC(load)/2 = 27\ mA/2 = 13.5\ mA$$

$$IDC(D3, D4) = IDC(load)/2 = 27\ mA/2 = 13.5\ mA$$

NOTE: If one or both diodes in a diode string were opened, halfwave rectification would occur at the load. However, the peak voltage and peak current remain the same (halfwave or full wave) but the DC current to the load is halved. Therefore, the DC current is supplied (shared) by both diode strings.

8. Solving for the RMS value of the AC Ripple Voltage at the load:

$$VAC(RMS) = Vp/3.25 = 42.42\ V/3.25 = 13.05\ V$$

9. Solving for the total V RMS at the output load:

$$V\ RMS = Vp/\sqrt{2} = 42.42\ V/1.414 \approx 30\ V$$

102

NOTE: V RMS can also be solved from:

$$V\ RMS = \sqrt{VDC^2 + VAC(RMS)^2} = \sqrt{27^2 + 13.05^2} \approx 30\ V$$

This equation was derived in the previous discussion on the center-tapped full wave circuit.

10. Solving for the PIV of the Diodes D1, D3 and D2, D4:

$$PIV = Vp = 42.42\ V$$

NOTE: The PIV of the diodes in the full wave bridge circuit is equal to the peak voltage. See Figures 3-18(a) and 3-18(b), where the forward biased diodes have been removed for clarity. In Figure 3-18(a), a positive-going pulse turns on diodes D1 and D3 to provide a reverse bias voltage across diodes D2 and D4 — equal to the voltage developed across the load resistor. Since the voltage at the load is Vp, then PIV of diodes D2 and D4 is Vp or 42.42 V. For the negative-going pulse, as shown in Figure 3-18(b), just the opposite occurs. The PIV of diodes D1 and D3 also equals the peak voltage of Vp. (The turned on diodes D1 and D3 of Figure 3-18(a) and diodes D2 and D4 of Figure 3-18(b) are represented by a "shorted" condition to illustrate PIV conditions.)

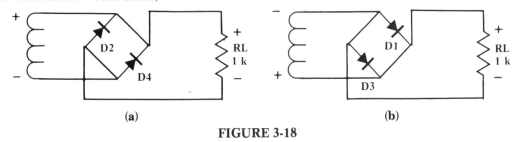

(a)   (b)

FIGURE 3-18

11. The percent ripple of the full wave bridge rectifier circuit is the same as that of the full wave center-tapped rectifier circuit. Therefore:

$$\%\ ripple = VAC(RMS)/VDC \times 100 = 13.05\ V/27\ V \times 100 = 48.3\%$$

NOTE: The waveshapes most commonly associated with both unfiltered and filtered "bulk" power supplies are the sine wave, halfwave, full wave, and sawtooth wave. Each of these waves are normally converted to RMS values and, since they have different shapes and areas, the converting (dividing) factor is slightly different for each wave. The wave shapes and their RMS converting factors are shown in Figure 3-19, where all ripple voltage amplitudes are considered peak-to-peak.

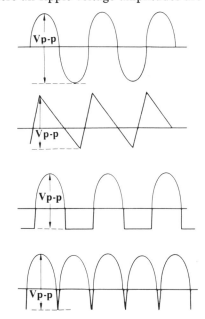

SINE WAVE

$$V\ RMS = \frac{Vp\text{-}p}{2\sqrt{2}} = \frac{Vp\text{-}p}{2.828}$$

SAWTOOTH WAVE

$$V\ RMS = \frac{Vp\text{-}p}{2\sqrt{3}} = \frac{Vp\text{-}p}{3.464}$$

HALFWAVE (RECTIFIED)

$$V\ RMS = \frac{Vp\text{-}p}{2.6}$$

FULL WAVE (RECTIFIED)

$$V\ RMS \approx \frac{Vp\text{-}p}{3.25}$$

FIGURE 3-19

103

EXPERIMENTAL OBJECTIVES

To investigate the unfiltered full wave bridge rectifier circuit.

LIST OF MATERIALS

1. Transformer: 120 V : 24 V or equivalent
2. Diodes: 1N4001 (four) or equivalents
3. Resistor: 1 kΩ (one).
4. Oscilloscope
5. Voltmeter
6. Ammeter

EXPERIMENTAL PROCEDURE

1. Connect the circuit as shown in Figure 3-20.

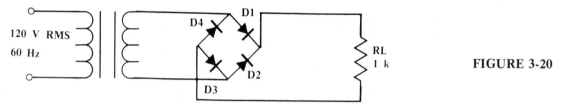

FIGURE 3-20

2. Across the secondary of the transformer:
   a) Measure the peak-to-peak voltage with a scope.
   b) Measure the RMS voltage with a voltmeter.
3. a) Calculate the Vp at the load from: $Vp = Vp\text{-}p / 2$
   b) Measure the Vp at the load
4. Calculate Ip from: $Ip = Vp/RL$
5. a) Calculate the VDC at the load from: $VDC = 2 \times Vp/\pi$. Then, measure VDC.
6. a) Calculate the DC curent flow through the load resistor from: $IDC = VDC/RL$
   b) Measure the DC current flow through the load. Measure with an ammeter or by monitoring the voltage drop across the known load resistor.
7. a) Calculate the DC current flow through the D1, D3 and D2, D4 diode strings. Calculate from:

$$I(D1, D3) = I(D2, D4) = IRL/2$$

   b) Measure the DC curent flow through each diode string, as shown in Figure 3-21, by placing an ammeter in series with diodes D1 or D3 and then with D2 or D4.

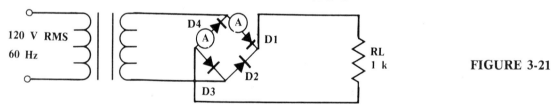

FIGURE 3-21

8. a) Calculate the VAC(RMS) ripple voltage, at the load, from: $VAC(RMS) = 0.3077$ or $Vp/3.25$.
   b) Measure the RMS value of the AC ripple voltage, at the load, using a standard voltmeter.
9 Calculate the total RMS voltage at the load from:

$$V\ RMS = Vp/\sqrt{2} \quad or \quad V\ RMS = \sqrt{VDC^2 + VAC(RMS)^2}$$

10 a) Calculate the PIV of (all) the diodes in the bridge circuit from:

$$PIV(D1) = PIV(D2) = PIV(D3) = PIV(D4) = Vp$$

   b) measure the PIV with a dual scope. Monitor the output with one scope probe and, alternately, monitor the anodes of D1, D2, D3, and D4 with the other probe.

11. Calculate the percent ripple from:

$$\% \text{ ripple} = \frac{\text{VAC(RMS)}}{\text{VDC}} \times 100$$

12. Insert the calculated and measured values, as indicated, into Table 3-3.

TABLE 3-3	Vp-p SEC	V RMS SEC	Vp LOAD	Ip LOAD	VDC LOAD
CALCULATED	////////	////////			
MEASURED				////////	

		IDC		VAC RMS	V RMS TOTAL	PIV				% RIPPLE
	LOAD	D1, D3	D2, D4			D1	D3	D2	D4	
CALCULATED										
MEASURED				////////						////////

PART 1B — FILTERED AC TO DC CONVERTERS

THE HALFWAVE RECTIFIER CIRCUIT

A large capacitor placed across the load, as shown in Figure 3-22, reduces the high ripple content associated with the output of the unfiltered half wave rectifier circuit. The capacitor repetitively charges up to the peak value of the positive-going pulses, 60 times per second, and then discharges on the negative-going pulses. Because the capacitor cannot discharge in 1/60 of a second under light loading, the amount of ripple will be relatively small. Since the ripple voltage ΔV rides on the DC voltage, the DC voltage will equal the peak voltage minus one half of the ripple voltage. If ΔV is small, the peak voltage approximately equals the DC voltage.

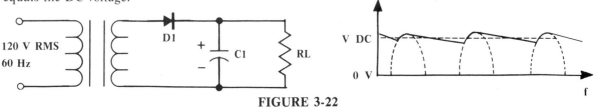

FIGURE 3-22

The ripple voltage has a sawtooth shape because the positive-going pulses (60 per second) charge the capacitor to peak value, and then the capacitor discharges into the load with the time constant $R_L C$ — a relatively long discharge time. The magnitude of the ripple voltage is a function of frequency, capacitance, and load current. The ripple voltage across the filter capacitor is solved from: $\Delta V_C = T/C \times VDC/RL$.

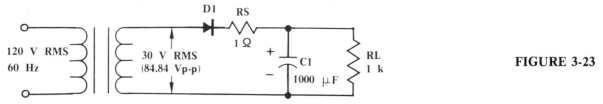

FIGURE 3-23

1. Solving for the peak-to-peak voltage at the secondary:

$$Vp\text{-}p = V\,RMS \times 2\sqrt{2} = 30\,V \times 2.828 = 84.84\,Vp\text{-}p$$

2. Solving for the peak voltage at the load:

$$Vpeak = Vp\text{-}p/2 = 84.84\,Vp\text{-}p/2 = 42.42\,Vpeak$$

3. Solving for the peak current through the load resistor:

$$IP = Vpeak/RL = 42.42\,Vp/1\,k\Omega = 42.42\,mA$$

4. Solving for the initial DC voltage at the load:

$$VDC \approx Vpeak \approx 42.4\,VDC$$

NOTE: While it is impossible to calculate, initially, the exact DC voltage across RL, an estimation of $VDC \approx Vp \approx 42.4\,V$ is acceptable, because the ripple voltage is rarely designed much above 1 Vp-p across the filter capactior. The actual VDC is solved only after the (ΔV) is known.

5. Solving for the ripple voltage magnitude across the filter capacitor from:

$$\Delta V_{C1} = \frac{T}{C} \times I_{RL} = \frac{T}{C} \times \frac{VDC}{RL} = \frac{1/60 \times 42.4\,V}{1000\,\mu F \times 1\,k\Omega} = \frac{16.6 \times 10^{-3} \times 42.4\,V}{10^{-3} \times 10^3} = 706\,mVp\text{-}p$$

NOTE: Because the ripple voltage is 706 mVp-p and rides on the DC voltage, the DC voltage equals, approximately 42.07 VDC and the ripple voltage is ± 0.35 volts. This is shown in **Figure 3.24** where the ripple voltage upper voltage limit is the peak voltage and the actual DC is solved from:

$$VDC = Vp - \Delta V_{C1}/2 = 42.42\,V - 0.7\,V/2 = 42.07\,VDC$$

106

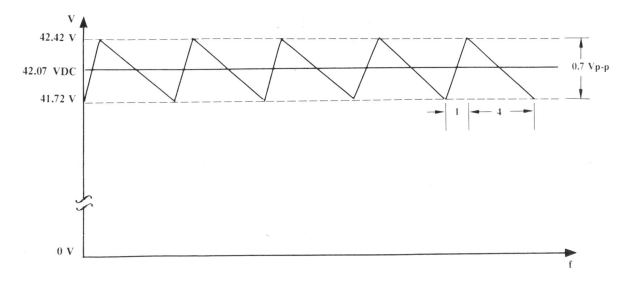

FIGURE 3-24

NOTE: ΔV across the capacitor carries the subscript ΔV_{C1}.

6. Solving for the DC current through the load from:

$$IDC = VDC/RL = 42.07\ VDC/\ 1\ k\Omega \approx 42\ VDC/1\ k\Omega = 42\ mA$$

7. Solving for the DC charging current through the diode where the ratio of charge time to discharge time is, for instance, 4 : 1. Therefore:

$$IDC(charging) = \frac{discharge\ time}{charge\ time} \times IDC = \frac{4}{1} \times 42\ mA = 168\ mA$$

NOTE: The advantages of filtering are reduced ripple and higher DC content to the load, which approaches Vpeak in magnitude. However, the disadvantage is increased current flow into the capacitor during the charging time and, hence, increased current flow through the series diode D1. This increased current occurs because the discharge time has a longer duration than the charge time Therefore, current flow drawn out of the capacitor during discharge time must be replenished in the shorter span of charge time. Fore instance, 42 mA delivered during dicharge will charge to approximately 168 mA if the charge time is one-fourth the discharge time. The charge-discharge sawtooth wave shape can be monitored across the load with an oscilloscope, and it is similar in shape to the 4 : 1 discharge-charge wave shape shown in Figure 3-24. Also, because the charge time occurs with each postive-going pulse, the 168 milliamps of DC charging current is recurring — that is, it occurs 60 time a second for the half-wave rectifier.

A second method for determining the DC charging current through the diodes is by monitoring the peak voltage across the series resistance RS, and then solving for the diode current from:

$$Ipeak(charging) = VRS(peak)/RS = 436.8\ mV/1\ \Omega = 436.8\ mA$$

Then, converting the "almost" half-wave shaped, peak-recurring, diode current to DC charging current from:

$$IDC(charging) = I_D(charging)/2.6 = 436.8\ mA/2.6 = 168\ mA$$

NOTE: The practical approach for solving the recurring current is through oscilloscope measurements. This is because the series charging resistance of the diode and the transformer can only be found through measurement techniques and the scope method is easily accomplished. For instance, if the scope monitors the output waveshape at the load and the charge-discharge ratio is monitored at 4 : 1, as shown in Figure 3-25, then the recurring DC charging current can be calculated rather easily.

Likewise, if the wave shape at the junction of the anode and the series resistor RS is monitored with the scope, the peak voltage (as shown in Figure 3-25) will be seen. Since the capacitance looks like a short during the charge time, the charging waveshape is the voltage developed across the series RS resistor. The

107

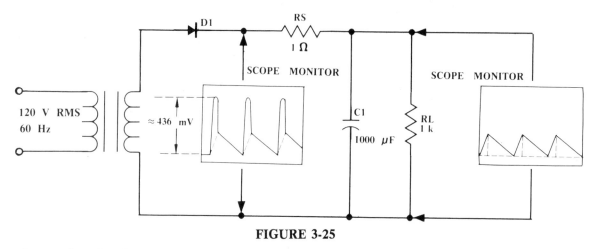

FIGURE 3-25

peak recurring charging current is solved from the ratio of peak charging voltage divided by the series resistance of RS, or VRS(peak)/RS. Again, for a half-wave rectifier, the recurring charging current occurs 60 times per second, and the diode must be able to handle the peak current conditions. Also, dividing the peak recurring current by 2.6 assumes that the peak pulse shape is half wave, but it could be triangular in shape, and the dividing factor would then be 2. Obviously, an exact comparison between peak and DC charging current can only be estimated until measurements are made:

8. Solving for the RMS value of the AC ripple voltage at the load:

$$VAC(RMS) = VAC/2\sqrt{3} = 706 \text{ mVp-p}/3.464 \approx 204 \text{ mV}$$

NOTE: The RMS value of a sawtooth ripple voltage wave is solved by dividing its peak-to-peak voltage value by $2\sqrt{3}$. This was shown in Figure 3-19.

9. Solving for the percent ripple from:

$$\% \text{ ripple} = (VAC[RMS]/VDC)100 = (204 \text{ mV}/41.05 \text{ V})100 \approx 0.496\%$$

10. Solving for the peak inverse voltage of diode D1 from: $PIV = 2 \text{ Vp} \approx 2 \times 42.42 \text{ V} = 84.84 \text{ V}$

NOTE: Since the charged filter capacitor remains at, or about, the Vp value of 42 volts, then the applied peak-to-peak voltage reaches − 42.42 volts, the reverse voltage across the diode is at ≈ 84 volts. This results in a PIV of 2 Vp. See Figure 3-26.

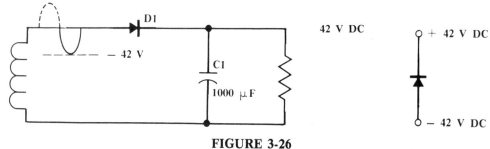

FIGURE 3-26

11. Solving for the percent regulation:

$$\% \text{ regulation} = \frac{V(\text{no load}) - V(\text{load})}{V(\text{no load})} \times 100 = \frac{42.42 \text{ V} - 41.07 \text{ V}}{42.42 \text{ V}} \times 100 = \frac{1.35 \text{ V}}{42.42 \text{ V}} \times 100 \approx 3.18\%$$

$$\% \text{ regulation} = \frac{V(\text{no load}) - V(\text{load})}{V(\text{load})} \times 100 = \frac{42.42 \text{ V} - 41.07 \text{ V}}{41.07 \text{ V}} \times 100 = \frac{1.35 \text{ V}}{41.07 \text{ V}} \times 100 \approx 3.29\%$$

NOTE: Two methods are used to show percent regulation because the Federal Communication Commission (FCC) uses the latter method and most of the "real world" uses the former method. Most

108

engineering standards are based on the normal 0 to 100% scale, while the FCC, in this instance, operates on a zero to infinity scale. The choice of using one method over the other is purely academic.

EFFECTS OF LOAD RESISTOR OR FILTER CAPACITOR VARIATIONS

Varying the load resistor or filter capactior will cause increased or decreased ripple voltage, VDC, and regulation. For instance, decreasing the load resistance to 100 ohms instead of 1 kΩ, or decreasing the filter capacitor to 100 μF instead of 1000 μF, will cause the ripple voltage to increase from approximately 0.7 Vp-p to 7 Vp-p and the VDC from 41.07 VDC to approximately 39.2 VDC. In calculating the initial DC voltage for a halfwave circuit having a load resistor of 1 kΩ and a filter capacitor of 100μF, the initial peak-to-peak ripple voltage across the filter capacitor is solved from:

$$\Delta VC1 = \frac{T}{C} \times I_{RL} = \frac{T}{C} \times \frac{VDC}{RL} = \frac{1/60}{100\ \mu F} \times \frac{42\ V}{1\ k\Omega} \approx \frac{0.7\ V}{10^{-1}} = 7\ Vp\text{-}p$$

However, the 7 Vp-p ripple voltage results in a "new" DC voltage of approximately:

$$VDC \approx Vp - \frac{\Delta VC_1}{2} = 42.4\ V - \frac{7\ V}{2} = 38.9\ VDC$$

which results in a ripple voltage at

$$\Delta VC1 = \frac{T}{C} \times \frac{VDC}{RL} = \frac{1/60}{100\ \mu F} \times \frac{38.9\ V}{1\ k\Omega} \approx 6.48 Vp\text{-}p$$

which raises the DC to

$$VDC \approx Vp - \Delta VC/2 = 42.4\ V - 6.48\ V/2 \approx 39.16\ V \approx 39.2\ V$$

Further refinements, obviously, are not needed. The circuit and approximate DC voltage and ripple voltage are shown in Figure 3.27.

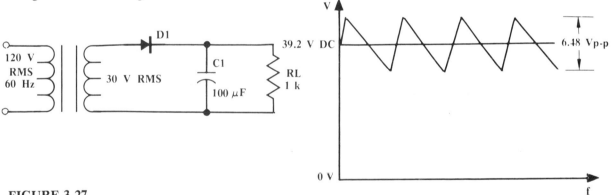

FIGURE 3-27

The % ripple for a VDC of 39.2 and a ripple voltage of 6.48 Vp-p is solved from:

$$\% \text{ ripple} = \frac{VAC(RMS) \times 100}{VDC} = \frac{6.48\ Vp\text{-}p/2\sqrt{3} \times 100}{39.2\ V} = 4.78\%$$

NOTE: Filtering of 4.78% ripple is considered intolerable. Also, relative accuracy can only be achieved for light loading when the ripple voltage magnitude is kept below 1 Vp-p.

CAPACTIVE FILTERING OF THE FULL WAVE RECTIFIER CIRCUIT

The full wave rectifier circuit is easier to filter than the half wave rectifier circuit because twice as many pulses (120 Hz) exist across the load. Therefore, the maginitude of the ripple voltage ΔV is cut in half and, in comparison to the halfwave rectifier, the full wave circuit needs only one half the capacitance or one half the value of load resistance in providing equal ripple voltage magnitude.

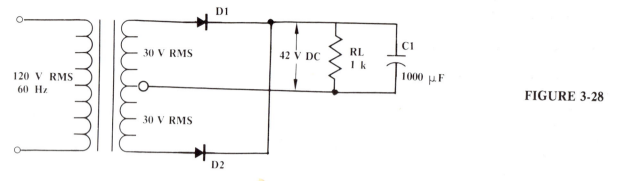

FIGURE 3-28

1. Solving for the peak-to-peak voltage at the anodes of D1 and D2:

$$Vp\text{-}p = VRMS \times 2\sqrt{2} = 30\ V \times 2.828 = 84.84\ Vp\text{-}p$$

NOTE: The center tapped full wave rectifier circuit is effectively two separate half wave circuits, operating 180° out of phase with each other. Therefore, each diode "sees" only one half of the transformer — 30 VRMS or 84.84 Vp-p.

2. Solving for the peak voltage at, and across, capacitor C1:

$$Vpeak = Vp\text{-}p/2 = 84.84\ Vp\text{-}p/2 = 42.42\ Vpeak$$

3. Solving for the peak current through the load resistor:

$$Ipeak = Vpeak/RL = 42.42\ V/1\ k\Omega = 42.42\ mA$$

4. Solving for the DC voltage at the load:

$$VDC \approx Vpeak, \text{ or more precisely:}$$

$$VDC = Vp - Vp\text{-}p/2 \approx 42.42\ V - 0.35\ V/2 = 42.42\ V - 0.175 = 42.245\ V \approx 42.25\ V$$

5. Solving for the magnitude of the ripple voltage:

$$\Delta V_{C1} = T/C \times I_{RL} = T/C \times VDC/RL = (1/120 \times 42.25\ V)/(1000\ \mu F \times 1\ k\Omega) \approx 0.35\ Vp\text{-}p$$

NOTE: For the full wave rectifier circuit, the pulse repetition rate is 120 Hertz per second.

6. The DC voltage, the peak voltage, and the ΔVC1 ripple voltage for the full wave rectifier circuit of Figure 3-28 is shown in Figure 3-29. The DC voltage is at ≈ 42.25 volts and the ripple voltage of ≈ 0.35 Vp-p rides on it. The peak excursion of the ripple voltage is 42.42 volts, the minimum excursion is 42.07 volts, and the VDC is 42.25 volts or, more precisely, 42.245 volts.

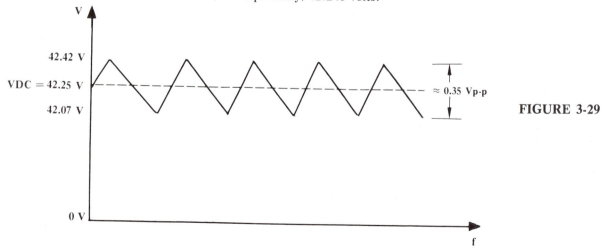

FIGURE 3-29

110

7. Solving for the DC current through the load:

$$IDC = VDC/RL = 42.25/1 \text{ k}\Omega = 42.25 \text{ mA}$$

8. Solving for the DC charging current through the diodes where, as an example, the ratio of discharge time to charge time is given at 4 : 1 — as shown in Figure 3-30. In this example, the capacitance is discharged into (by) the load over a period of time and it is recharged to peak value in one-fourth the discharge time. Therefore, the charge current into the capacitor must be four times larger than the discharge current.

$$IDC\text{(charging)} = \frac{IDC \times \text{discharge time}}{\text{charge time}} = \frac{42.25 \text{ mA} \times 4}{1} = 170 \text{ mA}$$

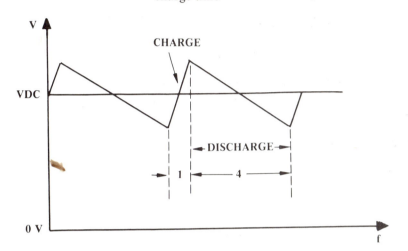

FIGURE 3-30

NOTE: Because of the discharge-charge ratio of 4 : 1, the reoccurring current flow through the diodes is four times greater than the DC current into the load.

9. Solving for the recurring charging current through each diode from:

$$I_{D1} = I_{D2} \approx IDC\text{(charging)} \times 2.6 = 170 \text{ mA} \times 2.6 = 442 \text{ mA}$$

NOTE: In the filtered full-wave, center-tapped rectifier circuit, each diode shares the total DC current to the load. However, each diode must be able to process the increased peak charging current ($\approx$ 442 mA).

10. Solving for the RMS value of the AC ripple voltage:

$$VAC\text{(RMS)} = \Delta V_C/2\sqrt{3} = 0.35 \text{ Vp-p}/4.464 = 0.101 \text{ V}$$

11. Solving for the percent ripple:

$$\% \text{ ripple} = (VAC[RMS]/VDC)100 = (0.101 \text{ V}/42.25 \text{ V})100 \approx 0.24\%$$

12. Solving for the PIV of diodes D1 and D2:

$$PIV = 2 \text{ Vpeak} = 2 \times 42.42 \text{ V} = 84.84 \text{ V}$$

NOTE: The PIV for the diodes in the full wave center-tapped rectifier circuit is the same, with or without a filter capacitor.

13. Solving for the percent regulation:

$$\% \text{ regulation} = (V[NL] - V[L]/V[NL])100 = ([42.22 \text{ V} - 42.25 \text{ V}]/42.42)100 \approx 0.40\%$$

$$\% \text{ regulation} = (V[NL] - V[L]/V[L])100 = ([42.42 \text{ V} - 42.25 \text{ V}]/42.25 \text{ V})100 \approx 0.414\%$$

111

NOTE: Obviously, either method of solution for percent regulation is valid (mathematically close) for low ripple voltage conditions. Again, approximate VDC values alter calculations slightly.

CAPACITIVE FILTERING OF THE FULL WAVE BRIDGE RECTIFIER CIRCUIT

The filtered full wave bridge rectifier circuit of Figure 3-31 has all the advantages of full wave rectification without the need for a center-tapped transformer secondary. Therefore, the transformer secondary can be 30 V RMS, instead of 60 V(CT) as required in the full wave center-tapped circuit configuration.

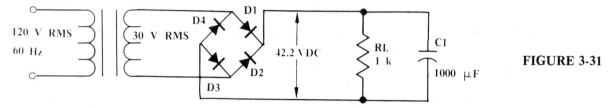

FIGURE 3-31

1. Solving for the peak-to-peak voltage at the anodes of D1 and D2:

$$Vp\text{-}p = V\ RMS \times 2\sqrt{2} = 30 \times 2.828 = 84.84\ Vp\text{-}p$$

2. Solving for the peak voltage at, and across, capacitor C1:

$$Vpeak = Vp\text{-}p/2 = 84.84\ Vp\text{-}p/2 = 42.42\ Vpeak$$

3. Solving for the peak current through the load resistor:

$$Ipeak = Vpeak/RL = 42.42\ V/1\ k\Omega = 42.42\ mA$$

4. Solving for the DC voltage at the load:

$$VDC \approx Vpeak \quad or,$$

$$VDC = Vp - (\Delta V_{C1}/2) = 42.42\ V - (0.35\ Vp\text{-}p/2) \approx 42.25\ V$$

5. Solving for the magnitude of the ripple voltage:

$$\Delta V_{C1} = T/C \times I_{RL} \approx T/C \times VDC/RL = (1/120 \times 42.42\ V)/(1000\ \mu F \times 1\ k\Omega) \approx 0.35\ Vp\text{-}p$$

6. The DC voltage, the peak voltage, and the ripple voltage are shown in Figure 3.32. The DC voltage is approximately 42.25 volts and the ripple voltage of 0.35 Vp-p rides on it. These values are the same as those for the full wave rectifier circuit.

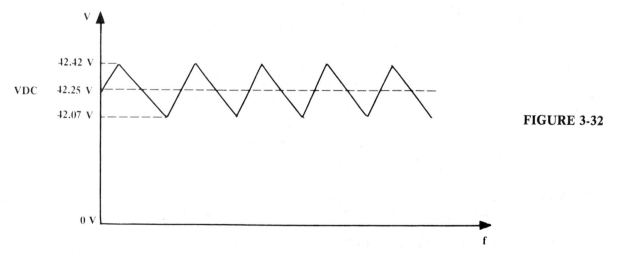

FIGURE 3-32

7. Solving for the DC current through the load:

$$IDC = VDC/RL \approx 42.25\ V/1\ k\Omega = 42.25\ mA$$

8. Solving for the recurrng DC charging current from the diode bridge when the ratio of charge time to discharge time is 4 : 1:

$$IDC(charging) = \frac{discharge\ time}{charge\ time} \times IDC = \frac{4}{1} \times 42.5\ mA = 170\ mA$$

9. Solving for the recurring peak charging current through series diodes D1, D3 and D2, D4:

$$I_p(charging) \approx IDC(charging) \times 2.6 = 170\ mA \times 2.6 = 442\ mA$$

NOTE: Each diode series string (D1, D3 and D2, D4) alternately contributes 442 mA of recurring peak charging current. Therefore, approximately the same amount of peak charging current (approximately 442 mA) flows through each diode.

10. Solving for the PIV of diodes D1, D2, D3, and D4:

$$PIV = Vp = 42.42\ V$$

NOTE: The PIV across the diodes of the unfiltered and filtered full wave bridge rectifier circuits is the same. The PIV for the bridge circuit was illustrated in Figure 3-18.

11. Solving for the percent ripple:

$$\%\ ripple = (VAC[RMS/VDC]100 = (\Delta VC/2\sqrt{3})/VDC \times 100$$

$$= (0.35\ Vp\text{-}p/3.464)/42.25\ V \times 100 = 0.24\%$$

EXPERIMENTAL OBJECTIVES

To investigate the full wave bridge rectifier circuit with capacitor filtering:

LIST OF MATERIALS

1. Transformer: 120 V : 24 V RMS or equivalent
2. Diodes: 1N4001 (four) or equivalents
3. Resistors: 1 kΩ (one)   1 Ω (one)
4. Capacitor: 1000 μF (one)
5. Oscilloscope
6. Voltmeter
7. Ammeter

EXPERIMENTAL PROCEDURES

1. Connect the circuit as shown in Figure 3-33.

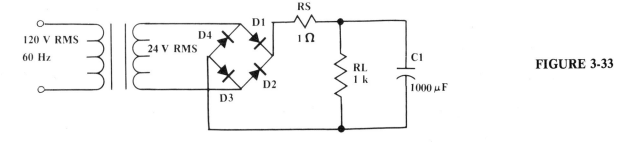

FIGURE 3-33

113

2. Across the secondary of the transformer:
   a) Measure the peak-to-peak voltage with an oscilloscope.
   b) Measure the RMS voltage with a voltmeter.
3. a) Calculate the Vp at the load from: $Vp\text{-}p/2 = VDC + (\Delta V_{C1}/2) \approx VDC(C1)$
   b) Measure the Vp at the load.

NOTE: Vp and VDC are approximately the same. Vp is slightly higher than VDC by $\Delta V_{C1}/2$. Vp is measured from the maximum peak excursion of ΔV_C with respect to ground. The measurement can be verified by removing the filter capacitor, momentarilly, and monitoring the unfiltered Vp wave. However, because it is the upper extremity of the ripple voltage, the AC-DC switch of the scope can be used in the reference and measurement of Vp. Also, VDC(C1) is the voltage developed across capacitor C1 and it equals the DC voltage at the load.

4. a) Calculate VDC from: $VDC \approx Vp$ and $VDC = Vp - \Delta V_{C1}/2$.
   b) Measure the DC voltage at the load.
5. a) Calculate the DC current through the load from: $IDC = VDC/RL$.
   b) Measure the DC current through the load with an ammeter.
6. a) Calculate the AC ripple voltage from: $\Delta VC1 = T/C \times I_{RL} = T/C \times VDC/RL$.
   b) Measure the AC ripple voltage at the load with an oscilloscope.
7. a) Calculate the DC and peak recurring charging current out of the diode bridge (refer to Figure 3-34):

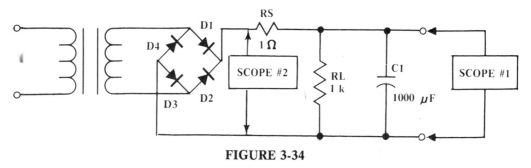

FIGURE 3-34

   1. Monitor the sawtooth ΔV_{C1} wave shape at the load. Note the discharge-charge ratio and calcu-ate from:

$$IDC(charging) = IDC \times discharge\ time/charge\ time \quad (Scope\ \#1)$$

   2. Monitor the wave shape on the bridge side of the rectifier — with respect to ground. Measure the peak recurring charging voltage and calculate the peak recurring charging current from:

$$Ip(charging = Vp(charging)/RS \quad (Scope\ \#2)$$

8. a) Calculate the peak inverse voltage of the bridge diodes from: $PIV = Vp$.
   b) Measure the PIV of diodes D1 and D2.

NOTE: It is only necessary to measure one diode in each of the diode strings. The measurement should be made with a dual scope.

9. Measure the ΔV_C at the load and calculate the RMS value of the AC content from:

$$VAC(RMS) = \Delta V_{C1}/2\sqrt{3}$$

10. Calculate the percent ripple from: $\%\ ripple = VAC(RMS)/VDC \times 100$.
11. Insert the calculated and measured values, as indicated, into Table 3-4.

114

TABLE 3-4	V_{p-p} SEC	V RMS SEC	V_p LOAD	VDC LOAD	IDC LOAD
CALCULATED	/////	/////			
MEASURED					

	I RECURRING		ΔV_{C1} Load	VAC(RMS) Load	PIV		% Ripple
	DC	PEAK			D1	D2	
CALCULATED			/////				
MEASURED	/////	/////		/////			/////

NOTE: The relationship of peak charging to DC charging current is about 2.6 : 1 if the waveshape of the peak pulse is half wave. However, in practice the 2.6 : 1 ratio is rarely achieved and the half-wave pulse wave shape estimation only serves to provide relative, not exact, calculations and measurements.

SECTION II: REGULATED POWER SUPPLIES

GENERAL DISCUSSION

Bulk supplies alone do not make very good regulators because, with increased loading, the DC voltage decreases and the ripple current increases. Therefore, additional circuitry, along with the bulk supply, is needed to insure against voltage output changes for varying degrees of load current changes. In solid state circuitry, the device most widely used as a voltage regulator is the zener diode.

PART 2A — ZENER DIODE REGULATORS

All diodes exhibit a linear breakdown region when the device is subject to reverse biased conditions. For regular diodes this region is the PIV voltage, where excessive power caused by the combination of high voltage and current can destroy the device. However, zener diodes are doped to take advantage of this zener or avalanche characteristic, where lower breakdown voltages at higher currents are normally used in low to medium power devices. Typical breakdown characteristics for both the regular diode and the zener diode are shown in Figure 3-35.

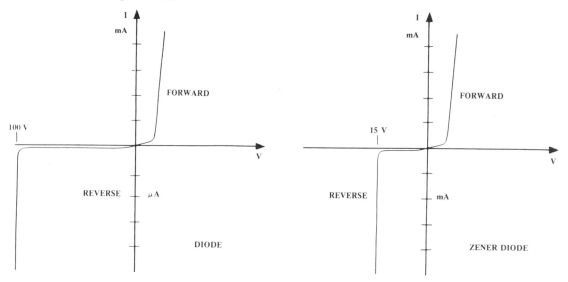

FIGURE 3-35

Zener or avalanche breakdown can occur when the zener diode is connected in a reverse biased condition. This is where the positive voltage is on the cathode and the negative voltage is on the anode, as shown in Figure 3-36 (a).   In this circuit configuration, various degrees of input voltages can be applied but the voltage across the zener diode will remain constant, with the excess voltage being dropped across the series resistor R1.   However, the current flow through the zener diode will increase with increased voltage across R1 — the maximum current being determined by the power dissipating capabilities of the device.

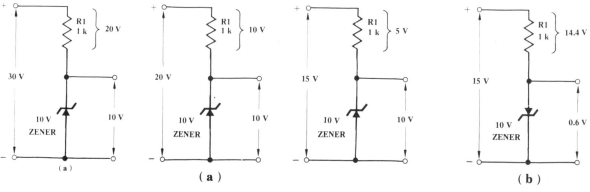

FIGURE 3-36

116

For instance, too low a current flow through the zener will cause the device to come out of regulation or become increasingly noisy as it approaches the knee of the curve. Therefore, the current flow through the zener diode is maintained at between 1 mA and 5 mA. When the zener diode is in a forward biased condition, as shown in Figure 3-36 (b), it acts like any normally forward biased diode. For the silicon zener diode, the forward bias voltage drop is approximately 0.6 volts — which is the same as that for the normal silicon (rectifier) diode.

ZENER DIODE LOADING

Placing a load resistor across the zener diode does not change the voltage drop across the zener if enough current is available to supply both the zener diode and the load resistor. This is because the zener needs only enough current through it to remain below the knee of the curve and in zener condition. Therefore, from 1 mA to 5 mA of current must always flow through the zener or else it will come out of regulation and revert to a normally reverse biased diode, exhibiting a high (greater than 100 kΩ) resistance.

For instance, if 30 V DC is applied to a series 1 kΩ resistor and a 10 V zener diode, 20 volts is dropped across the 1 kΩ resistor and 10 volts is dropped across the zener diode. However, when a load resistor is placed across the zener, a portion of the available R1 current (20 mA) flows into the load resistor and the remaining current flows through the zener. This is shown in Figure 3-37 for varying degrees of no load and loaded conditions.

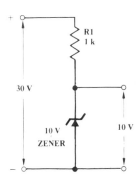

$$V_Z = 10 \text{ V}$$

$$V_{R1} = Vin - V_Z = 30 \text{ V} - 10 \text{ V} = 20 \text{ V}$$

$$i_{R1} = V_{R1}/R1 = 20 \text{ V}/1 \text{ k}\Omega = 20 \text{ mA}$$

$$I_Z = I_{R1} = 20 \text{ mA}$$

FIGURE 3-37 (a)

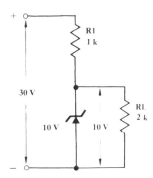

$$V_Z = 10 \text{ V}$$

$$V_{R1} = Vin - V_Z = 30 \text{ V} - 10 \text{ V} = 20 \text{ V}$$

$$I_{R1} = V_{R1}/R1 = 20 \text{ V}/1 \text{ k}\Omega = 20 \text{ mA}$$

$$I(load) = V_{RL}/RL = V_Z/RL = 10 \text{ V}/2 \text{ k}\Omega = 5 \text{ mA}$$

$$I_Z = I_{R1} - I(load) = 20 \text{ mA} - 5 \text{ mA} = 15 \text{ mA}$$

FIGURE 3-37 (b)

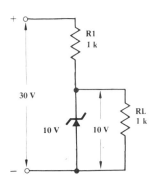

$$V_Z = 10 \text{ V}$$

$$V_{R1} = Vin - V_Z = 30 \text{ V} - 10 \text{ V} = 20 \text{ V}$$

$$I_{R1} = V_{R1}/R1 = 20 \text{ V}/1 \text{ k}\Omega = 20 \text{ mA}$$

$$I(load) = V_{RL}/RL = V_Z/RL = 10 \text{ V}/1 \text{ k}\Omega = 10 \text{ mA}$$

$$I_Z = I_{R1} - I(load) = 20 \text{ mA} - 10 \text{ mA} = 10 \text{ mA}$$

FIGURE 3-37 (c)

117

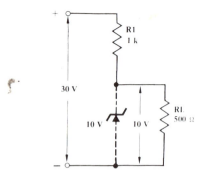

$$V_Z = 10 \text{ V}$$

$$V_{R1} = \text{Vin} - V_Z = 30 \text{ V} - 10 \text{ V} = 20 \text{ V}$$

$$I_{R1} = V_{R1}/R1 = 20 \text{ V}/1 \text{ k}\Omega = 20 \text{ mA}$$

$$I(\text{load}) = V_{RL}/RL = V_Z/RL = 10 \text{ V}/500\Omega = 20 \text{ mA}$$

$$I_Z = I_{R1} - I(\text{load}) = 20 \text{ mA} - 20 \text{ mA} = 0 \text{ mA}$$

FIGURE 3-37 (d)

NOTE: Therefore, in this latter example, the zener diode is out of regulation and has reverted to the high resistance of approximately 100 kΩ. This is a normally reverse biased diode.

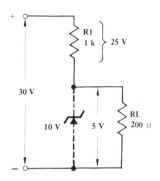

$$V_{RL} = \frac{\text{Vin} \times RL}{R1 + RL} = \frac{30 \text{ V} \times 200 \text{ }\Omega}{1 \text{ k}\Omega + 200 \text{ }\Omega} = 5 \text{ V}$$

$$V_{R1} = \frac{\text{Vin} \times R1}{R1 + RL} = \frac{30 \text{ V} \times 1000 \text{ }\Omega}{1 \text{ k}\Omega + 200 \text{ }\Omega} = 25 \text{ V}$$

FIGURE 3-37 (e)

NOTE: Placing a lower 200 Ω RL resistance across the zener increases the amount of current to the load resistor and provides a 5 V drop across it. Therefore, the zener voltage is at approximately 5 V, out of regulation, and is at a high enough resistance (approximately 100 kΩ) to be ignored. Hence, the series R1 and RL are solved as two resistors in series.

THE ZENER DIODE REGULATOR CIRCUIT

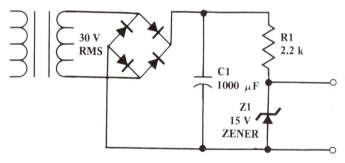

FIGURE 3-38

1. Solving for the peak-to-peak voltage at the secondary:

$$\text{Vp-p} = (\text{V RMS}) 2\sqrt{2} = 30 \text{ V} \times 2.828 = 84.84 \text{ Vp-p}$$

2. Solving for the peak voltage at the load:

$$\text{Vpeak} = \text{Vp-p}/2 = 84.84 \text{ Vp-p}/2 = 42.42 \text{ Vp}$$

3. Solving for the DC voltage across Capacitor C1:

$$\text{VDC(C1)} = \text{Vpeak} = 42.42 \text{ V, or } \text{Vp} - (\Delta V_{C_1}/2) = 42.42 \text{ V} - 0.1 \text{ V}/2 = 42.32 \text{ VDC}$$

118

4. Solving for the voltage drop across resistor R1:

$$V_{R1} = VDC(C1) - V_{Z1} = 42.3\text{ V} - 15\text{ V} = 27.3\text{ V}$$

5. Solving for the current through R1:

$$I_{R1} = V_{R1}/R1 = 27.3\text{ V}/2.2\text{ k}\Omega \approx 12.4\text{ mA}$$

NOTE: For the unloaded condition, the full current of R1 flows through the zener diode.

6. Solving for the AC ripple voltage across the filter capacitor C1:

$$\Delta V_{C1} = T/C \times I_{RL} = \frac{T}{C} \times \frac{V_{R1}}{RL} = \frac{1/120}{1000\ \mu\text{F}} \times \frac{27.3\text{ V}}{2.2\text{ k}\Omega} = 8.33 \times 12.4\text{ mA} = 103.3\text{ mVp-p}$$

NOTE: The current flow through R1 represents the load current or $I_{R1} = I_{RL}$.

7. Therefore, a ripple voltage of 103.3 mVp-p exists across the filter capacitor. However, the ripple voltage across the zener will be considerably lower because it is stepped down by the effective ratio of Z1 to R1.

For instance, the resistance of most zeners, found in the data sheets, is between 5 ohms and 30 ohms. However, a "ball park" estimate is that the dynamic resistance of a zener is approximately equal to the zener voltage: 20 V = 20 Ω, 5 V = 5 Ω, and so on. Therefore, if data is not available on the zener, this kind of estimate is valid until measurements can be made.

Hence, assuming that the 15 V zener is 15 ohms, then the ripple voltage across the zener will be:

$$\Delta V_{C1} = \frac{\Delta V_C \times R_{Z1}}{R_{Z1} + R1} = \frac{103.3\text{ mVp-p} \times 15\ \Omega}{15\ \Omega + 2200\ \Omega} = 0.7\text{ mVp-p}$$

Therefore, the DC voltage across the zener, and to the load, is 15 V DC and the ripple voltage is less than 1 mVp-p or 0.7 mVp-p.

ZENER DIODE LOADED CONDITIONS

The zener will continue to provide a constant 15 volts of DC and less than 1 mVp-p of ripple voltage, for varying degrees of loading, as long as zener regulation is maintained. However, in order to make sure the zener does not come out of regulation, a minimum of 1 mA to 5 mA of DC current flow is provided to the zener. Also, a resistor must be chosen to insure against too much current flow through the device when the load is removed. And the no load condition means that all the current flows through the zener so the zener must be able to dissipate the power safely.

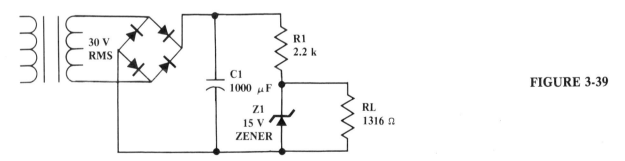

FIGURE 3-39

1. Solving for the maximum current flow into the load, maintaining 1 mA of zener regulation current:

$$I_{R1} = \frac{V\,DC_{(C1)} - V_{Z1}}{R1} = \frac{42.3\text{ V} - 15\text{ V}}{2.2\text{ k}\Omega} = 12.4\text{ ma}$$

$$I_{RL} = I_{R1} - I_{Z1} = 12.4\text{ mA} - 1\text{ mA} = 11.4\text{ mA}$$

2. Solving for the minimum load resistance possible for the 15 V zener, maintaining the 1 mA of regulation current:

$$RL = V_{Z1}/I_{RL} = 15 \text{ V}/11.4 \text{ mA} = 1315.8 \text{ } \Omega$$

3. Solving for the power dissipated by the zener:
   a. For full load condition where only 1 mA of zener current flows:

$$P_{Z1} = I_{Z1} \times V_{Z1} = 1 \text{ mA} \times 15 \text{ V} = 15 \text{ mW}$$

   b. No load condition where 12.4 mA of zener current flows:

$$P_{Z1} = I_{Z1} \times V_{Z1} = 12.4 \text{ mA} \times 15 \text{ V} = 186 \text{ mW}$$

NOTE: The "rule of thumb" ratio for choosing the power capability of a device is a minimum of 2 : 1 over the actual power dissipation conditions. Therefore, the zener diode (device) dissipation parameter should be 400 mW.

EXPERIMENTAL OBJECTIVES

To investigate the zener diode regulated circuit.

LIST OF MATERIALS

1. Transformer: 120 V : 24 V
2. Diodes: 1N4001 or equivalents (four)
3. Resistors: 1 Ω   470 Ω   3.3 kΩ   4.7 kΩ   10 kΩ potentiometer   (one each)
4. Capacitor: 1000 μF   (one)
5. Zener Diode: 15 V

EXPERIMENTAL PROCEDURES

1. Connect the circuit as shown in Figure 3-40.

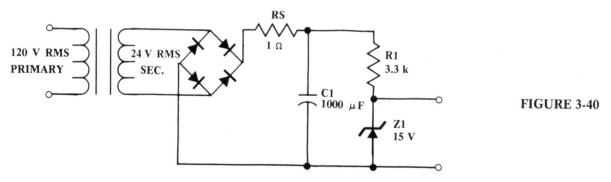

FIGURE 3-40

2. Across the secondary of the transformer:
   a. Measure the Vp-p with an oscilloscope.
   b. Measure the RMS voltage with a voltmeter.
3. Calculate and measure the Vp across the filter capacitor.

NOTE: Vp is approximately equal to V DC, but it is slightly higher by $\Delta V(C1)/2$. VDC is measured from the center of the AC ripple voltage with respect to ground while Vp is from the maximum peak excursion of the ripple voltage with respect to ground.

4. Calculate the DC voltage across the filter capacitor from:

$$VDC(C1) \approx Vp \quad \text{or} \quad VDC(C1) = Vp - (\Delta V_{C1}/2)$$

Measure the DC voltage across the filter capacitor.

5. Calculate the voltage drop across the current limiting resistor R1 from:

$$V_{R1} = VDC_{C1} - V_{Z1}$$

Measure the voltage drop across R1.
Measure the voltage drop across the zener.

6. Calculate the AC ripple voltage across the filter capacitor from:

$$\Delta V_{(C1)} = T/C \times I_{RL} = T/C \times I_{R1} \quad \text{where } I_{R1} = V_{R1}/R1$$

NOTE: The load current is I_{R1}, where $V_{R1} = VDC_{C1} - V_{Z1}$.

Measure the AC ripple voltage across the filter capacitor.

7. Calcute the AC ripple voltage across the zener from:

$$\Delta V_{(Z1)} = \frac{\Delta V_{C1} \times R_{Z1}}{R_{Z1} + R1}$$

NOTE: R_{Z1} can be obtained from the device's data sheet. However, if this information is not available the "rule of thumb" voltage-resistance, equivalent approximation is acceptable until actual measurements can be made.

Measure the AC ripple (ΔV) across the zener.

8. Monitor the sawtooth ΔV_{C1} discharge-charge ratio across the filter capacitor.
Calculate the recurring DC charging current of the bridge from:

$$IDC\text{(charging)} = \frac{I_{RL} \times \text{discharge time}}{\text{charge time}} \quad \text{where: } I_{RL} = I_{R1}$$

Monitor the peak recurring charging current across RS, the series 1 ohm resistor —bridge side of RS with respect to ground.
Calculate the peak recurring charging current out of the bridge from:

$$Ip\text{(charging)} = \frac{Vp \text{ recurring}}{RS}$$

9. Calculate the RMS value of the AC ripple voltage across the filter capacitor from:

$$V_{AC(RMS)C1} = \Delta V_{C1}/2\sqrt{3}$$

Calculate the RMS value of the AC ripple voltage across the zener from: $V_{AC(RMS)Z1} = \dfrac{\Delta V_{Z1}}{2\sqrt{3}}$

10. Calculate the percent ripple across the Zener from: % ripple $= \dfrac{V_{AC(RMS)Z1} \times 100}{V_{Z1}}$

11. Insert the calculated and measured value, as indicated, into Table 3-5.

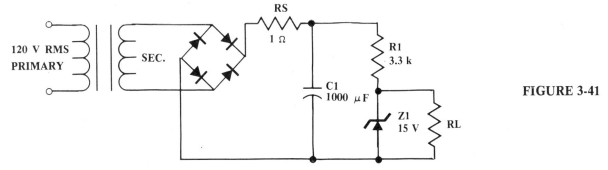

FIGURE 3-41

12. Connect the 4.7 k load resistor across the zener.
    a. Measure the output voltage across the zener.

b. Measure the ΔV across the zener.

c. Measure the V DC across the filter capacitor.

d. Measure the ΔV across the filter capacitor.

13. Insert the measured values into Table 3-5 and note for any significant changes.

14. Connect the 470 Ω load resistor across the zener.

a. Measure the output voltage across the zener.

b. Measure the ΔV across the zener.

c. Measure the V DC across the filter capacitior.

d. Measure the ΔV across the filter capacitor.

15. Insert the measured values into Table 3-5 and note the significant changes.

16. Connect the variable 10 kΩ potentiometer across the zener.

a. Monitor the regulated output voltage (across the zener) with the potentiometer and set to approximately 10 kΩ.

b. Vary the value of the pot until the output voltage changes. Just before it changes, or the zener comes out of regulation, note the rise in "noise".

NOTE: Zener diodes are extremely "noisy" devices and the noise increases dramatically just before it comes out of regulation.   Step 16 demonstrated this fact.

TABLE 3-5		V_{p-p} SEC.	V RMS SEC.	V_p	V DC C1	V_Z	V_{R1}	I_{R1}
N.L.	CALC.	////	////					
	MEAS.							
4.7 kΩ	MEAS.	////	////	////			////	////
470 Ω	MEAS.	////	////	////			////	////

		ΔV C1	ΔV Z1	I recurring DC	I recurring PEAK	VAC(RMS C1	VAC(RMS Z1	% Ripple
N.L.	CALC.							
	MEAS.					////	////	////
4.7 kΩ	MEAS.			////	////	////	////	////
470 Ω	MEAS.			////	////	////	////	////

PART 2B: THE SERIES PASS VOLTAGE REGULATOR

A power supply should be capable of delivering a constant voltage to a variety of loads with minimal ripple and with excellent regulation. The zener diode alone connot provide all of these specifications, but a zener diode combined with power transistors can and with (almost) zero ohms of output impedance approached.

In the circuit of Figure 3-42 (a), for instance, the zener diode provides 99 mA to a load and 1 mA to the zener, for a total of 100 mA of current through resistor R1. However, if the value of R1 is relatively low the ripple voltage across Z1 and, hence across the load, is relatively high. Additionally, and more importantly, if the load is removed, the zener must absorb the entire 100 mA of current and the device will be forced to dissipate 1.5 watts.

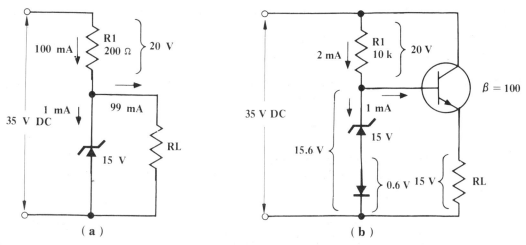

FIGURE 3-42

A solution to this problem is the circuit of Figure 3-42 (b) where the emitter follower provides 15 V of regulated voltage and approximately 100 mA of current to the load. And, in addition, the ripple content across the zener has been decreased. This is because only about 1 mA of base current is required to deliver 100 mA of current to the load for a transistor with a beta of approximately 100. Therefore, R1 can be increased from 200 ohms to 10 kohms , which causes the ripple voltage to the base and hence the load to be decreased. Also, the maximum zener current is now only 2 mA with Q1 disconnected and this insures that the power dissipated by the reference zener Z1 is low (30 milliwatts) and drift (instability) due to temperature minimized. Additionally, the output impedance has been reduced from a nominal 15 ohms to 50 ohms range to less than one ohm.

THE DARLINGTON PAIR COMPOUND

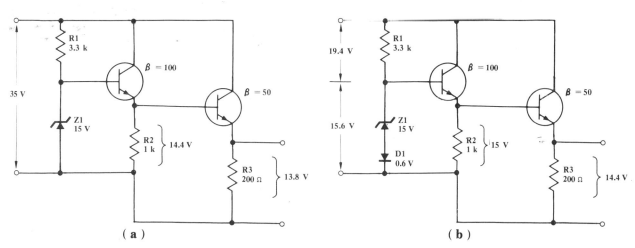

FIGURE 3-43

The darlington connection of Figure 3-43 is widely used to enhance the overall current gain of a circuit so that both ripple content to the load can be further attenuated and the output impedance can be further decreased.

123

The circuit of Figure 3-43 (b) also make use of the forward biased diode in series with the zener to further increase stability due to temperature. This is because the zener diode has a positive temperature coefficient which causes the current through the zener to rise with increased temperature, while the forward biased diode has a negative temperature coefficient which causes the current flow to decrease with increased temperature. Therefore, with the two devices in series, the net effect is a minimization of drift due to temperature change. Occasionally, two diodes, instead of one, are used to accomplish the same effect. Therefore, one and sometimes two forward biased diode are used in series with the zener diode. Also, single package zener diode-diode combinations can be utililized where the package combination can approach an optimum zero temperature coefficient, if manufacturer specifications and suggestions are followed.

SHORT CIRCUIT PROTECTION

The power supply is the one circuit in the system that must be protected from failures by other circuits within the system. Bench supplies also need some form of short circuit protection to insure against inadverten short circuit conditions that can exist during external breadboarding and test conditions. Two methods of protection used widely are the current sampling resistor-transistor technique and the current sampling resistor series diode technique. Both methods provide current limiting with the desired maximum which is established through the choice of sampling resistors and where even under short circuit conditions the current flow cannot be exceeded.

1. DIODE SHORT CIRCUIT PROTECTION

The addition of three diodes and a resistor (D2, D3, D4, and R3) is shown in Figure 3-44 where the diodes are connected in parallel with the series base-emitter junction of Q1, Q2, and resistor R3 to provide current limiting. This is because the diode string (D2, D3, and D4) remains ineffective in an open circuit condition until 1.8 V is dropped across them, and this only happens when approximately 0.6 V is dropped across the sampling resistor R3 — and this occurs when a minimum 300 mA flows through the 2 ohm resistor R3. Therefore, under the condition of 300 mA, 1.8 V is dropped across both the diode string of D2, D3, and D4 and the parallel base emitter junction of Q1 $\approx$ 0.6 V, Q2 $\approx$ 0.6 V, and R3 $\approx$ 0.6 V. Hence, a maximum current conditions exists (300 mA) because the diode string will not drop much greater than 1.8 V, and this, in turn, limits the voltage drop across the 2 ohm resistor R3 at 0.6 V. The

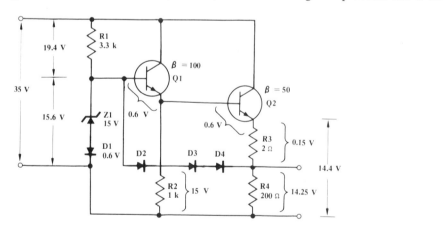

FIGURE 3-44

maximum current limit of 300 mA is illustrated in the following example for no load condition, loaded conditions of 48 ohms, and short circuit conditions. See Figure 3-45.

Under no load conditions, as shown in Figure 3-44, approximately 0.15 volts are dropped across R3 and the diode string has no effect on the circuit. However, when a load resistance of 48 ohms is connected across the output terminals, as shown in Figure 3-45 (a), approximately 300 mA is drawn through resistor R3, turning on the parallel diode string D2, D3, and D4. Decreasing the load resistance further, or even placing a short circuit across the output terminals, as shown in Figure 3-45 (b), still causes 300 mA to flow through R3, since the diode string still remains at 1.8 volts and V_{R3} at 0.6 volts. Therefore, the current is limited at 300 mA, under short circuit conditions. Also note, that during short circuit conditions, 1.8 volts exist at the base of Q1, with respect to ground, and zener diode Z1 is (obviously) out of regulation. See Figure 3-45(b).

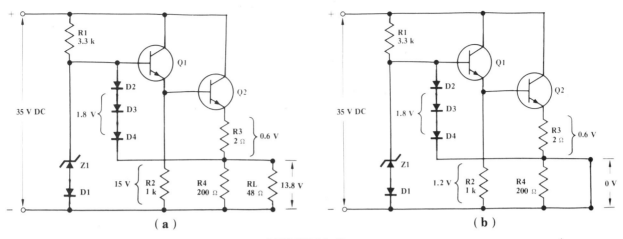

FIGURE 3-45

2. TRANSISTOR SHORT CIRCUIT PROTECTON

The addition of the sampling resistor and the transistor, as shown in Figure 3-46, provides essentially the same current limiting technique as does the series diode method. However, in this case, 0.6 volts dropped across the current sampling resistor R3 causes 0.6 volts to be dropped across the base emitter junction of Q3, causing Q3 to saturate and act as an electronic switch. Both the current flow through the transistor and the minimum voltage drop across it, even under worst case conditions (1.8 V) warrant no more than a low power device. The power dissipated by Q3 is solved in the sample problem. Also, the value of the base resistor R5 limits the base current to approximately 5 mA, which is more than enough current to insure adequate saturation of the Q3 transistor.

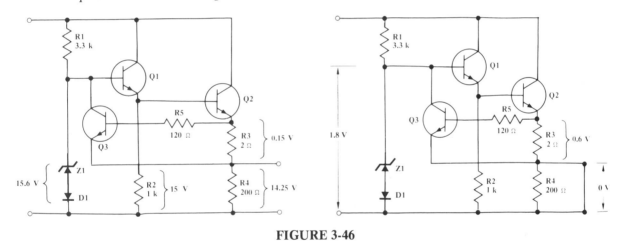

FIGURE 3-46

NOTE: Under short or circuit or heavy load conditions ($\approx$ 300 mA), the voltage at the input will drop from 35 volts to a lower value that is determined by the filter capacitor and the current drawn out of the filter capacitor. Therefore, the voltage drop across the R1 resistor can be estimated at $\approx$ 33 V (worst case) and the current through transistor Q3 at 10 mA, from $V_{R1}/R1 = 33 \text{ V}/3.3 \text{ k}\Omega$. Note, also, that 10 mA is about the same current that will flow through diodes D2, D3, and D4 during the short circuit condition as shown in Figure 3-45(b).

THE SERIES PASS REGULATOR

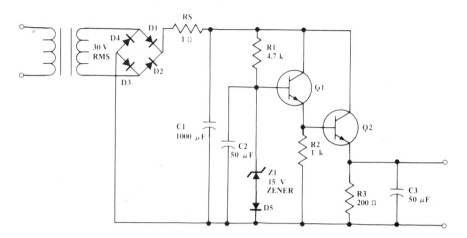

FIGURE 3-47

1. Solving for the peak voltage across the capacitor:

$$\text{Vpeak} = \sqrt{2} \times V \text{ RMS} = 1.414 \times 30 \text{ V} \approx 42.42 \text{ V.}$$

2. Solving for:
    a. The VDC across the C1 capacitor (where $\text{VDC}_{C1} = V_{C1}$):

$$V_{C1} \approx \text{Vpeak or, more correctly,}$$

$$V_{C1} = \text{Vp} - \Delta V_{C1}/2 = 42.42 \text{ V} - 783.5 \text{ mV}/2 \approx 42.03 \text{ V DC}$$

   b. The V DC at the base of Q1:

$$V_{B1} = V_{Z1} + V_{D5} = 15 \text{ V} + 0.6 \text{ V} = 15.6 \text{ V}$$

   c. The V DC across the R1 series resistor:

$$V_{R1} = V_{C1} - V_{B1} = 42.03 \text{ V} - 15.6 \text{ V} = 26.42 \text{ V}$$

   d. The current flow through resistor R1:

$$I_{R1} = V_{R1}/R1 = 26.42 \text{ V}/4.7 \text{ k}\Omega = 5.62 \text{ mA}$$

   e. The V DC at the emitter of Q1:

$$V_{E1} = V_{B1} - V_{BE1} = 15.6 \text{ V} - 0.6 \text{ V} = 15 \text{ V}$$

   f. The V DC at the emitter of Q2:

$$V_{E2} = V_{B2} - V_{BE2} = 15 \text{ V} - 0.6 \text{ V} = 14.4 \text{ V} \qquad \text{where: } V_{B2} = V_{E1}$$

   g. The current flow thourgh the emitter resistor of Q1:

$$I_{R2} \approx V_{R2}/R2 = 15 \text{ V}/1 \text{ k}\Omega = 15 \text{ mA}$$

   h. The current flow through the emitter resistor of Q2:

$$I_{E2} = V_{R3}/R3 = 14.4 \text{ V}/200\Omega = 72 \text{ mA}$$

   i. The voltage drop across the collector emitter of Q1:

$$V_{CE1} = V_{C1} - V_{E1} = 42.03 \text{ V} - 15 \text{ V} = 27.03 \text{ V} \qquad \text{where: } V_{C1} = \text{VDC}_{C1}$$

126

j. The voltage drop across the collector emitter of Q2:

$$V_{CE2} = V_{C2} - V_{E2} = 42.03 \text{ V} - 14.4 \text{ V} = 27.63 \text{ V} \qquad \text{where:} \quad V_{C2} = V_{C1} = VDC_{C1}$$

3. Solving for the power dissipated by Q1:

$$P_{Q1} \approx V_{CE1} \times (I_{E1} + I_{B2}) = 27.63 \text{ V} \times (15 \text{ mA} + 1.44 \text{ mA}) = 454.2 \text{ mW}$$

Solving for the power dissipated by Q2:

$$P_{Q2} = V_{CE2} \times I_{E2} = 27.03 \text{ V} \times 72 \text{ mA} \approx 1.946 \text{ watts}$$

4. Solving for the current demand from the zener for no load conditions:
   a. Solving for the current demand at the base of Q2 by virtue of the emitter current flow of 72 mA:

$$I_{B2} = I_{E2}/(\beta + 1) \approx 72 \text{ mA}/50 = 1.44 \text{ mA} \qquad \text{where:} \quad \text{beta of Q2} \approx 50$$

   b. Solving for the current demand out of the emitter of Q1 by virtue of the current flow into the base of Q2 and into the emitter resistor R2:

$$I_{E1} = I_{B2} + I_{R2} = 1.44 \text{ mA} + 15 \text{ mA} = 16.44 \text{ mA}$$

   c. Solving for the current demand out of the base of Q1 by virtue of the emitter current flow of 16.44 mA:

$$I_{B1} = I_{E1}/(\beta + 1) = 16.44 \text{ mA}/100 \approx 164 \text{ } \mu A \qquad \text{where:} \quad \text{the beta of Q1} \approx 100$$

NOTE: Since approximately 5.62 mA flows through the R1 resistor, then 169 μA flows into the base of Q1 and the balance of the current, or 55.36 mA, flows into the zener diode.

5. Solving for:
   a. The ΔV_{C1} ripple voltage at the filter capacitor for no load conditions:

$$\Delta V_{C1} = T/C \times I_T = \frac{T}{C} \times \left(\frac{V_{R1}}{R1} + \frac{V_{R2}}{R2} + \frac{V_{R3}}{R3} \right)$$

$$= \frac{1/120}{1000 \text{ } \mu F} \times (5.62 \text{ mA} + 16.44 \text{ mA} + 72 \text{ mA}) = 8.33 \times 94.06 \text{ mA} = 783.5 \text{ mA}$$

NOTE: The ripple voltage is solved by multiplying the T/C factor by the total current flow out of the capacitor. Therefore, the currents in the parallel branches are additive and they provide the total current out of C1. I_T is the total current.

   b. Solving for the ΔV ripple voltage across the zener diode where $RZ \approx 20$ ohms:

$$\Delta V_{Z1} = \frac{\Delta V_{C1} \times R_Z}{R_Z + R1} = \frac{783.5 \text{ mV} \times 20 \text{ } \Omega}{20 \text{ } \Omega + 4700 \text{ } \Omega} = 3.32 \text{ mVp-p}$$

   c. Solving for the ΔV ripple voltage at the emitter of Q1:

$$\Delta V \text{(emitter Q1)} = \frac{\Delta V_Z \times R2}{r_{e_1} + R2} = \frac{3.32 \text{ mVp-p} \times 1000 \text{ } \Omega}{2 \text{ } \Omega + 1 \text{ k}\Omega} \approx 3.31 \text{ mVp-p}$$

where: $r_{e_1} = 26 \text{ mV}/15 \text{ mA} \approx 2 \text{ } \Omega$

   d. Solving for the ΔV ripple voltage at the emitter of Q2:

$$\Delta V \text{(emitter Q2)} = \frac{\Delta V_{E2} \times R3}{r_{e_2} + R3} = \frac{3.31 \text{ mVp-p} \times 200 \text{ } \Omega}{1 \text{ } \Omega + 200 \text{ } \Omega} \approx 3.29 \text{ mVp-p} \qquad \text{where:} \quad r_{e_2} \approx 1 \text{ } \Omega$$

127

NOTE: Therefore, approximately the same ripple voltage across the zener diode is present at the output. However, ripple voltage can come from other sources. For instance, ripple voltage can filter through the normally high resistance of the reverse biased base-collector junctions of Q2 and possibly Q1. This is because the power transistors sometimes have base-collector junction resistances of less than 10 kΩ, which causes a portion of the ripple voltage at the collector to "leak" through and be developed at the emitter of Q2 and, hence, the output.

A second source of ripple can develop from a load being switched on and off at a given rate. This is because rapidly changing load causes rapidly changing currents which, when developed across the output impedance, can cause a ripple voltage to develop at the switching frequency.

Both of these types of ripple can be minimized by placing an additional capacitor (C3) in parallel with the output emitter resistor R3. Further improvement can be made by adding a 0.1 μF to 0.01 μF capacitor in parallel with C3 to minimize the output inductance of the C3 electrolytic capacitor.

6. Solving for the output impedance at the emitter of Q2:

$$Z_o = r_{e_2} \,/\!/\, R3 + \left[\frac{R2 \,/\!/ \left(\dfrac{RZ \,/\!/\, R1}{\beta Q1} \right)}{\beta Q2} \right] \approx r_{e_2} + \frac{RZ}{\beta Q1 \times \beta Q2} \approx 0.28\ \Omega + \frac{20\ \Omega}{100 \times 50} = 0.32\ \Omega$$

where: $r_{e_2} = 26\ mV/I_E = 26\ mV/92.7\ mA \approx 0.28\ \Omega$

where: $\beta Q1 = 100$ and $\beta Q2 = 50$

NOTE: Therefore, the output impedance of the darlington pair (Q1 and Q2) provides an output impedance of approximately 0.28 ohms by reflecting the impedance at the base of Q1 into the emitter of Q1 and then into the emitter of Q2 and the output. However, bulk resistance of the base emitter junction of Q2 rarely allows the output impedance (resistance) to go much below 0.5 Ω — the nominal resistance being between 0.5 Ω and 1 Ω. For this reason, power supplies that need lower output impedance use feedback to accomplish this end.

7. In Figure 3-48, a 50 Ω load resistance has been connected across the output terminals of the circuit of under discussion.

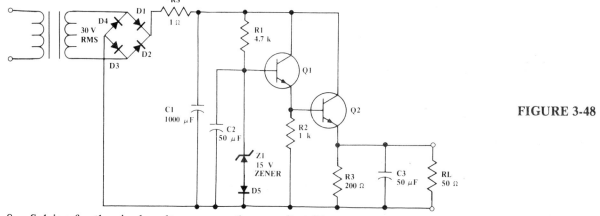

FIGURE 3-48

8. Solving for the ripple voltage across the capacitor C1:

$$\Delta V_{C1} = T/C \times (I_{R1} + I_{R2} + I_{R3} + I_{RL})$$

$$\approx (1/120)/1000\ \mu F \times (4.25\ mA + 15\ mA + 72\ mA + 288\ mA) \approx 3.14\ V\text{p-p}$$

where: $I_{R1} = V_{R1}/R1 = 20\ V/4.7\ k\Omega \approx 4.25\ mA$

$I_{R2} = V_{E1}/R2 = 15\ V/1\ k\Omega = 15\ mA$

$I_{R3} = V_o/R3 = 14.4\ V/200\Omega = 72\ mA$

$I_{RL} = V_o/RL = 14.4\ V/50\ \Omega = 288\ mA$

128

NOTE: Because of the heavy current demand out of capacitor C1, the voltage across C1 drops, causing less voltage drop across R1 and less current available to both the zener diode Z1 and the base of Q1. For this reason, the value of R1 is critical.

To insure against the possibility of not having enough current available (to keep the circuit in regulation) through proper selection of the R1 value, a test of the filter capacitor can be made. Connect a load resistance (chosen for the approximate maximum current condition) directly across the filter capacitor C1 and monitor the DC voltage across this capacitor. This will be the low DC value which will provide "guidance" in choosing the R1 resistor. For instance, if the DC voltage drops from 42.4 v to 35 V, then only 20 volts exist across R1 and approximately 4 mA flow through R1.

Another method is to solve for $\Delta V(C1)$, knowing either RL or the total current, and, hence, the DC level can be calculated at one-half the ripple magnitude level from the peak voltage level.

Solving for the ripple voltage across the zener diode for $R_Z \approx 20$ ohms:

$$\Delta V_{Z1} = \frac{\Delta V_{C1} \times R_Z}{R_Z + R1} = \frac{3.22 \text{ Vp-p} \times 20 \text{ } \Omega}{20 \text{ } \Omega + 4700 \text{ } \Omega} \approx 13.6 \text{ mVp-p}$$

NOTE: The ripple at the output (emitter of Q2) is approximately equal to the ripple across the zener diode Z1.

9. Solving for the current demand from the zener for loaded conditions:
    a. Solving for the current demand at the base of Q2 due to the emetter current flow of Q2:

$$I_{B2} = I_{E2}/(\beta + 1) \approx 360 \text{ mA}/50 = 7.2 \text{ mA}$$

$$\text{where: } I_{E2} = \frac{Vo}{RL} + \frac{V_{R2}}{R3} \approx \frac{14.4 \text{ V}}{50 \text{ } \Omega} + \frac{14.4 \text{ V}}{200 \text{ } \Omega} = 288 \text{ mA} + 72 \text{ mA} = 360 \text{ mA}$$

NOTE: The DC voltage across the output will drop slightly from 14.4 V DC under loaded 50 ohm conditions because of the signal loss across r_{e_2}.

b. Solving for the current demand out of the emitter of Q1 due to the current flow into the base of Q2 and the emitter resistor R2:

$$I_{E1} = I_{B2} + I_{R2} = 7.2 \text{ mA} + 15 \text{ mA} = 22.2 \text{ mA}$$

c. Solving for the current demand out of the base of Q1 due to the emitter current flow of 22.2 mA:

$$I_{BI} = I_{E1}/(\beta + 1) = 22.2 \text{ mA}/100 = 222 \text{ } \mu A$$

NOTE: Therefore, approximately 4 mA flow through the zener diodes and approximately 0.22 mA flow through to the base of Q1 for loaded conditions of 50 ohms.

10. Solving for the power dissipated by transistor Q1:

$$P_{Q1} = V_{CE1} \times I_E \approx 25.6 \text{ V} \times 22.2 \text{ mA} = 568.3 \text{ mW}$$

11. Solving for the power dissipated by Q2:

$$P_{Q2} = V_{CE2} \times I_{E2} \approx 26.4 \text{ V} \times 360 \text{ mA} = 9.5 \text{ watts}$$

NOTE: Under loaded conditions of 50 ohms, the voltage across Q1 can drop to 40.8 V from $Vp - \Delta V_C/2 = 42.42 \text{ V} - 3.19 \text{ Vp-p}/2$. This lessens the effective power dissipated by Q1 and Q2 with respect to a condition of no external load effect to capacitor C1. Therefore, $V_{CE1} = 40.8 \text{ V} - 15 \text{ V} = 25.8 \text{ V}$ and $V_{CE2} = 40.8 \text{ V} - 14.4 \text{ V} = 26.4 \text{ V}$.

12. Solving for the percent regulation using a 50 ohm load where the voltage out for a 50 ohm load is:

$$V_L = \frac{Vo(\text{no load}) \times R(\text{load})}{Zo + R(\text{load})} = \frac{14.4 \text{ V} \times 50 \text{ } \Omega}{0.50 \text{ } \Omega + 50 \text{ } \Omega} = \frac{720}{50.5} \approx 14.25 \text{ V}$$

where: Zo is calculated at 0.28 ohms but is rarely less than 0.5 ohms.

$$\% \text{ regulation} = \frac{V(NL) - V(L)}{V(NL)} \times 100 = \frac{14.4 \text{ V} - 14.25 \text{ V}}{14.4 \text{ V}} \times 100 = \frac{0.15 \text{ V}}{14.4 \text{ V}} \times 100 = 1.04\%$$

The above formula is the industry standard and the FCC formula solves percent regulation at:

$$\% \text{ regulation} = \frac{V(NL) - V(L)}{V(L)} \times 100 = \frac{14.4 \text{ V} - 14.25 \text{ V}}{14.25} \times 100 = 1.05\%$$

In Figure 3-49, additional components (three diodes and a resistor) have been added to the circuit to provide a short circuit proof condition by limiting the flow of current through the series pass transistor Q1.

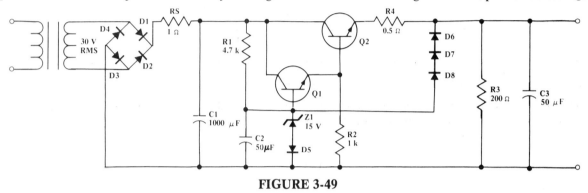

FIGURE 3-49

13. Solving for the output impedance, which includes the series R4 resistor:

$$Zo \approx r_{e_2} + (R_Z/[\beta_1 \times \beta_2]) + R4 = 0.50 \text{ }\Omega + (20 \text{ }\Omega/5000) + 0.5 \text{ }\Omega \approx 1 \text{ }\Omega$$

14. **SHORT CIRCUIT CONDITIONS:**
   a. Solve for the maximum current to current limiting resistor R4:

$$I_{SC} = \frac{V_{R4}}{R4} = \frac{V_{(D1 + D2 + D3)} - (V_{BE1} + V_{BE2})}{R4}$$

$$= \frac{(0.6 \text{ V} + 0.6 \text{ V} + 0.6 \text{ V}) - (0.6 \text{ V} + 0.6 \text{ V})}{0.5 \text{ }\Omega} = \frac{0.6 \text{ V}}{0.5 \text{ }\Omega} = 1.2 \text{ A}$$

   b. Solving for the power dissipated by Q2:

$$P_{Q2} = V_{CE2} \times I_{E2} = 36.8 \text{ V} \times 1.2 \text{ A} = 44.16 \text{ watts}$$

where: $V_{CE2} = VDC_{C1} - V_{R4} = 37.4 \text{ V} - 0.6 \text{ V} = 36.8 \text{ V}$

NOTE: $V_{C1} = Vp - \Delta V_{C1}/2 = 42.42 \text{ V} - 10 \text{ V}/2 \approx 37.4 \text{ V DC}$

   c. Solving for the power dissipated by Q1:

$$P_{Q1} = V_{CE1} \times I_{E1} \approx 36.2 \text{ V} \times 25.2 \text{ mA} = 912.24 \text{ mW}$$

where: $I_{E1} \approx I_{E2}/\beta_{Q1} + V_{R2}/R2 = 1.2 \text{ A}/50 + 1.2 \text{ V}/1 \text{ k}\Omega = 24 \text{ mA} + 1.2 \text{ mA} = 25.2 \text{ mA}$

NOTE: Under short circuit conditions approximately 1.2 amperes are flowing out of capacitor C1 and this could force the DC voltage across the filter capacitor C1 to drop to approximately 37.4 V DC. Therefore, the transistor Q2 must be able to withstand approximately 45 watts of power dissipation and Q1 approximately 900 mW for as long as the short circuit condition exists. Remember that a shorted condition places a ground at the normally positive output terminal and the current through the diode string is at maximum condition.

   d. Solving for the current through the diode string.

$$I(\text{diode string}) = \frac{V_{C1} - V_{(D6 + D7 + D8)}}{R1} = \frac{37.4 \text{ V} - 1.8 \text{ V}}{4.7 \text{ k}\Omega} = 7.57 \text{ mA}$$

130

15. Reconnecting the circuit, as shown in Figure 3-50, takes advantage of a second method of short circuit protection, where the current is again limited by series resistor R4. However, this time, R4 is used to turn on Q3, when $V_{BE3} = 0.6$ volts.

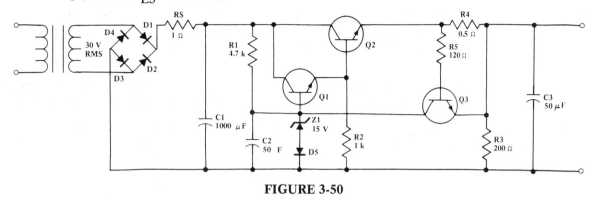

FIGURE 3-50

16. With the output impedance approximately 1 ohm and the short circuit condition approximately 1.2 amperes, as solved in the previous circuit, solve for:

a. The power dissipated by Q1:

$$P_{Q1} = V_{CE1} \times I_{E1} \approx 36.2 \text{ V} \times 25.2 \text{ mA} = 912.24 \text{ mW}$$

b. The power dissipated by Q2:

$$P_{Q2} = V_{CE2} \times I_{E2} = 36.8 \text{ V} \times 1.2 \text{ A} = 44.16 \text{ watts} \quad \text{where:} \quad I_{E2} = I_{SC}$$

c. The power dissipated by Q3:

$$P_{Q3} = V_{CE3} \times I_{E3} \approx 1.8 \text{ V} \times 7.57 \text{ mA} = 13.62 \text{ mW}$$

$$\text{where:} \quad I_{E3} \approx (V_{C1} - V_{CE3})/R1 = (37.4 \text{ V} - 1.8 \text{ V})/4.7 \text{ k}\Omega \approx 7.57 \text{ mA}$$

$$\text{where:} \quad V_{CE3} = V_{BE1} + V_{BE2} + V_{R4} = 0.6 \text{ V} + 0.6 \text{ V} + 0.6 \text{ V} = 1.8 \text{ V}$$

NOTE: Therefore, Q3 need only be a low power transistor. This is because the collector-emitter is connected across the series base emitter junctions of Q1 and Q2 and the current limiting resistor R4. The voltage drop across Q3, under normal operation conditions, is approximately 1.2 V to 1.4 V and, under short circuit conditions, approximately 1. 8 V — with about 0.6 V dropped across R4 and about 0.01 V across R5. Resistor R5 is used to prevent Q3 from going to a fully saturated condition where $Q3 \approx 0.4$ volts — a condition that would negate the short circuit protection circuit.

EXPERIMENTAL OBJECTIVES

To investigate the series pass voltage regulated power supply.

LIST OF MATERIALS

1. Transformer: 120 V : 24 V
2. Transistors: 2N3055 (one), 2N3053 (one), or equivalents
3. Diodes: 1N4001 (three) or equivalents
4. Resistors: 1 Ω (two)   1 kΩ (one)   2.2 kΩ (one)   3.3 kΩ (one)
5. Capacitors: 100 µF (one)   35 µF (one)
6. Zener Diode: 15 volt

EXPERIMENTAL PROCEDURES

1. Connect the circuit as shown in Figure 3-51.

131

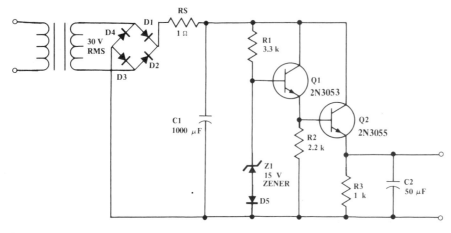

FIGURE 3-51

2. Across the secondary of the transformer:
   a. Measure the Vp-p with an oscilloscope.
   b. Measure the RMS volatage with a voltmeter.
3. Calculate the Vp across the filter capacitor C1.
   Measure the Vp across the filter capacitor C1 where: $V_p = VDC_{C1} + \Delta V_{C1}/2$
4. Calculate the DC voltage across the filter capacitor from:

$$V_{C1} = V_p - \Delta V_{C1}/2 \approx V_p \quad \text{where:} \quad VDC_{C1} = V_{C1}$$

   Measure the DC voltage across the filter capacitor.
5. Calculate the DC voltage:
   a. At the base of Q1 from: $V_{B1} = V_{Z1} + V_{D5}$
   b. At the base of Q2 from: $V_{B2} = V_{B1} - V_{BE1}$

   c. At the emitter of Q2 from: $V_{E2} = V_{B2} - V_{BE2}$
   Measure the DC voltage:
   a. At the base of Q1.
   b. At the base of Q2 — emitter of Q1.
   c. At the emitter of Q2 — across the output.
6. Calculate the voltage drop across:
   a. Transistor Q1 from: $V_{CE1} = V_{C1} - V_{E1}$ where: $V_{C1} = VDC_{C1}$
   b. Transistor Q2 from: $V_{CE2} = V_{C2} - V_{E2}$ where: $V_{C2} = VDC_{C1}$

   Measure the voltage drop across transistors Q1 and Q2.
7. Calculate the ripple voltage:
   a. Across the filter capacitor from:

$$\Delta V_{C1} = T/C \times I_L$$

   where: $I_L = V_{R1}/R1 + V_{R2}/R2 + V_{R3}/R3$

   b. Across the zener diode from:

$$\Delta Z1 = \frac{\Delta V_{C1} \times R_{Z1}}{R1 + R_{Z1}}$$

NOTE: Unless the dynamic impedance of the zener diode is known (from data sheets), use the "rule of thumb" $R_Z = V_Z$ until measurements can be made. Also, ignore the dynamic resistance of D5.

   c. Across the R2 resistor (emitter of Q1) from: $\Delta V_{R2} = \dfrac{\Delta V_{Z1} \times R2}{r_{e_1} + R2}$

   d. Across the R3 resistor (emitter of Q2) from: $\Delta V_{R3} = \dfrac{\Delta V_{R2} \times R3}{r_{e_2} + R3}$

132

Measure the ripple voltage across:
 a.   Filter capacitor C1.
 b.   The series zener diode Z1 and Diode D5 — base of Q1.
 c.   Resistor R2 — base of Q2.
 d.   Resistor R3 — output.
8. Calculate the power dissipated by:
 a.   Transistor Q1 from: $P_{Q1} = V_{CE1} \times I_{E1}$ where: $I_{E1} = I_{R2} + I_{BQ2}$
 b.   Transistor Q2 from: $P_{Q2} = V_{CE2} \times I_{E2}$ where: $I_{E2} = I_L$

9. Insert the calculated and measure values, as indicated, into Table 3-6.

TABLE 3-6	Vp-p SEC	V RMS SEC	Vp C1	VDC C1	VB Q1	VB Q2	VE Q2
CALCULATED	/////	/////					
MEASURED							
	V_{CE1}	V_{CE2}	P_{Q1}	P_{Q2}	ΔV_{C1}	ΔV_{B1}	ΔV_{R2}	ΔV_{R3}
CALCULATED								
MEASURED		/////	/////					

10. Connect the circuit as shown in Figure 3-52, which utilizes short circuit proof diode protection.

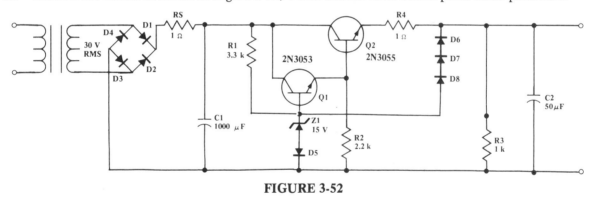

FIGURE 3-52

11. Monitor the DC voltages around the circuit to make sure the circuit is working. Connect a 100 Ω load resistor across the output and calculate the ΔV_{C1} from:

$$\Delta V_{C1} = \frac{T}{C} \times I_L \quad \text{where:} \quad I_L = \frac{V_{R1}}{R1} + \frac{V_{R2}}{R2} + \frac{V_{R3}}{R3} + \frac{Vo}{RL} \quad \text{(under load)}$$

NOTE: $V_{R3} = V_{E2} - V_{R4}$, but V_{R4} is only a few tenths of a volt and can be ignored in the calculations. Also, ΔV_{C1} is monitored under no load conditions to provide a loaded condition comparison.

Measure the ripple voltage for a loaded condition of 100 Ω across:
 a.   Filter capacitor C1.
 b.   The series zener diode Z1 and diode D5 — base of Q1 and at the emitter of Q1.
 c.   Resistor R3 — 100 Ω output resistor.
12. Measure the DC voltages for the 100 Ω loaded conditions:
 a.   Across the filter capacitor C1 — collectors of Q1 and Q2.
 b.   Emitters of Q1 and Q2.
 Calculate the power dissipated by:
 a.   Transistor Q1 from: $P_{Q1} = V_{CE1} \times I_{E1}$
 b.   Transistor Q2 from: $P_{Q2} = V_{CE2} \times I_{E2}$

133

13. Remove the 100 Ω load, momentarily, and measure the voltage across the output. Then, reconnect the 100 Ω load and remeasure the voltage across the output. Calculate the percent regulation from:

$$\% \text{ regulation} = \frac{V(NL) - VL}{V(NL)} \times 100$$

14. Calculate the short circuit current of the circuit from:

$$I_{SC} = \frac{V_{(D6 + D7 + D8)} - (V_{BE1} + V_{BE2})}{R4}$$

$$I_{SC} = \frac{V_{R4}}{R4} \quad \text{where } V_{R4} \text{ can be monitored or estimated at about 0.6 V.}$$

Momentarily short out the output terminals — long enought to make the following measurements:
a.   Measure the actual voltage drop across R4.
b.   Or, measure the short circuit current with an ammeter.
c.   Measure the DC voltage across the filter capacitor C1.
Calculate I_{SC} from $V_{R4}/R4$, and compare to the measured value taken. If R4 is a known value, the current meter reading in unnecessary.
Calculate the power dissipated by Transistors Q1 and Q2 from:

$$P_{Q2} = V_{CE2} \times I_{SC} \quad \text{and} \quad P_{Q1} = V_{CE1} \times I([I_{SC}/\beta_{Q2}] + I_{R2})$$

15. Insert the calculated and measured values, as indicated, into Table 3-7.

TABLE 3-7		V DC					V_{CE}	
		C1	B1	B2	E2	Vo	Q1	Q2
NO LOAD	CALCULATED	/////	/////	/////	/////	/////	/////	/////
	MEASURED							
100 Ω LOAD	CALCULATED	/////	/////	/////	/////	/////	/////	/////
	MEASURED							
SHORT CKT.	CALCULATED	/////	/////	/////	/////	/////	/////	/////
	MEASURED							

		ΔV				I_{SC}	P_{Q1}	P_{Q2}	% REG
		C1	B1	E2	Vo				
NO LOAD	CALCULATED	/////	/////	/////	/////	/////	/////	/////	/////
	MEASURED					/////	/////	/////	/////
100 Ω LOAD	CALCULATED		/////	/////	/////				
	MEASURED					/////	/////	/////	/////
SHORT CKT.	CALCULATED	/////	/////	/////	/////				/////
	MEASURED			/////	/////		/////	/////	/////

16.   Connect the circuit as shown in Figure 3-53, which obtains short circuit protection by utilizing transistor Q3 and Resistors R4 and R5.

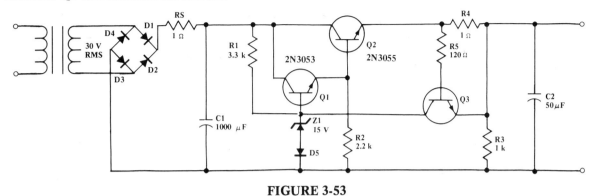

FIGURE 3-53

17.   Monitor the DC voltages around the circuit to make sure the circuit is working. Connect a 100 Ω load resistor across the output.
   a.   Monitor the DC voltage across capacitor C1, at the bases of Q1 and Q2 and at the emitter of Q2.
   b.   Monitor the DC voltage across the output and across transistors Q1 and Q2.

NOTE: The DC voltages should be close in value to those of the previous (three diode short circuit protected) circuit and the above measurements should confirm the similar values.

   Measure the 100 Ω loaded circuit for ripple voltage at:
   a.   The filter capacitor C1, at the base of Q1 and the emitter of Q2.
   b.   Resistor R3 — 100 Ω output resistor.
   Measure the no load circuit conditions and the 100 Ω loaded circuit conditions and calculate percent regulation from:

$$\% \text{ regulation} = \frac{V(NL) - V(L)}{V(NL)} \times 100$$

18.   Calculate the short circuit conditions of the circuit from:

$$I_{SC} = \frac{V_{BE3} + V_{R5}}{R4} \approx \frac{V_{BE3}}{R4}$$

   Momentarily short out the output terminals — long enough to make the following measurements:
   a.   The voltage drop across R4.
   b.   The DC voltage across the filter capacitor C1.
   Calculated the I_{SC} from:

$$I_{SC} = V_{R4}/R4, \quad \text{where R4 is known}$$

NOTE: V_{R4} will be slightly higher than that of the three diode circuit because of the added voltage drop across resistor R5.

   Calculate for the current flow through R3 from:

$$IE3 = \frac{V_{C1} - V_{CE3}}{R1} \quad \text{where:} \quad V_{CE3} \approx V_{BE1} + V_{BE2} + V_{R4}$$

   Monitor V_{CE3}, during short circuit conditions, and test the value of R5 by momentarily shorting it out. Monitor both V_{CE3} and I_{SC} conditions.

3.   Calculate the power dissipated by transistors Q2 and Q3 from:

$$P_{Q2} = V_{CE2} \times I_{SC} \quad \text{and} \quad P_{Q1} = V_{CE1} \times I_{SC}/\beta_{Q2}$$

$$PQ3 \approx V_{CE3} \times I_{R1} \quad \text{where:} \quad V_{R1} = VDC_{C1} - V_{CE3}$$

135

19. Insert the calculated and measured values, as indicated, into Table 3-8.

TABLE 3-8		V DC					V CE	
		C1	B1	B2	E2	Vo	Q1	Q2
NO LOAD	CALCULATED	▨	▨	▨	▨	▨	▨	▨
	MEASURED							
100 Ω LOAD	CALCULATED	▨	▨	▨	▨	▨	▨	▨
	MEASURED							
SHORT CKT.	CALCULATED	▨	▨	▨	▨	▨	▨	▨
	MEASURED							

		ΔV							
		C1	B1	E2	Vo	ISC	P_{Q2}	P_{Q3}	% REG
NO LOAD	CALCULATED	▨	▨	▨	▨	▨	▨	▨	
	MEASURED					▨	▨	▨	▨
100 Ω LOAD	CALCULATED					▨	▨	▨	
	MEASURED					▨	▨	▨	▨
SHORT CKT.	CALCULATED	▨	▨	▨	▨				▨
	MEASURED						▨	▨	▨

136

SECTION III: THE VOLTAGE ADJUSTABLE VARIABLE POWER SUPPLY

Most bench power supplies are voltage variable over a wide voltage range, usually starting at zero volts and extending upwards. For solid state applications, a supply that ranges from 0 — 30 V at 1 Amp is more than adequate. Voltage adjustable power supplies, on the other hand, vary over limited voltage ranges of about 5 volts and normally have screw driver adjust. The voltage adjustable supply is used in all systems where optimum V DC is a requirement. A major advantage of the voltage adjustable, or voltage variable, power supply is that low output impedance, approaching zero ohms, is obtained by negative feedback techniques. By using negative feedback, a constant output voltage is obtained while providing whatever current that the load requires.

PART 3A — THE VOLTAGE ADJUSTABLE FEEDBACK REGULATOR

The voltage adjustable feedback circuit shown in Figure 3-54 provides two immediate improvements over the previously discussed non-feedback series pass regulators. The first improvement is that the voltage can be adjusted (over a limited range) and the second improvement is that Q3 provides 180° feedback, which effectively lowers the output impedance to zero ohms. Both of these improvements will be explained in detail. The main drawback of the circuit of Figure 3-54 is that improved ripple attenuation is needed, and this attenuation will be accomplished by using a high impedance in place of resistor R1. For this purpose, a constant current source will be used.

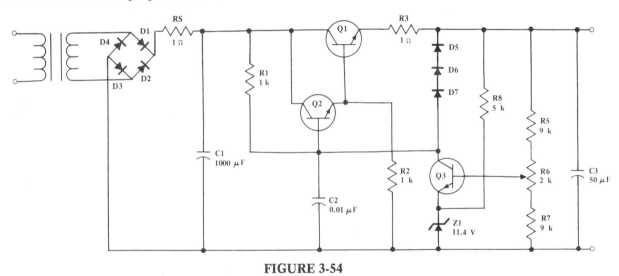

FIGURE 3-54

THE ADJUSTABLE OUTPUT VOLTAGE

The ouput voltage of the circuit of Figure 3-54 can be adjusted from 21.8 volts to 26.6 volts. When the wiper of variable resistor R6 is set to mid-rotation, the output voltage is 24 volts. The voltage output for the mid, bottom, and top rotations of the variable resistor R6 can be determined as follows:

1. A mid-rotation condition exists when the variable resistor R6 (potentiometer) is set so that the 2 kΩ resistor is divided into equal 1 kΩ values and, effectively, 9 kΩ + 1 kΩ exist on each half of the voltage divider circuit — as shown in Figures 3-55 (a) and 3-55 (b).

The output voltage for mid-rotation of variable resistor R6 can be solved by two methods:

a. The reverse voltage divider equation, where the voltage across the sampling resistance is compared to the total of the series resistance, can be used; and like the voltage divider equation, it provides a comparison of the two resistances.

$$V_O = \frac{V_{B3} \times (R5 + R6 + R7)}{R7 + R6/2} = \frac{12\,V \times (9\,k\Omega + 2\,k\Omega + 9\,k\Omega)}{9\,k\Omega + 1\,k\Omega} = \frac{12\,V \times 20\,k\Omega}{10\,k\Omega} = 24\,V$$

where: $V_{B3} = V_{Z1} + V_{BE3} = 11.4\,V + 0.6\,V = 12$ volts

b. The second method uses ohms law to solve for the current through resistors R7 and R6/2, and it

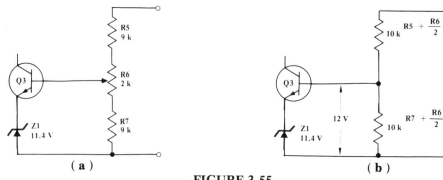

FIGURE 3-55

assumes that approximately the same current flows through the remaining R5 + R6/2 resistance, with little current flow into the Base of Q3.

$$I(R7 + R6/2) = \frac{V(R7 + R6/2)}{R7 + R6/2} = \frac{V_{B3}}{R7 + R6/2} = \frac{12 \text{ V}}{9 \text{ k}\Omega + 1 \text{ k}\Omega} = 1.2 \text{ mA}$$

$$Vo = I(R7 + R6/2)(R5 + R6 + R7) = 1.2 \text{ mA}(9 \text{ k}\Omega + 2 \text{ k}\Omega + 9 \text{ k}\Omega) = 24 \text{ V}$$

where: $I(R7 + R6/2) \approx I(R5 + R6 + R7)$, which is sometimes referred to as I(bleed)

2. Bottom rotation is obtained when the wiper of variable resistor R6 is adjusted so that the 12 volts of the base of Q3 are developed across resistor R7 only. This condition is solved from:

a. $$Vo = \frac{V_{B3} \times (R5 + R6 + R7)}{R7} = \frac{12 \text{ V} \times 20 \text{ k}\Omega}{9 \text{ k}\Omega} \approx 26.6 \text{ V} \quad \text{(See Figure 3-56)}$$

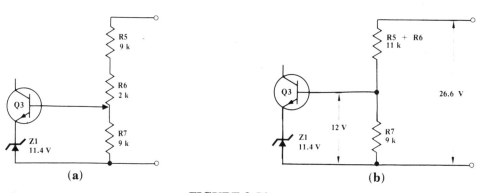

FIGURE 3-56

b. The second method is to divide the Q3 base voltage, which is developed across R7, by the resistance of R7 and multiply the developed current by the total series resistance. Therefore:

$$I_{R7} = I(\text{bleed}) = V_{B3}/R7 = 12 \text{ V}/9 \text{ k}\Omega \approx 1.33 \text{ mA, and}$$

$$Vo = (R5 + R6 + R7)I(\text{bleed}) = 20 \text{ k}\Omega \times 1.33 \text{ mA} \approx 26.6 \text{ V}$$

3. Top rotation is obtained when the wiper of the variable resistor is adjusted so that the 12 V at the base of Q3 is developed across the series R6 and R7 resistors. For this condition, the output voltage is solved from:

a. $$Vo = \frac{V_{B3}(R5 + R6 + R7)}{R6 + R7} = \frac{12 \text{ V} \times 20 \text{ k}\Omega}{2 \text{ k}\Omega + 9 \text{ k}\Omega} \approx 21.8 \text{ V} \quad \text{(See Figure 3-57)}$$

b. The second method of solution is to divide the base voltage of Q3, which is developed across series resistors R6 and R7, by R6 + R7 and multiply this current by the total series resistance. Therefore:

138

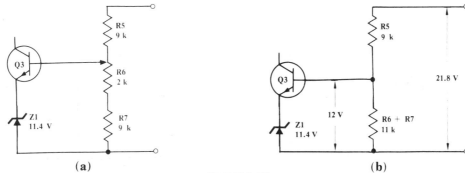

(a)   **(b)**

FIGURE 3-57

$$V_O = \frac{V_{B3}}{R6 + R7} \times (R5 + R6 + R7) = \frac{12\ V}{2\ k\Omega + 9\ k\Omega} \times (9\ k\Omega + 2\ k\Omega + 9\ k\Omega)$$

$$\approx 1.09\ mA \times 20\ k\Omega = 21.8\ V.$$

180° COMPENSATING FEEDBACK

The 180° (negative) feedback performs the function of monitoring the output voltage for changes in voltage and then corrects for these changes through the 180° phase shifting of Q2. These corrections, in effect, provide a constant output voltage at the output under loaded or no load conditons. This compensating feedback is shown in the simplified circuit of Figure 3-58 where, because of the zener diode and the R3 and R4 voltage divider resistors, 24 volts exist at the output and 24.6 volts exist at the base of Q1. Therefore, the voltage drop across the R1 resistor is 5.4 volts, where $V_{C1} = 30$ volts and where 2 mA flow through R1 because $V_{R1}/R1 = 5.4\ V/2.7\ k\Omega$. Hence, if the voltage at the output rises or falls because of loading or unloading, then the initial 2 mA of current through R1 also rises and falls, which causes the voltage at the base of Q1 to fall and rise in opposition to, or 180° out of phase with, the output voltage. The net effect is as follows:

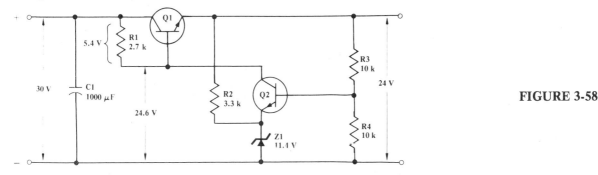

FIGURE 3-58

1. Assume that the voltage across the output rises, momentarily, from 24 V DC to 24.4 V DC. When this happens, the voltage at the junction of resistor R3 and R4 rise from 12 volts to 12.2 volts and the base-emitter junction voltage rises from 0.6 volts to 0.8 volts. Therefore, the base current of Q2 rises, the collector current of Q2 rises, and the voltage drop across the resistor R1 rises. If the collector current of Q2, and hence the R1 resistor current, rises from 2 mA to 2.2 mA, for example, then the voltage drop across R1 increase from 5.4 volts to 5.94 volts and the voltage at the base of Q1 drops from 24.6 volts to 24.06 volts. Therefore, the output voltage drops to 23.46 volts, which is below the initial 24 V DC, and this will cause the V DC at the output to correct for itself.
2. Therefore, assuming the voltage across the output drops to 23.46 volts, then the voltage across R4 drops to 11.73 volts and the voltage across the base-emitter junction of Q2 drops to 0.33 volts — considerably below the initial 0.6 volts. As a result, the base current decreases, the collector current decreases, the current through resistor R1 decreases, the voltage drop across resistor R1 decreases, and the voltage at the base of Q1 increases. For instance, if the collector current of Q2 drops to 1.8 mA, the V_{R1} is 4.8 volts and the voltage at the base of Q1 is 25.14 volts and the output voltage is 24.54 volts. Obviously, further correction is needed to re-establish initial voltage requirements at 24 volts.
3. Overall, the feedback corrects for voltage rise and fall by providing opposite voltage change to voltage

139

changes at the output. This forces the output to attempt to remain constant and to provide an output impedance that, theoretically, approaches the "optimum" zero ohm condition.

RIPPLE ATTENUATION USING THE CONSTANT CURRENT SOURCE

The main disadvantage of the feedback circuit of Figure 3-59(a) is that ripple content developed across C1 is not attenuated enough by resistor R1. This is because the resistance (impedance) at the base of Q1 is high, which means that either a filter capacitance must be placed at the base of Q2 to provide a lower impedance point or the value of R1 must be increased. However, placing a large capacitor at the base of Q1 would negate the advantage of the feedback circuit by slowing down the reaction time, and physically increasing the value of R1 is limited to the available distributed DC voltage, since too large a resistance will make circuit operation fail. A circuit that provides reduced ripple, without the associated high voltage drop, is the constant current source connection of Figure 3-59(b). Resistor R4 and zener Z2 provide the (Voltage divider) ripple voltage attenuation and Q3 provides the constant current source and further attenuation.

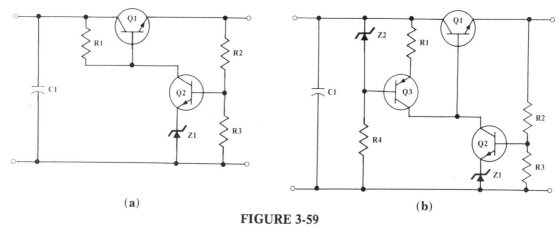

(a)   (b)

FIGURE 3-59

RIPPLE ATTENUATION USING A SEPARATE POWER SUPPLY

Another techniques for ripple attenuation is to provide a separate power supply to power the base of Q1, as shown in Figure 3-60. Here, when a load across the output causes current drain out of the filter

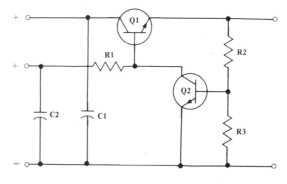

FIGURE 3-60

capacitor C1, the ripple content across C1 cannot flow through the high (reverse biased) collector resistance of Q1. Also, the ripple developed across filter capacitor C2 is at least beta times less than the loaded or unloaded conditions of C1 and, hence, no appreciable ripple voltage appears at the output. The only drawbacks to this connection is the need for a multiple secondary transformer and additional rectification and filter capacitors. The idea of a separate supply to power the regulating circuit, however, is widely used in commercial power supplies.

Other techniques used for ripple attenuation involving the single secondary winding, but using FET and constant current drives, are discussed in the differential amplifier driven power supplies in the next section.

ZERO OHM CONDITIONS

A circuit that provides 10 volts of output voltage under loaded or no load conditions can be considered to have, theoretically, zero ohms of output resistance and, hence, zero percent regulation. In practice, this can only be approached, but power supplies having 0.01% load and line regulation are common.

To illustrate the zero ohm concept further, consider the simplistic circuits of Figure 3-61, where the output resistances of the circuits are one ohm and zero ohms. Under load or no load conditions, the voltages drop from 10 volts no load to 9 volts loaded for the one ohm output resistance circuit, and the voltage remains at 10 volts for the zero ohm output resistance circuit. Obviously, zero voltage change between no load and loaded conditions implies a zero ohm output condition.

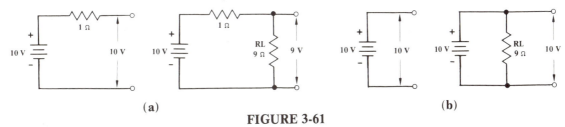

FIGURE 3-61

1. Percent regulation for the circuit of Figure 3-61 (a):

$$\% \text{ regulation} = \frac{V(NL) - V(L)}{V(NL)} \times 100 = \frac{10\,V - 9\,V}{10\,V} \times 100 = 10\%$$

2. Percent regulation for the circuit of Figure 3-61 (b):

$$\% \text{ regulation} = \frac{V(NL) - V(L)}{V(NL)} \times 100 = \frac{10\,V - 10\,V}{10\,V} \times 100 = 0.0\%$$

Since feedback networks attempt to maintain the output voltage at a constant value under loaded or no load conditions, the circuit of Figure 3-58 attempts to simulate the conditions of the Figure 3-61 (b) circuit.

THE VOLTAGE ADJUSTABLE POWER SUPPLY WITH 180° FEEDBACK — CIRCUIT ANALYSIS

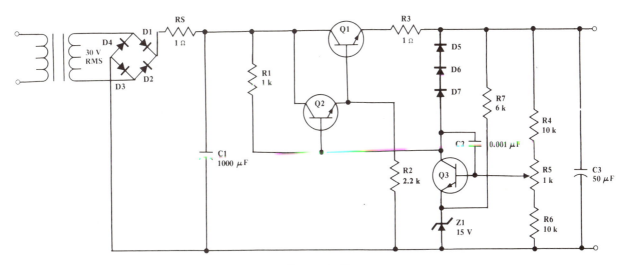

FIGURE 3-62

1. Solving for the peak voltage across filter capacitor C1:

$$V_{peak} = \sqrt{2} \times RMS = 1.414 \times 30\,V \approx 42.42\,V$$

2. Solving for:

a. The V DC across capacitor C1:

$$VDC_{C1} \approx Vpeak, \text{ or } VDC_{C1} = V_p - (\Delta V_{C1}/2) = 42.42 \text{ V} - (2.38 \text{ mV}/2 \approx 42.3 \text{ V DC}$$

b. The VDC_{C1} at the base of Q3:

$$V_{B3} = V_{Z1} + V_{BE3} = 15 \text{ V} + 0.6 \text{ V} = 15.6 \text{ V}$$

c. The voltage at the output for R5 mid-rotation conditions:

$$V_o = \frac{V_{B3} \times (R4 + R5 + R6)}{R6 + R5/2} = \frac{15.6 \text{ V} \times (10 \text{ k}\Omega + 1 \text{ k}\Omega + 10 \text{ k}\Omega)}{10 \text{ k}\Omega + 1 \text{ k}\Omega/2} = 31.2 \text{ V}$$

d. The voltage at the base of Q1:

$$V_{B1} = V_{E1} + V_{BE1} \approx 31.2 \text{ V} + 0.6 \text{ V} = 31.8 \text{ V} \qquad \text{where: } V_{R3} \text{ is small and can be initially ignored.}$$

e. The voltage at the base of Q2:

$$V_{B2} = V_{B1} + V_{BE2} = 31.8 \text{ V} + 0.6 \text{ V} = 32.4 \text{ V}$$

f. The voltage drop across resistor R1:

$$V_{R1} = V_{C1} - V_{B2} = 42.3 \text{ V} - 32.4 \text{ V} = 9.9 \text{ V}$$

g. The voltage drop across the collector-emitter of Q1:

$$V_{CE1} = V_{C1} - V_{E1} = 42.3 \text{ V} - 31.2 \text{ V} = 11.1 \text{ V}$$

$$\text{where: } V_{C1} = V_{C2} = VDC_{C1} \text{ and } V_{E1} \approx V_o$$

h. The voltage drop across the collector-emitter of Q2:

$$V_{CE2} = V_{C2} - V_{E2} = 42.3 \text{ V} = 31.8 \text{ V} = 10.5 \text{ V}$$

$$\text{where: } V_{C2} = V_{C1} = VDC_{C1} \text{ and } V_{E2} = V_{B1}$$

i. The voltage drop across the collector-emitter of Q3:

$$V_{CE3} = V_{C3} - V_{E3} = 32.4 \text{ V} - 15 \text{ V} = 17.4 \text{ V}$$

$$\text{where: } V_{C3} = V_{B2} \text{ and } V_{E3} = V_{Z1}$$

3. Solving for the voltage at the output where the wiper of the potentiometer R5 is at full bottom rotation.

$$V_o = \frac{V_{B3}(R4 + R5 + R6)}{R6} = \frac{15.6 \text{ V}(10 \text{ k}\Omega + 1 \text{ k}\Omega + 10 \text{ k}\Omega)}{10 \text{ k}\Omega} = 32.76 \text{ V}$$

NOTE: Therefore, the voltage distribution around the circuit changes, accordingly, to reflect the new output voltage. For instance, now $V_{CE1} = 9.54 \text{ V}$, $V_{CE2} = 8.94 \text{ V}$, and $V_{CE3} = 18.96 \text{ V}$.

4. Solving for the voltage at the output where the wiper of the potentiometer is at full top rotation:

$$V_o = \frac{V_{B3}(R4 + R5 + R6)}{R6 + R5} = \frac{15.6 \text{ V}(10 \text{ k}\Omega + 1 \text{ k}\Omega + 10 \text{ k}\Omega)}{10 \text{ k}\Omega + 1 \text{ k}\Omega} = 29.78 \text{ V}$$

NOTE: Therefore, the voltage distributions changes accordingly and the new $V_{CE2} = 12.52$ V, $V_{CE1} = 11.92$ V, and $V_{CE3} = 15.98$ V.

5. Solving for the ripple voltage across filter capacitor C1 for mid-rotation conditions of R5:

$$\Delta V_{C1} = \frac{T}{C} \times I_L = \frac{1/120}{1000 \ \mu F} \times 28.54 \text{ mA} \approx 238 \text{ mV}$$

where: $I_L = I_{R1} + I_{R2} + I_{R7} + I_{(R4 \ + \ R5 \ + \ R6)}$

$$I_{R1} = V_{R1}/R1 = (V_{C1} - V_{B2})/R1 \approx (42.3 \text{ V} - 32.4 \text{ V})/1 \text{ k}\Omega = 9.9 \text{ mA}$$

$$I_{R2} = V_{E2}/R2 = 31.8 \text{ V}/2.2 \text{ k}\Omega = 14.45 \text{ mA}$$

$$I_{R7} = (V_o - V_{Z1})/R7 = (31.2 \text{ V} - 15 \text{ V})/6 \text{ k}\Omega = 2.7 \text{ mA}$$

$$I(R4 + R5 + R6) = V_o/(R4 + R5 + R6) = 31.2 \text{ V}/21 \text{ k}\Omega \approx 1.48 \text{ mA}$$

RIPPLE ATTENUATION USING CONSTANT CURRENT SOURCE TECHNIQUES

The ripple voltage across the filter capacitor C1 can be solved by the ΔV_{C1} formula, but this ripple voltage cannot be attenuated adequately by the existing circuitry of Figure 3-62. This is because the resistance at the base of Q2 is higher than 10 kΩ because of the resistance at the emitters of Q2 and Q1 being reflected back into the base of Q2. Therefore, in Figure 3-63, additional circuitry is provided, where a constant current source in the form of Q3 (and associated components) has replaced resistor R1. Hence, an effective high resistance, without the associated high DC voltage drop, is achieved. Additionally, forward biased diodes D8 and D9 are used instead of a low voltage zener diode.

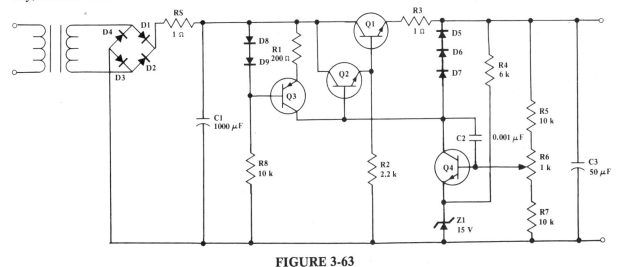

FIGURE 3-63

1. Mid-Rotation condition of R6:
   a. Solving for the V DC at the base of Q4:

$$V_{B4} = V_{Z1} + V_{BE4} = 15 \text{ V} + 0.6 \text{ V} = 15.6 \text{ V}$$

   b. Solving for the voltage at the output:

$$V_o = \frac{V_{B4}(R5 + R6 + R7)}{R7 + R6/2} = \frac{15.6 \text{ V}(10 \text{ k}\Omega + 1 \text{ k}\Omega + 10 \text{ k}\Omega)}{10 \text{ k}\Omega + 1 \text{ k}\Omega/2} = 31.2 \text{ V}$$

   c. Solving for the voltage at the base of Q1:

$$V_{B1} = V_{E1} + V_{BE1} = 31.2 \text{ V} + 0.6 \text{ V} = 31.8 \text{ V} \qquad \text{where: } V_{E2} = V_{B1}$$

d. Solving for the voltage at the base of Q2:

$$V_{B2} = V_{B1} + V_{BE2} = 31.8 \text{ V} + 0.6 \text{ V} = 32.4 \text{ V}$$

e. Solving for the voltage drop across resistor R8:

$$V_{R8} = V_{C1} - V(D8 + D9) = 42.3 \text{ V} - 1.4 \text{ V} = 40.9 \text{ V}$$

NOTE: Series diodes D8 and D9 make excellent low voltage references, without the zener diode drawback of noise. Also, notice that voltage drops of 0.7 volts, instead of 0.6 volts, are now being used, where earlier in the text 0.6 volts were used. The reason for this abrupt change is because both 0.6 volts and 0.7 volts are used in literature, and one should get used to seeing either. Too, the practical voltage difference between the diode and base-emitter of transistors is about 0.1 volts. Therefore, the diode voltage at 0.7 volts and the base-emitter-diode junction at 0.6 volts are more acceptable.

f. Solving for the voltage drop across R1:

$$V_{R1} = V_{(D8 + D9)} - V_{BE3} = 1.4 \text{ V} - 0.6 \text{ V} = 0.8 \text{ V}$$

g. Solving for the voltage drop across the transistor collector-emitters:

$$V_{CE1} = V_{C1} - V_{E1} \approx 42.3 \text{ V} - 31.2 \text{ V} = 11.1 \text{ V}$$

$$V_{CE2} = V_{C2} - V_{E2} \approx 42.3 \text{ V} - 31.8 \text{ V} = 10.5 \text{ V}$$

$$V_{CE3} = V_{E3} - V_{C3} = 41.5 \text{ V} - 32.4 \text{ V} = 9.1 \text{ V}$$

$$V_{CE4} = V_{C4} - V_{E4} = 32.4 \text{ V} - 15 \text{ V} = 17.4 \text{ V}$$

2. Solving for the ripple voltage — no load, mid-rotation condition:

a. Across the filter capacitor from:

$$\Delta V_{C1} = \frac{T}{C} \times I_L = \frac{1/120}{1000 \ \mu F} \times 26.72 \text{ mA} = 223 \text{ mV}$$

$$\text{where:} \quad I_L = \frac{V_{R8}}{R8} + \frac{V_{R1}}{R1} + \frac{V_{R2}}{R2} + \frac{V_{R4}}{R4} + \frac{V_o}{R5 + R6 + R7}$$

$$= \frac{40.9 \text{ V}}{10 \text{ k}\Omega} + \frac{0.8 \text{ V}}{200 \ \Omega} + \frac{31.8 \text{ V}}{2.2 \text{ k}\Omega} + \frac{16.2 \text{ V}}{6 \text{ k}\Omega} + \frac{31.2 \text{ V}}{21 \text{ k}\Omega}$$

$$= 4.09 \text{ mA} + 4 \text{ mA} + 14.45 \text{ mA} + 2.7 \text{ mA} + 1.48 \text{ mA} = 26.72 \text{ mA}$$

b. Across the reference diodes (D8 + D9):

$$\Delta V_{ref} = \frac{\Delta V_{C1} \times R_{(D8 + D9)}}{R_{(D8 + D9)} + R8} = \frac{223 \text{ mV} \times 12.4 \ \Omega}{12.4 \ \Omega + 10 \text{ k}\Omega} = 0.277 \text{ mV}$$

NOTE: The effective resistance of series diodes D8 and D9 can be calculated at:

$$re_1 + re_2 = (26 \text{ mV}/4.09 \text{ mA}) + (26 \text{ mv}/4.09 \text{ mA}) = 12.4 \ \Omega$$

$$\text{where:} \quad I_{re_1, \ re_2} = I_{R8} = V_{R8}/R8 = 4.09 \text{ mA}$$

c. To the load:

The ripple content at the base of Q3 is stepped down slighlty across the base-emitter diode junction of Q3 and, since the collector of Q3 "looks into" a high impedance (base of Q2 — collector of Q4), little

144

attenutation is expected to the base of Q2. Also, further attenuation across the base emitter diode junctions of Q2 and Q1 is slight. Therefore, the ripple voltage developed at the base of Q3 is, essentially, the ripple voltage at the output.

NOISE AND OSCILLATORY CONDITIONS

Both zener noise and possible oscillatory conditions because of the feedback circuitry can be minimized through low pass capacitive filtering. Therefore, with reference to the circuit of Figure 3-63, capacitor C2, connected across the base-collector junction of Q4, or some other form of high frequency filtering, needs to be used to protect against parasitic oscillations. Likewise, Q1 and Q2 may require similar filtering.

ZENER SUBSTITUTIONS

Reference voltages can be provided by zeners, forward biased diodes, or modified transistors. Both zeners and forward biased diodes are familiar by now, but modified transistors have not been discussed. A 2N3638 (PNP), a 2N3906 (PNP), or a 2N3904 (NPN) — for instance — can be transformed into "zeners", with adequate dynamic resistance, when the base lead is tied to the emitter and the DC voltage is applied to the collector-emitter leads. Allowing the base lead to float provides a sightly higher (0.6 V) voltage breakdown, but the dynamic resistance (breakdown slope) is poorer. $V(ref) \approx 8$ volts.

PROTECTIVE CIRCUITRY

Short circuit protection will take care of extreme output voltage losses across the output terminals because of shorted or low resistance conditions across the output, but it provides no protection against extreme voltage rise or reverse voltage conditions across the output terminals. These conditions must be handled with additional protective circuitry.

REVERSE VOLTAGE PROTECTION

In the circuit of Figure 3-64, a diode has been added across the series pass transistor Q1 and another has been added across the output terminals. The diode across the output terminals protects the supply against reverse voltage conditions that could easily occur if the supply is being used in tandem with another supply or if the supply is powering an active load. This diode protects both the output capacitor and the series pass transistor.

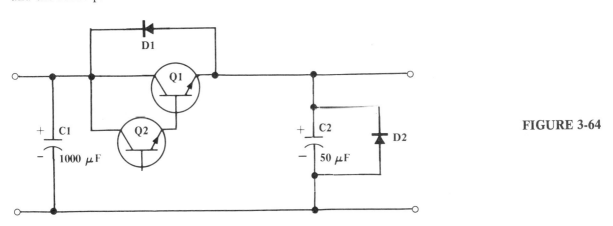

FIGURE 3-64

Additionally, reverse voltage protection is provided to the series pass device and driver transistor by diode D1. This diode prevents the output capacitor of the supply from becoming charged before filter capacitor C1 becomes charged, during parallel power supply conditions. Without the diode, the unenergized capacitor would have the output filter capacitor charged, the C1 filter capacitor not charged, and a reverse voltage condition across Q1 and Q2.

The current rating of Diode D1 is not too critical but the current rating for diode D2 is critical. This is because diode D2 must be able to withstand a higher current rating than the supply itself. However, if it does short, the cost of replacing a single diode is considerably less than the cost of replacing the series pass transistor, the driver, and possibly associated circuitry.

OVERVOLTAGE PROTECTION

If the voltage across the output terminals were to rise significantly above the maximum voltage output level, it could cause damage to the load. This rise could occur if the series pass transistor were to fail, or conversely, damage to the power supply could occur if extreme external voltages are applied to the output terminals by a failure in the load because of some external power source. To prevent these occurances by providing a safe limit at some preset voltage value, the "crowbar" circuit can be used. Then, if the preset voltage is exceeded by an overvoltage condition, an SCR is "turned on" or shorted, causing the voltage across the output terminals to drop immediately. The SCR can be connected across the output terminals or across the input terminals (filter capacitor C1), but the sensing circuit is normally connected across the output terminals.

In the simplest of "crowbar" circuits, as shown in Figure 3-65, variable resistor R2 is adjusted to provide firing of the SCR at just about the maximum output voltage level. Therefore, when the maximum voltage is exceeded, the SCR "turns on" placing a shorted condition across the output terminals. A nominal gate trigger voltage for the GE-C106 SCR is about 0.5 volts.

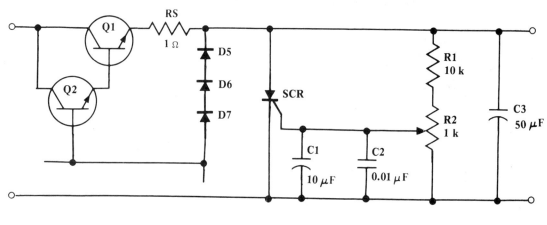

FIGURE 3-65

When the SCR is connected across the input terminals, it simply causes the fuse to blow. Either method is valid, but the blown fuse method (SCR across the input terminals as shown in Figure 3-66) has a built in flag when the fuse blows. Capacitor C1 and C2 (in both circuits) prevent false triggering in the form of "spikes", because capacitors react slowly to rapid increases and decreases of voltages.

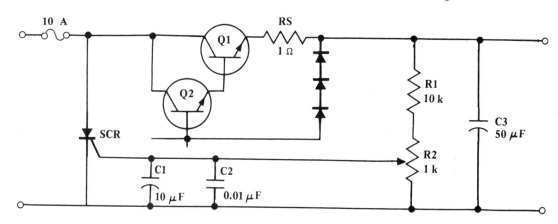

FIGURE 3-66

146

EXPERIMENTAL OBJECTIVES

To investigate the voltage adjustable regulated power supply with 180° feeback.

LIST OF MATERIALS

1. Transformer: 120 V : 24 V
2. Diodes: 1N4001 or equivalents (nine)
3. Zener Diode: 12 V (one)
4. Transistors: 2N3055 (one) 2N6178 (one) 2N3904 (one) 2N3906 (one) or equivalents
5. Resistors: 1 Ω (two) 100 Ω (two) 1 kΩ (potentiometer) 2.2 kΩ (two) 4.7 kΩ (one) 10 kΩ (three)
6. Capacitors: 1000 μF (one) 50 μF (one) 0.01 μF (one)

EXPERIMENTAL PROCEDURE (Part I):

1. Connect the circuit as shown in Figure 3-67. Adjust R6 to mid-rotation.

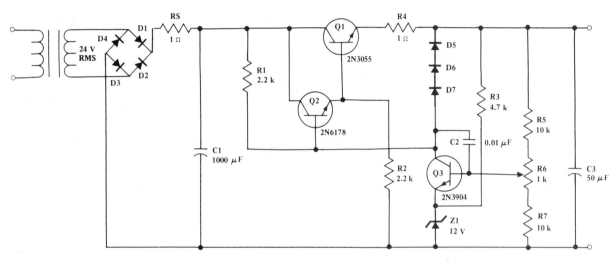

FIGURE 3-67

2. Across the secondary of the transformer, perform either measurement.
    a. Measure the Vp-p with an oscilloscope.
    b. Measure the RMS voltage with a voltmeter.
3. Calculate the Vp across the filter capacitor C1.
   Measure the Vp across the filter capacitor C1.
4. Calculate the DC voltage across the filter capacitor from:

$$V\,DC = Vp - \Delta V_{C1}/2 \approx Vp$$

   Measure the DC voltage across the filter capactior C1.
5. For mid-rotation conditions of variable resistor R6:
   a. Calculate the DC voltage:

      1. At the base of Q3 from: $V_{B3} = V_{Z1} + V_{BE3}$

      2. At the output from: $Vout = \dfrac{V_{B3} \times (R5 + R6 + R7)}{R7 + R6/2}$

      3. At the base of Q1 from: $V_{B1} = Vo + V_{BE1}$

      4. At the base of Q2 from: $V_{B2} = V_{B1} + V_{BE2}$

147

b. Measure the DC voltage:
  1. Across the zener.
  2. At the base of Q3.
  3. Across the output (Vo).
  4. At the base of Q1 — emitter of Q2.
  5. At the base of Q2 — collector of Q3.
6. Calculate the DC voltage across:

a. Transistor Q1 from: $V_{CE1} = V_{C1} - V_{E1}$

b. Transistor Q2 from: $V_{CE2} = V_{C2} - V_{E2}$

c. Transistor Q3 from: $V_{CE3} = V_{C3} - V_{E3}$

Measure the voltage drops across transistor Q1, Q2, and Q3.

7. Calculate the ripple voltage:

a. Across the filter capacitor from: $\Delta V_{C1} = T/C \times I_L$

Where: $I_L = \dfrac{V_{R1}}{R1} + \dfrac{V_{R2}}{R2} + \dfrac{V_{R3}}{R3} + \dfrac{Vo}{R5 + R6 + R7}$

Measure the ripple voltage:
a. Across the filter capacitor.
b. At the Base of Q1.
c. At the base of Q2.
d. Across the output.
8. For top-rotation conditions of variable resistor R6 :
  a. Calculate the DC voltage:

  1. At the output from $Vout = \dfrac{V_{B3} \times (R5 + R6 + R7)}{R7 + R6}$

  2. At the base of Q1.
  3. At the base of Q2.
  b. Measure the DC voltage:
  1. Across the zener diode.
  2. At the base of Q3.
  3. Across the output (Vo).
  4. At the base of Q1 — emitter of Q2.
  5. At the base of Q2 — collecor of Q3.
9. Calculate the DC voltage drops across:
  a. Transistor Q1.
  b. Transistor Q2.
  c. Transistor Q3.
Measure the voltage drops across transistors Q1, Q2, and Q3.
10. For bottom-rotation conditions of variable resistor R6:
  a. Calculate the DC voltage:

  1. At the output from: $Vout = \dfrac{V_{B3} \times (R5 + R6 + R7)}{R7}$

  2. At the base of Q1 and at the base of Q2.
  3. Across transistors Q1, Q2, and Q3.
  b. Measure the DC voltages:
  1. At the bases of Q1, Q2, and Q3.
  2. Across the zener diode, output (Vo), and transistors Q1, Q2, and Q3.
11. Insert the calculated and measured values, as indicated, into Table 3-9.

12. For 100 Ω loaded conditions:
   a. Measure Vp and VDC across filter capacitor C1.
   b. Measure the DC voltages:
      1. At the bases of Q1, Q2, and Q3.
      2. Across the zener diode and across the output load resistor.
13. Calculate and measure the DC voltage drops across Q1, Q2, and Q3.
14. Calculate the ripple voltage across the filter capacitor from:

$$\Delta V_{C1} = T/C \times I_L$$

where; $\quad I_L = \dfrac{V_{R1}}{R1} + \dfrac{V_{R2}}{R2} + \dfrac{V_{R3}}{R3} + \dfrac{Vo}{R5 + R6 + R7} + \dfrac{Vo}{R_L}$

where: $\quad RL = 100\ \Omega$

15. Measure the ripple voltage across the filter capactior and the output load resistor.
16. Calculate the power dissipated by Q1 and Q2.
17. Vary the voltage variable resistor R6 (under loaded conditions):
   a. To top rotation and to bottom rotation.
   b. Measure the Vo and ΔVo for top rotation and, then, for bottom rotation.
18. Insert the calculated and measured values, as indicated, into Table 3-10.

TABLE 3-9	Vp-p SEC	V RMS SEC	Vp C1	V DC C1	V DC				
					V_{Z1}	V_{B3}	V_{B2}	V_{B1}	Vo
Mid-Roation CALCULATED	////	////			////				
Mid-Rotation MEASURED									
Top-Rotation CALCULATED	////	////	////	////	////	////			
Top-Rotation MEASURED	////	////	////	////					
Bottom-Rotation CALCULATED	////	////	////	////	////	////			
Bottom-Rotation MEASURED	////	////	////	////					

	V_{CE}			ΔV			
	Q1	Q2	Q3	C1	B_{Q1}	B_{Q2}	Vo
Mid-Rotation CALCULATED					////	////	////
Mid-Rotation MEASURED							
Top-Rotation CALCULATED				////	////	////	////
Top-Rotation MEASURED				////	////	////	////
Bottom-Rotation CALCULATED				////	////	////	////
Bottom-Rotation MEASURED				////	////	////	////

TABLE 3-10	V_p C1	V DC C1	Mid-Rotation				V_{CE1}	V_{CE2}	V_{CE3}
			V_{B3}	V_{B2}	V_{B1}	V_o			
100 Ω Load CALCULATED	////	////	////	////	////	////			
100 Ω Load MEASURED									

	Mid-Rotation				Top-Rotation		Bottom-Rotation	
	ΔV_{C1}	ΔV_o	P_{Q1}	P_{Q2}	V_o	ΔV_o	V_o	ΔV_o
100 Ω Load CALCULATED		////			////	////	////	////
100 Ω Load MEASURED			////	////				

EXPERIMENTAL PROCEDURES (Part II):

1. Connect the circuit as hown in Figure 3-68. Adjust R6 to mid-rotation. Resistor R9 has been added to improve feedback sensitivity.

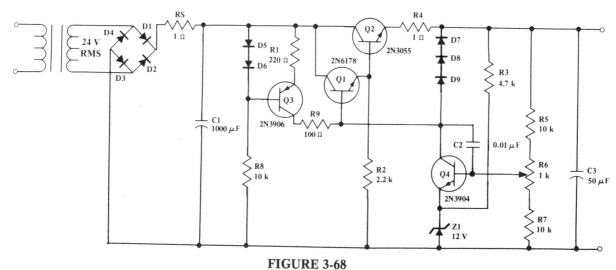

FIGURE 3-68

2. Calculate the DC voltage:
   a. Across the filter capacitor C1 (previously calculated).
   b. At the base of Q3 from: $V_{B3} = V_{C1} - V_{(D5 + D6)}$

   c. At the emitter of Q3 from: $V_{E3} = V_{B3} + V_{BE3}$

   d. Across resistor R1 from: $V_{R1} = V_{(D5 + D6)} - V_{BE3}$

   e. At the base of Q4 from: $V_{B4} = V_{Z1} + V_{BE4}$

   f. At the output for mid-rotation conditions.
   g. At the base of Q1.
   h. At the base of Q2.
   Measure the DC voltages that were calculated.
3. Calculate the DC voltage drops across transistor Q1, Q2, Q3, and Q4.
   Measure the DC voltage drops across transistor Q1, Q2, Q3, and Q4.

150

4. Calculate the Ripple voltage across filter capacitor C1 from: $\Delta V_{C1} = T/C \times I_L$

$$\text{where:} \quad I_L = \frac{V_{R1}}{R1} + \frac{V_{R2}}{R2} + \frac{V_{R3}}{R3} + \frac{V_{R8}}{R8} + \frac{V_o}{R5 + R6 + R7}$$

Measure the ripple voltage across the filter capacitor, at the base of Q3, at the emitter of Q3, at the base of Q3, and across the output terminals.

NOTE: Feedback circuits are susceptible to oscillatory conditions and capacitor C2 is added to the circuit to minimize these oscillations. Too, since emmiter follower circuits are susceptible to oscillatory conditions, also, further capacitor additions ($\approx$ 100 pF) can be added across the base-collector junctions of either Q1 or Q2.

5. LOADED CONDITIONS: With the exception of the added Q3 stage, the circuit operation and measured DC values should be close to those of the **Figure 3-67** circuit. Therefore, measure the DC voltages, as indicated, in Table 3-11.
  a. Connect the 100 Ω load, vary the variable R6 resistor from top, to mid, to bottom rotation. Measure the mid, maximum, and minimum rotation output voltages.
  b. At mid-rotation, monitor the ripple voltage:
  1. Across the filter capacitor.
  2. At the base and the emitter of Q3.
  3. At the base of Q1.
  4. Across the output terminals.
6. At mid-rotation, monitor the output volage under no load conditons and, then, under 100 Ω loaded conditions. Calculate the percent regulation from:

$$\% \text{ regulation} = \frac{V(NL) - V(L)}{V(NL)} \times 100$$

7. Insert the calculated and measured values, as indicated, into Table 3-11.

TABLE 3-11	\multicolumn{10}{c}{V DC — Mid-Rotation}
	V_{C1}	V_{B3}	V_{E3}	V_{B1}	V_{B2}	V_{B4}	V_{CE1}	V_{CE2}	V_{CE3}	V_{CE4}
No Load CALCULATED										
No Load MEASURED										
100 Ω Load CALCULATED	///	///	///	///	///	///	///	///	///	///
100 Ω Load MEASURED										

	\multicolumn{5}{c}{ΔV — Mid-Rotation}	\multicolumn{3}{c}{V_o}	% Regulation Mid-Rotation
	V_{C1}	VD3	V_{E3}	V_{B1}	V_o	Mid	Max	Min	
No Load CALCULATED		///	///	///	///	///	///	///	///
No Load MEASURED						///	///	///	///
100 Ω Load CALCULATED	///	///	///	///	///	///	///	///	
100 Ω Load MEASURED									///

PART 3B — THE VOLTAGE ADJUSTABLE POWER SUPPLY UTILIZING THE DIFFERENTIAL AMPLIFIER

GENERAL DISCUSSION

Further improvement of the voltage stability of power supplies, to overcome temperature instability, can be obtained by using the differential amplifier — as shown in Figure 3-69 — because current variations are cancelled, theoretically, in the differential configuration when matched pair transistors are used. Secondly, the voltage at the base of Q5 follows the voltage at the base of Q4, in the differential mode, and provides a condition that allows voltage variable supplies to adjust down to reference voltage. Thirdly, the Q5 transistor of the differential amplifier provides 180° feedback in obtaining the extremely low output impedance necessary in well regulated power supplies.

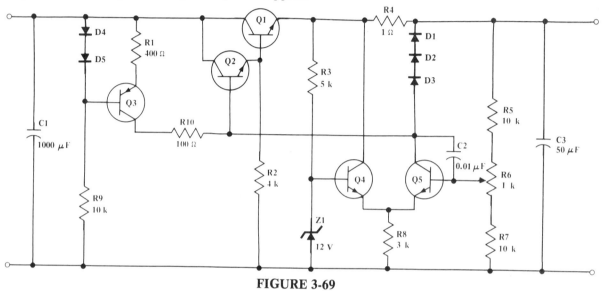

FIGURE 3-69

The analysis of the differential amplifier begins at the 12 volt zener, where 12 volts exist at the base of Q4, 11.4 volts at the emitters of Q4 and Q5, and 12 volts at the base of Q5. Therefore, 12 volts exist at the wiper of the variable resistor R6 and, for mid-rotation conditions, 24 volts exist at the output terminals. This is exactly the procedure used in the previous example problem.

Other similar techniques used in this circuit are diodes D4 and D5, in place of a zener diode, and the inclusion of R10 to insure immediate reaction to 180° feedback. This is because transistor Q3 acts as a constant current source, but any change in current out of feedback transistor Q5 will be sensed by resistor R10. The feedback path using transistor Q5, the sensing resistor using resistor R10, and the low voltage referencing and ripple-attenuation circuit are illustrated in Figure 3-70.

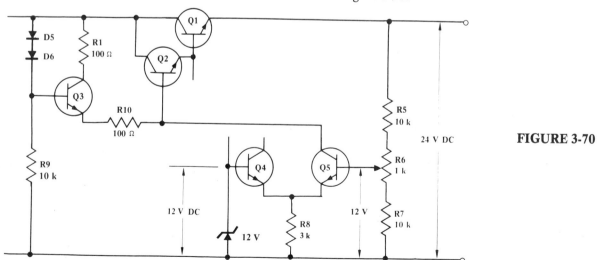

FIGURE 3-70

Another technique that can be used to provide an effective constant current source is the field effect transistor as shown in Figure 3-71. The only limitation with this circuit is that the minimum voltage across the drain-to-source of the device must be greater than the pinch-off voltage Vp, to insure a constant current condition. The maximum current condition occurs when $R1 = 0\ \Omega$ and $I_D = I_{DSS}$.

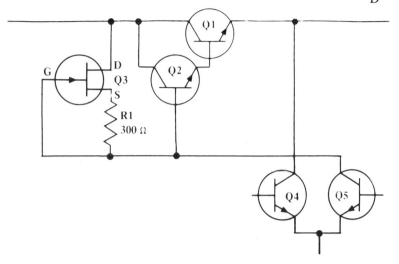

FIGURE 3-71

A third technique that can be used is the constant current diode. These diodes are essentially a FET connection in a single diode package and they come in a variety of constant current (I_f) values. The device equivalent, the symbol, and the circuit connection are shown in Figure 3-72.

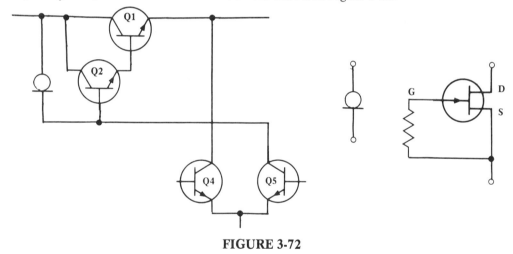

FIGURE 3-72

ADDITIONAL CIRCUIT CONFIGURATIONS

A circuit configuration which allows relatively large output voltages, without having to resort to devices with large breakdown voltages is shown in Figure 3-73. And a simple technique to accomplish this is to reverse the poitions of Resistors R4 and the 12 volt zener. Therefore, if the output voltage is 96 volts and the input voltage is 120 volts, then the large voltage drops are occuring across resistors instead of the circuit devices.

1. $V_{Z1} = 12\ V$

2. $V_{CE4} = V_{Z1} + V_{BE4} = 12\ V + 0.6\ V = 12.6\ V$

3. $V_{R6} = V_{Z1} = 12\ V$

153

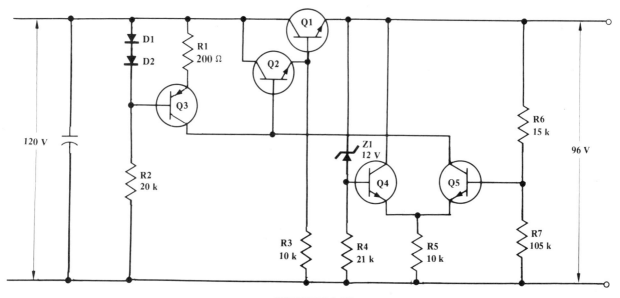

FIGURE 3-73

4. $V_o = \dfrac{V_{R6}\,(R6 + R7)}{R6} = \dfrac{12\ V\,(15k\Omega + 105\ k\Omega)}{15\ k\Omega} = 96\ V$

5. $V_{R7} = V_o - V_{R6} = 96\ V - 12\ V = 84\ V$

6. $V_{R5} = V_{R7} - V_{BE5} = 84\ V - 0.6\ V = 83.4\ V$

7. $V_{R4} = V_{R5} + V_{BE4} = 83.4\ V + 0.6\ V = 84\ V$

8. $I_{R4} = V_{R4}/R4 = 84\ V/21\ k\Omega = 4\ mA$

9. $I_{R5} = V_{R5}/R5 = 83.4\ V/10\ k\Omega = 8.34\ mA$

10. $V_{B1} = V_o + V_{BE1} = 96\ V + 0.6\ V = 96.6\ V$

11. $V_{B2} = V_{B1} + V_{BE2} = 96.6\ V + 0.6\ V = 97.2\ V$

12. $V_{B3} = V_{C1} - V_{(D1 + D2)} = 120\ V - 1.4\ V = 118.6\ V$

13. $V_{E3} = V_{B3} + V_{BE3} = 118.6\ V + 0.6\ V = 119.2\ V$

14. $V_{CE3} = V_{E3} - V_{C3} = 119.2\ V - 97.2\ V = 22\ V$

15. $V_{CE1} = V_{C1} - V_o = 120\ V - 96\ V = 24\ V$

16. $V_{CE2} = V_{C1} - V_{B1} = 120\ V - 96.6\ V = 23.4\ V$

17. $V_{CE4} = V_o - V_{E4} = 96\ V - 83.4\ V = 12.6\ V$

18. $V_{CE5} = V_{B2} - V_{E5} = 97.2\ V - 83.4\ V = 13.8\ V$

Another circuit that provides high output voltages without using devices with high breakdown voltages is shown in Figure 3-74. This circuit configuration does not use the differential amplifier and, therefore, it must use PNP devices to insure negative feedback. However, the PNP configuration has an advantage in that ripple content attenuation by preregulators, such as used in the NPN series pass regulators, are not required. Feedback by way of Q3 and output capacitor C2 insures a low output impedance, which normally would be high because the load is fed from the collector of Q1. The DC analysis is as follows:

1. $V_{Z1} = 12.6\ V$

154

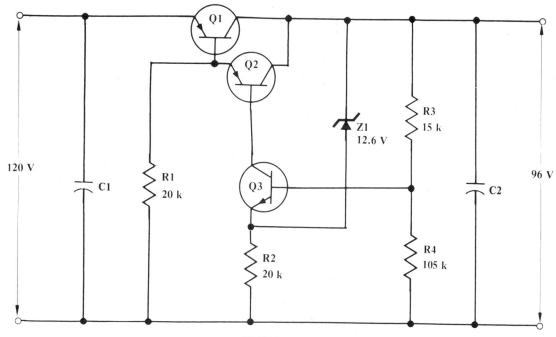

FIGURE 3-74

2. $V_{R3} = V_{Z1} - V_{BE3} = 12.6 \text{ V} - 0.6 \text{ V} = 12 \text{ V}$

3. $Vo = \dfrac{V_{R3}(R3 + R4)}{R3} = \dfrac{12 \text{ V}(15 \text{ k}\Omega + 105 \text{ k}\Omega)}{15 \text{ k}\Omega} = 96 \text{ V}$

4. $V_{R4} = Vo - V_{R3} = 96 \text{ V} - 12 \text{ V} = 84 \text{ V}$

5. $V_{R1} = V_{E2} = VDC_{C1} - V_{BE1} = 120 \text{ V} - 0.6 \text{ V} = 119.4 \text{ V}$

6. $V_{B2} = V_{E2} - V_{BE2} = 119.4 \text{ V} - 0.6 \text{ V} = 118.8 \text{ V}$

7. $V_{R2} = V_{R4} - V_{BE3} = 84 \text{ V} - 0.6 \text{ V} = 83.4 \text{ V}$

8. $V_{CE1} = VDC_{C1} - Vo = 120 \text{ V} - 96 \text{ V} = 24 \text{ V}$

9. $V_{CE2} = V_{E2} - Vo = 119.4 \text{ V} - 96 \text{ V} = 23.4 \text{ V}$

10. $V_{CE3} = V_{B2} - V_{E3} = 118.8 \text{ V} - 83.4 \text{ V} - 35.4 \text{ V}$

There are several other circuit configurations that can be used to advantage in providing either a constant current source output or a low voltage output without having to resort to high voltage breakdown devices.

The circuit of Figure 3-75 provides a constant current source where the resistive load determines the output voltage. This type of supply can be used in the precise measurement of unknown resistance, or in any other system requiring a constant current source. The DC analysis is as follows:

1. $V_{Z1} = 13.2 \text{ V}$

2. $V_{R1} = V_{Z1} - (V_{BE1} + V_{BE2}) = 13.2 \text{ V} - 1.2 \text{ V} = 12 \text{ V}$

3. $I_{R1} = V_{R1}/R1 = 12 \text{ V}/3 \text{ k}\Omega = 4 \text{ mA}$

4. $I_{RL} \approx I_{R1} = 4 \text{ mA}$

5. $Vo = I_{RL} \times RL = 4 \text{ mA} \times 4 \text{ k}\Omega = 16 \text{ V}, \quad \text{where } RL = 4 \text{ k}\Omega$

155

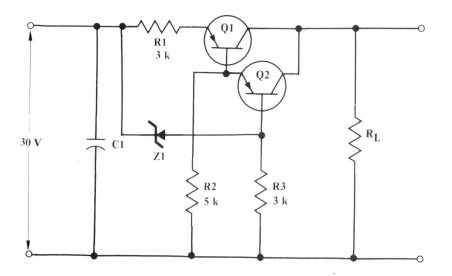

FIGURE 3-75

$$V_0 = I_{RL} \times RL = 4 \text{ mA} \times 3 \text{ k}\Omega = 12 \text{ V} \quad \text{where:} \quad RL = 3 \text{ k}\Omega$$

$$V_0 = I_{RL} \times RL = 4 \text{ mA} \times 2 \text{ k}\Omega = 8 \text{ V} \quad \text{where:} \quad RL = 2$$

The circuit of Figure 3-76 provides a low voltage output from a high voltage source by dropping the major portion of the voltage across a zener diode. Therefore, the circuit requires a zener with a high breakdown voltage, and it uses an NPN driving a PNP series pass configuration. The NPN driving a PNP effectively provides an NPN stage.

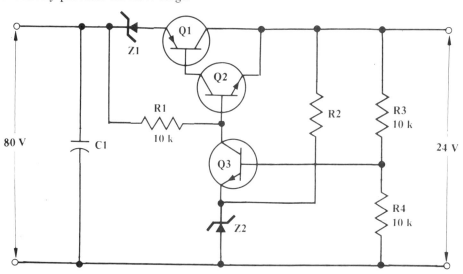

FIGURE 3-76

1. $V_{Z2} = 11.4 \text{ V}$

2. $V_{B3} = 12 \text{ V}$

3. $V_0 = \dfrac{V_{B3}(R3 + R4)}{R3} = \dfrac{12 \text{ V} (10 \text{ k}\Omega + 10 \text{ k}\Omega)}{10 \text{ k}\Omega} = 24 \text{ V}$

4. $V_{B2} = V_0 + V_{BE2} = 24 \text{ V} + 0.6 \text{ V} = 24.6 \text{ V}$

5. $V_{E1} = V_{cap_1} - V_{Z1} = 80 \text{ V} - 46 \text{ V} = 34 \text{ V}$

6. $V_{B1} = V_{E1} - V_{BE1} = 34 \text{ V} - 0.6 \text{ V} = 33.4 \text{ V}$

7. $V_{R1} = VDC_{C1} - V_{B2} = 80 \text{ V} - 24.6 \text{ V} = 55.4 \text{ V}$

156

8. $V_{R2} = V_o - V_{Z2} = 24\ V - 11.4\ V = 12.6\ V$

9. $V_{CE1} = V_{E1} - V_o = 34\ V - 24\ V = 10\ V$

10. $V_{CE2} = V_{B1} - V_o = 33.4\ V - 24\ V = 9.6\ V$

11. $V_{CE3} = V_{B2} - V_{Z2} = 24.6\ V - 11.4\ V = 13.2\ V$

VOLTAGE ADJUSTABLE-DIFFERENTIAL AMPLIFIER STABILIZED POWER SUPPLY — CIRCUIT ANALYSIS

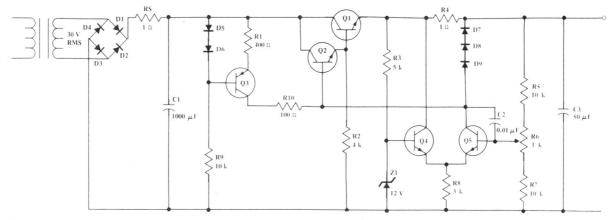

FIGURE 3-77

1. Solving for the peak voltage across the capacitor C1:

   $$V_{peak} = \sqrt{2} \times V\ RMS = 1.414 \times 30\ V = 42.42\ V$$

2. Solving for the:

   a. V DC across the C1 capacitor from:

   $$VDC_{C1} \approx V_{peak}, \quad or \quad VDC_{C1} = V_p - (\Delta V_{C1}/2) = 42.42\ V - (166.6\ mV/2) \approx 42.33\ V$$

   b. Voltage at the base of Q4 from: $\quad V_{B4} = V_{Z1} = 12\ V$

   c. Voltage at the emitters of Q4 and Q5 from:

   $$V_{E4} = V_{E5} = V_{B4} - V_{BE4} = 12\ V - 0.6\ V = 11.4\ V$$

   d. Voltage at the base of Q5 from: $\quad V_{B5} = V_{E5} + V_{BE5} = 11.4\ V + 0.6\ V = 12\ V$

   e. The voltage at the output for mid-rotation conditions from:

   $$V_o = \frac{V_{B5}(R5 + R6 + R7)}{R7 + (R6/2)} = \frac{12\ V(10\ k\Omega + 1\ k\Omega + 10\ k\Omega)}{10\ k\Omega + (1\ k\Omega/2)} = \frac{12\ V \times 21\ k\Omega}{10.5\ k\Omega} = 24\ V$$

   f. Voltage at the base of Q1 from: $\quad V_{B1} = V_{E1} + V_{BE1} = 24\ V + 0.6\ V = 24.6\ V$

NOTE: $V_{E1} \approx V_o$, since the voltage drop across R4 for no load conditions is minimal.

   g. Voltage at the base of Q2 from: $\quad V_{B2} = V_{B1} + V_{BE2} = 24.6\ V + 0.6\ V = 25.2\ V$

   h. Voltage drop across resistor R1 from: $\quad V_{R1} = V_{(D5 + D6)} - V_{BE3} = 1.4\ V - 0.6\ V = 0.8\ V$

157

Current flow through resistor R1 from: $I_{R1} = V_{R1}/R1 = 0.8 \text{ V}/400 \ \Omega = 2 \text{ mA}$

    i.   Voltage drop across resistor R8 from: $V_{R8} = V_{Z1} - V_{BE4} = 12 \text{ V} - 0.6 \text{ V} = 11.4 \text{ V}$

Current flow through resistor R8 from: $I_{R8} = V_{R8}/R8 = 11.4 \text{ V}/3 \text{ k}\Omega = 3.8 \text{ mA}$

NOTE: The current flow in R8 is shared by both Q4 and Q5. However, the current flow of R1 also flows through Q5 and is developed across resistor R8. Therefore, as I_{R1} is increased less current is available to transistor Q4, and if enough excess I_{R1} current is available it can cut off the Q4 transistor completely, forcing the entire power supply out of regulation. Consequently, the current flow in R8 is purposely maintained at about twice the current flow of I_{R1} to prevent this from occurring.

3. Solving for the voltage drop across:

    a. Transistor Q1 from: $V_{CE1} = V_{C1} - V_{E1} = 42.33 \text{ V} - 24 \text{ V} = 18.33 \text{ V}$

    b. Transistor Q2 from: $V_{CE2} = V_{C2} - V_{E2} = 42.33 \text{ V} - 24.6 \text{ V} = 17.73 \text{ V}$

    c. Transistor Q3 from: $V_{CE3} = V_{E3} - V_{C3} = 41.53 \text{ V} - 25.4 \text{ V} = 16.13 \text{ V}$

          where: $V_{E3} = V_{C1} - V_{R1} = 41.53 \text{ V}$ and $V_{C3} = V_{B2} + V_{R10} = 25.4 \text{ V}$

    d. Transistor Q4 from: $V_{CE4} = V_{C4} - V_{E4} \approx 24 \text{ V} - 11.4 \text{ V} = 12.6 \text{ V}$

    e. Transistor Q5 from: $V_{CE5} = V_{C5} - V_{E5} = 25.2 \text{ V} - 11.4 \text{ V} = 13.8 \text{ V}$

4. Solving for the voltage at the output for:

    a. Mid-rotation conditions:

$$Vo = \frac{V_{B5}(R5 + R6 + R7)}{R7 + (R6/2)} = \frac{12 \text{ V}(10 \text{ k}\Omega + 1 \text{ k}\Omega + 10 \text{ k}\Omega)}{10 \text{ k}\Omega + (1 \text{ k}\Omega/2)} = \frac{12 \text{ V} \times 21 \text{ k}\Omega}{10.5 \text{ k}\Omega} = 24 \text{ V}$$

    b. Full bottom-rotation conditions:

$$Vo = \frac{V_{B5}(R5 + R6 + R7)}{R7} = \frac{12 \text{ V} \times 21 \text{ k}\Omega}{10 \text{ k}\Omega} = 25.2 \text{ V}$$

    c. Full top-rotation conditions:

$$Vo = \frac{V_{B5}(R5 + R6 + R7)}{R6 + R7} = \frac{12 \text{ V} \times 21 \text{ k}\Omega}{1 \text{ k}\Omega + 10 \text{ k}\Omega} = 22.9 \text{ V}$$

NOTE: Voltage distribution around the circuit changes to accomodate the voltage changes at the output. Care should be taken to make sure that no transistor either goes into saturation or cutoff as the output voltage is varied. The voltage drops are solved exactly as they were for mid-rotation conditions.

5. Ripple voltage to the load is usually less than 10 mV, as was solved in the previous voltage feedback circuit using the constant current source.

The ripple voltage across the filter capacitor C1 is similar to that of the previous circuit where the ripple voltage magnitude is solved from: $\Delta V_{C1} = T/C \times I_L$, where I_L is the total current delivered by the capacitor to the circuit. It is solved:

    a. For no load, mid-rotation conditions from:

$$\Delta V_{C1} = T/C \times I_L = \frac{1/120}{1000 \ \mu F} \times 20 \text{ mA} = 166.6 \text{ mVp-p}$$

where: $I_L = \dfrac{V_{R9}}{R9} + \dfrac{V_{R2}}{R2} + \dfrac{V_{R3}}{R3} + \dfrac{V_{R8}}{R8} + \dfrac{Vo}{R5+R6+R7} = 20$ mA

$$\approx \dfrac{40.9 \text{ V}}{10 \text{ k}\Omega} + \dfrac{24.6 \text{ V}}{4 \text{ k}\Omega} + \dfrac{24 \text{ V}}{5 \text{ k}\Omega} + \dfrac{11.4 \text{ V}}{3 \text{ k}\Omega} + \dfrac{24 \text{ V}}{21 \text{ k}\Omega}$$

b. For loaded conditions of 100 ohms at mid-rotation:

$$\Delta V_{C1} = T/C \times I_L = T/C \times (I_L[NL] + Vo/RL) = \dfrac{1/120}{1000 \ \mu F} \times (20 \text{ mA} + 240 \text{ mA}) = 2.167 \text{ Vp-p}$$

where: Vo/RL = 24 V/100 Ω = 240 mA

EXPERIMENTAL OBJECTIVES

To investigate the voltage adjustable regulated power supply utilizing the differential amplifier and 180° feedback compensation.

LIST OF MATERIALS

1. Transformer: 120 V : 24 V
2. Diodes: 1N4001 (nine) or equivalent
3. Resistors: 1 Ω(two) 100 Ω(one) 220 Ω(one) 330 Ω(one) 1 kΩ potentiometer (one) 1.5 kΩ(two)
        2.2 kΩ(two) 4.7 kΩ(one) 10 kΩ(three)
4. Capacitors: 1000 μF (one) 50 μF (one) 0.01 μF (one)
5. Zener: 12 V (one) or equivalent
6. Transistors: 2N3055(one) 2N6178(one) 2N3904(two) 2N3906(one) 2N3819(one) or equivalents

EXPERIMENTAL PROCEDURE (PART 1):

1. Connect the circuit as shown in Figure 3-78. Adjust resistor R6 to mid-rotation.

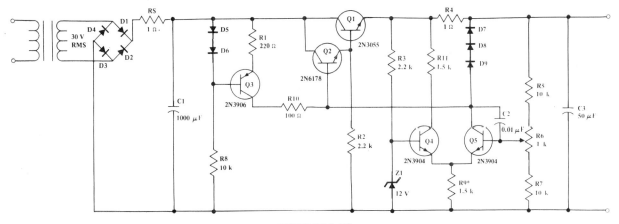

FIGURE 3-78

2. Across the secondary of the transformer, perform either measurement:
    a. Measure the Vp-p with an oscilloscope.
    b. Measure the RMS with a voltmeter.
3. Calculate the Vp across the filter capacitor C1.
    Measure the Vp across the filter capacitor.
4. Calculate the DC voltage across the filter capacitor from:

$$V \ DC = Vp - \Delta \ V_{C1}/2 \approx Vp$$

Measure the DC voltage across the filter capacitor C1.

159

5. Mid-rotation condition:
   a. Calculate the DC voltages:

   1. At the base of Q4 from: $V_{B4} = V_{Z1}$

   2. At the emitters of Q4 and Q5 from: $V_{E4} = V_{E5} = V_{B4} - V_{BE4}$

   3. At the base of Q5 from: $V_{B5} = V_{E5} + V_{BE5}$

   4. Across the output (V_o) from: $Vo = \dfrac{V_{B5}(R5 + R6 + R7)}{R7 + R6/2}$

   5. At the Base of Q1 from: $V_{B1} = Vo + V_{BE1}$

   6. At the base of Q2 from: $V_{B2} = V_{B1} + V_{BE2}$

   7. At the collector of Q3 from: $VC3 = VB2 + VR10$

   8. At the base of Q3 from: $V_{B3} = V_{C1} - V_{(D5 + D6)}$

   b. Measure the DC voltages as indicated (Step 1 through Step 8).

NOTE: The value of R9 can be critical and, ideally, I_{R9} should be $\approx 2I_{R1}$.

6. Calculate the DC voltage drops across:

   a. Transistor Q1 from: $V_{CE1} = V_{C1} - V_{E1}$

   b. Transistor Q3 from: $V_{CE3} = V_{E3} - V_{C3}$

   c. Transistor Q5 from: $V_{CE5} = V_{C5} - V_{E5}$

   d. Measure the voltage drops across transistors Q1, Q3, and Q5.

7. Repeat Steps 5 and 6 for a loaded condition of 100 Ω, but take measurements only. Make a comparison between loaded and unloaded conditions.

8. Calculate the ripple voltage across the filter capacitor:

   a. For no load conditions from: $\Delta V_{C1} = T/C \times I_L$

   where: $I_L = \dfrac{V_{R8}}{R8} + \dfrac{V_{R2}}{R2} + \dfrac{V_{R3}}{R3} + \dfrac{V_{R9}}{R9} + \dfrac{Vo}{R5 + R6 + R7}$

   b. For a loaded condition of 100 ohms from: $\Delta V_{C1} = T/C \times I_L$

   where: $I_L = \dfrac{V_{R8}}{R8} + \dfrac{V_{R2}}{R2} + \dfrac{V_{R3}}{R3} + \dfrac{V_{R9}}{R9} + \dfrac{Vo}{R5 + R6 + R7} + \dfrac{Vo}{RL}$

NOTE: The I_{R1} current flow is included in the I_{R8} current total and **R8** may have to be changed to assure minimum Q4 current condition.

Measure the ripple voltage for both unloaded and loaded conditions.
   a. Across the filter capacitor.
   b. At the base of Q3.
   c. At the base of Q2.
   d. Across the output.

9. For top and bottom rotation of resistor R6 (for loaded and unloaded conditions):
  a. Measure the DC voltage across the output and at the base of Q2.
  b. Measure ΔVo at the output terminals.
  Measure the DC voltage around the circuit to make sure the circuit is working.

NOTE: Remember resistor R8 can be critical and must be capable of delivering enough current so that Q4 is kept out of cutoff condition. However, too much current can also create power dissipation problems for Q4.

10. Test the short circuit condition at mid-rotation.
  a. Calculate I_{SC} from $V_{R4}/R4$.
  b. Measure I_{SC}.
11. Measure and calculate the percent regulation at mid-rotation from:

$$\% \text{ regulation} = \frac{V(NL) - V(L)}{V(NL)} \times 100 \quad \text{where:} \quad RL = 100$$

12. Insert the calculated and measured values, as indicated, into Table 3-12 and Table 3-13.

TABLE 3-12	Vp-p SEC	V RMS SEC	Vp C1	V DC C1	V DC — Mid-Roation			
					V_{B4}	V_{E4}	V_{B5}	Vo
NO LOAD CALCULATED	///////	///////						
NO LOAD MEASURED								
100 Ω LOAD CALCULATED	///////	///////	///////	///////	///////	///////	///////	///////
100 Ω LOAD MEASURED								
	VDC — Mid-Rotation				V_{CE} — Mid-Rot.		
	V_{B1}	V_{B2}	V_{C3}	V_{B3}	Q1	Q3	Q5
NO LOAD CALCULATED							
NO LOAD MEASURED							
100 Ω LOAD CALCULATED	///////	///////	///////	///////	///////	///////	///////
100 Ω LOAD MEASURED							

TABLE 3-13	ΔV Mid-Rotation			Top Rotation			Bottom R.			Mid-Rotation	
	C1	B_{Q2}	B_{Q3} Vo	Vo	B_{Q2}	ΔVo	Vo	B_{Q2}	ΔVo	I_{SC}	% Reg.
NO LOAD CALCULATED		///////	/////// ///////	///////	///////	///////	///////	///////	///////		///////
NO LOAD MEASURED											///////
100 Ω LOAD CALCULATED		///////	/////// ///////	///////	///////	///////	///////	///////	///////		
100 Ω LOAD MEASURED										///////	///////

EXPERIMENTAL PROCEDURE (PART 2):   FET CONSTANT CURRENT SOURCE

1.   Connect the circuit as shown in Figure 3-79.   Again, the resistance value of R8 is critical. If the resistance value is too high, it can force Q4 into cutoff. Too low a resistance value will cause excessive power dissipation of Q4. Also critical is I_{R1}, which is determined by the I_{DSS} of Q1.

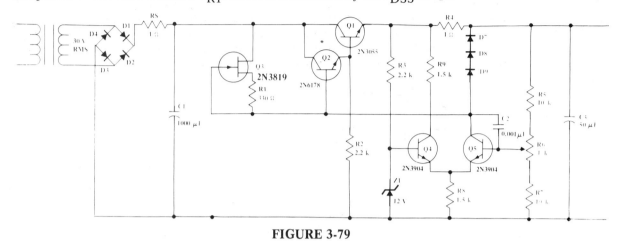

FIGURE 3-79

2.   Monitor the voltage drops around the circuit to make sure the circuit is working.

NOTE: This circuit is quite similar to the previous circuit with the exception of the FET constant current source. However, the voltage drop across Q3 will be different and this slightly alters the DC distribution. V_{DS} can be solved is Vp and I_{DSS} of the device are known.

3.   Insert the calculated and measured values, as indicated, into Table 3-14.

TABLE 3-14	V DC — Mid-Rotation							
	V_{C1}	V_{B4}	V_{E4}	V_{B5}	Vo	V_{B1}	V_{B2}	V_{S3}
NO LOAD CALCULATED	/////	/////	/////	/////	/////	/////	/////	/////
NO LOAD MEASURED								
100 Ω LOAD CALCULATED	/////	/////	/////	/////	/////	/////	/////	/////
100 Ω LOAD MEASURED								

	ΔV Mid-Rotation			Top Rotation		Bottom Rotation		Mid-Rotation	
	C1	B_{Q1}	Vo	Vo	ΔVo	Vo	ΔVo	I_{SC}	% Reg.
NO LOAD CALCULATED	/////	/////	/////	/////	/////	/////	/////	/////	
NO LOAD MEASURED									/////
100 Ω LOAD CALCULATED	/////	/////	/////	/////	/////	/////	/////		
100 Ω LOAD MEASURED								/////	/////

162

PART 3C — THE VOLTAGE VARIABLE POWER SUPPLY

GENERAL DISCUSSION

A voltage variable power supply should be capable of varying over a wide DC output voltage range with a minimum condition of zero output volts. However, in order to accomplish a zero output voltage condition, a negative power supply along with the positive unregulated power supply is needed. This is necessary so that the control circuitry of the power supply can remain "active" during zero and low voltage conditions. Two methods used to accomplish the plus and minus unregulated supplies are shown in Figure 3-80.

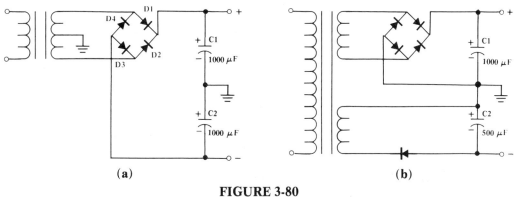

(a)　　　　　　　　　　　　　　**(b)**

FIGURE 3-80

Of the two circuits shown in Figure 3-80, the "b" circuit is more efficient because the secondary winding driving the bridge circuit supplies the voltage for the series pass "power half" of the circuit, while the transformer secondary driving the half wave rectifier circuit need only supply current to the reference and differential amplifier circuitry. Therefore, filter capacitor C1 must be able to handle the load current so it should be larger than capacitor C2.

The circuit of Figure 3-80(a) also provides plus and minus voltage. However, since the current demands of C1 are normally much higher thant the current demands of C2, power capability of the transformer secondary is wasted. This is because the wire size and current demand capabilities are the same but only half powering the load is used efficiently.

Another interesting feature about the circuit of Figure 3-80(a) is that it is not a full wave bridge circuit, but two full wave center-tapped circuits — one positive and one negative. Diodes D1 and D2, with respect to ground, provide the positive DC voltage developed across capacitor C1, and Diodes D3 and D4, with respect to ground, provide the DC voltage developed across capacitor C2. This is illustrated in Figure 3-81.

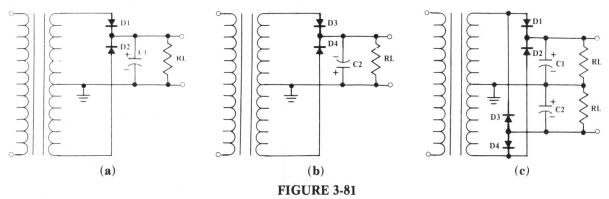

(a)　　　　　　　　　　**(b)**　　　　　　　　　　**(c)**

FIGURE 3-81

THE VARIABLE OUTPUT VOLTAGE

The advantage of the — 15 volt supply is that a ground condition exists at the base of Q4, and hence at the base of Q5. Therefore, with regard to the voltage developed across series resistors R6 + R7, it is 45 volts. However, across R7 alone it is 15 volts and across R6 alone it is 30 volts. And since the ground condition exists at the junction of R6 and R7, with respect to ground, a plus 30 volts exists at the top of

R6 and a minus 15 volts exists at the bottom of R7. This condition is illustrated in Figure 3-82, and details are analyzed in a sample problem.

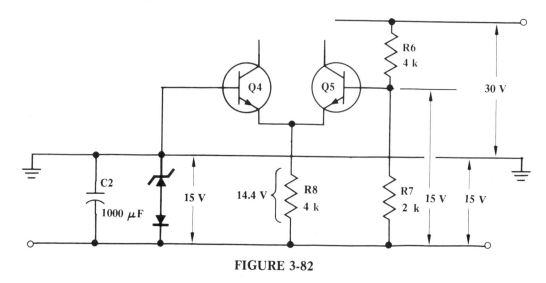

FIGURE 3-82

VARIABLE DIRECT CURRENT POWER SUPPLY ANALYSIS — 0-30 V DC

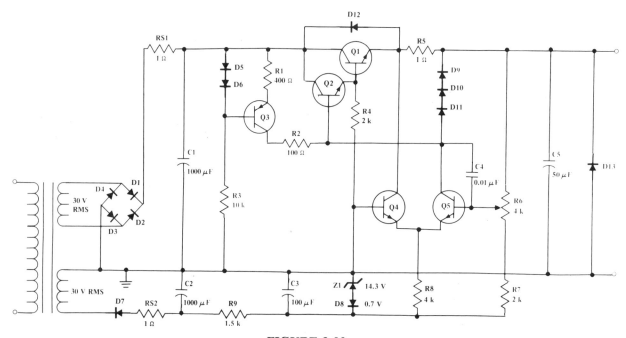

FIGURE 3-83

1. Solving for the peak voltage across filter capacitors C1 and C2:

$$\text{Vpeak} = \sqrt{2} \times \text{V RMS} = 1.414 \times 30 \text{ V} = 42.42 \text{ V}$$

2. Solving for the DC voltage developed across the filter capacitors:

   C1 from: $\text{V DC} = \text{Vp} - \Delta V_{C1}/2 = 42.42 \text{ V} - 254 \text{ mV}/2 \approx 42.3 \text{ V}$

   C2 from: $\text{V DC} = \text{Vp} - \Delta V_{C2}/2 = 42.42 \text{ V} - 151 \text{ mV}/2 \approx 42.35 \text{ V}$

3. Solving for:

a. The voltage across the reference diodes D5 and D6:

$$V(D5 + D6) = 0.7 \text{ V} + 0.7 \text{ V} = 1.4 \text{ V}$$

NOTE: These are normally forward biased diodes that are being used as part of the Q3 constant current source.

b. The voltage drop across R1 from: $V_{R1} = V(D5 + D6) - V_{BE3} = 1.4 \text{ V} - 0.6 \text{ V} = 0.8 \text{ V}$

c. The current flow through R1: $I_{R1} = V_{R1}/R1 = 0.8 \text{ V}/400 \text{ }\Omega = 2 \text{ mA}$

d. The voltage drop across R2: $V_{R2} = I_{R1} \times R2 = 2 \text{ mA} \times 100 \text{ }\Omega = 0.2 \text{ V}$

e. The voltage drop across R3 from: $V_{R3} = VDC_{C1} - V(D5 + D6) = 42.3 \text{ V} - 1.4 \text{ V} = 40.9 \text{ V}$

f. The voltage at the base of Q4: $V_{B4} = \text{ground} = 0 \text{ V}$

g. The voltage at the emitters of Q4 and Q5:

$$V_{E4} = V_{E5} = V_{B4} - V_{BE4} = 0 \text{ V} - 0.6 \text{ V} = -0.6 \text{ V}$$

h. The voltage at the base of Q5: $V_{B5} = V_{E5} + V_{BE5} = -0.6 \text{ V} + 0.6 \text{ V} = 0 \text{ V}$

NOTE: The base of Q5 follows the base of Q4 and a virtual short between the two exists. That is to say, if the base of Q4 is tied to ground, the base of Q5 is also at zero volts or "ground" condition.

i. The voltage of the negative reference supply: $\text{Vref} = V_{Z1} + V_{D3} = 14.3 \text{ V} + 0.7 \text{ V} = 15 \text{ V}$

NOTE: Since the base of Q4 is ground and Vref is tied between the base of Q4 and the junction of R7, R8, and R9, then minus 15 volts exists at this junction. Additionally, it is this minus 15 volts that allows transistors Q4 and Q5 to remain in the active region when the bases of both devices are at ground.

j. The voltage drop across R8: $V_{R8} = \text{Vref} - V_{BE4} = 15 \text{ V} - 0.6 \text{ V} = 14.4 \text{ V}$

k. The current flow through resistor R8: $I_{R8} = V_{R8}/R8 = 14.4 \text{ V}/4 \text{ k}\Omega = 3.6 \text{ mA}$

NOTE: The current flow in R8 must be approximately equal to twice the current flow of R1. This is to make sure there is an adequate current flow through transistor Q4. (This is similar to the previous differential amplifier circuit.) Therefore, $I_{Q4} \approx I_{R8} - I_{R4} = 3.6 \text{ mA} - 2 \text{ mA} = 1.6 \text{ mA}$. This is a condition that is prevalent when more than one source of current exists in one circuit.

4. With full bottom rotation of variable resistor R6.
   a. Solving for the voltage drop across R7:

$$V_{R7} = V_{B5} - (-[V_{Z1} + V_{D8}]) = 0 \text{ V} - (-[14.3 \text{ V} + 0.7 \text{ V}]) = 15 \text{ V}$$

NOTE: Although the base of Q5 is at ground condition, the voltage drop across R7 is 15 volts for full bottom rotation. This is because 15 volts is developed across the zener and diode reference, 14.4 volts across resistor R8, and 15 volts across resistor R7 which, with respect to ground, is 15 volts.

b. Solving for the current flow through resistor R7: $I_{R7} = V_{R7}/R7 = 15 \text{ V}/2 \text{ k}\Omega = 7.5 \text{ mA}$

c. Solving for the voltage drop across resistor R6 from: $V_{R6} = I_{R7} \times R6 = 7.5 \text{ mA} \times 4 \text{ k}\Omega = 30 \text{ V}$

NOTE: The current flow through R7 flows mostly through R6 and a slight amount flows into the base of Q5. However, if the current through the series R6 and R7 resistors is large enough, then the current drain into the base of Q5 can be and is ignored. Therefore, $I_{R6} \approx I_{R7} = 7.5 \text{ mA}$. The voltage developed across R6 is also the voltage out since "ground" condition exists at the base of Q5. However, real ground is referenced so that voltage developed across R6 is not altered.

5. The most direct method of solving for the voltage out, when variable resistor R6 is at full bottom rotation, as illustrated in Figure 3-84, is solved from:

a. $V_0 = \dfrac{V_{B5} \times R6}{R7} = \dfrac{15\ V \times 4\ k\Omega}{2\ k\Omega} = 30\ V$

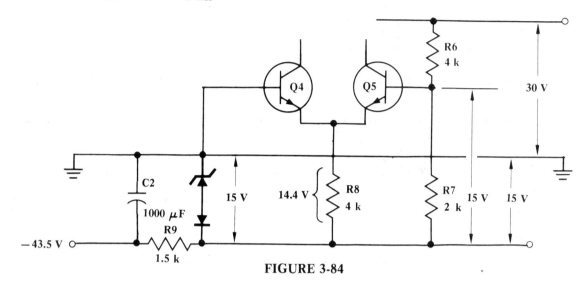

FIGURE 3-84

b. Solving for the remaining DC voltage drops around the circuit:

1. $V_{B1} \approx V_0 + V_{BE1} = 30\ V + 0.6\ V = 30.6\ V$

2. $V_{B2} = V_{B1} + V_{BE2} = 30.6\ V + 0.6\ V = 31.2\ V$

3. $V_{C3} = V_{B2} + V_{R2} = 31.2\ V + 0.2\ V = 31.4\ V$

4. $V_{E3} = VDC_{C1} - V_{R1} = 42.3\ V - 0.8\ V = 41.5\ V$

5. $V_{R9} = VDC_{C2} - Vref = 42.35 - 15\ V = 27.35\ V$

NOTE: The value of R9 can be critical because it must be chosen so that it delivers enough current to the R7, R8, and reference diode paths. Since a worse case condition for R7 is 7.5 mA, R8 $\approx$3.6 mA and the reference diode at approximately 5 mA, then R9 can be solved from $V_{R9}/I_T = 27.35\ V/16.1\ mA \approx 1698.7$ ohms. Therefore, a nominal 1.5 kΩ resistor is used.

c. Solving for the voltage drops across the circuit transistors:

1. $V_{CE1} = V_{C1} - V_{E1} \approx 42.3\ V - 30\ V = 12.3\ V$

2. $V_{CE2} = V_{C2} - V_{E2} = 42.3\ V - 30.6\ V = 11.7\ V$

3. $V_{CE3} = V_{E3} - V_{C3} = 41.5\ V - 31.4\ V = 10.1\ V$

4. $V_{CE4} = V_{C4} - V_{E4} \approx 30\ V - (-0.6\ V) = 30.6\ V$

5. $V_{CE5} = V_{C5} - V_{E5} = 31.2\ V - (-0.6\ V) = 31.8\ V$

d. Solving for the ripple voltage:
1. Across Filter capacitor C1 from:

$$\Delta V_{C1} = \frac{T}{C} \times \left(\frac{V_{R3}}{R3} + \frac{V_{R4}}{R4} + \frac{V_{R8}}{R8} + \frac{V_0}{V_{R6}} \right)$$

166

$$= \frac{1/120}{1000 \ \mu F} \times \left(\frac{40.9 \ V}{10 \ k\Omega} + \frac{30.6 \ V}{2 \ k\Omega} + \frac{14.4 \ V}{4 \ k\Omega} + \frac{30 \ V}{4 \ k\Omega} \right) \approx 254 \ mV$$

2. Across filter capacitor C2 from:

$$\Delta V_{C2} \approx T/C \times I_{R9} = \frac{1/120}{1000 \ \mu F} \times 18.2 \ mA \approx 151 \ mV$$

where: $V_{R9}/R9 = 27.35 \ V / 1.5 \ k\Omega = I_{R9}$

NOTE: Again worse case is noted. However, the full 3.6 mA of current demanded by R8 is not supplied wholely from filter capacitor C2.

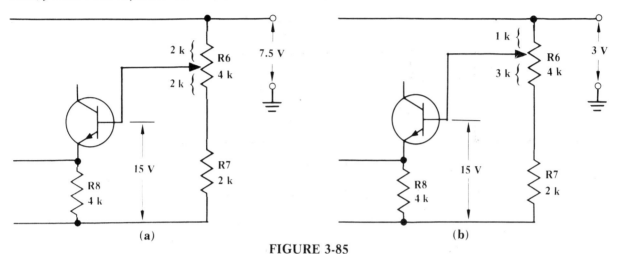

(a)   **(b)**

FIGURE 3-85

6. Mid-rotation or R6:

a. Solving for the voltage drop across R7 and one-half of the resistance of R6 when the wiper of the variable resistor R6 is at mid-rotation, as shown in Figure 3-85(a).

$$V(R7 + R6/2) = V_{B5} - (-[V_{Z1} + V_{D8}]) = 0 \ V - (-15 \ V) = 15 \ V$$

NOTE: It should be obvious by now that the base voltage, with respect to the minus reference voltage, always remains at 15 volts regardless of the wiper position of variable resistance R6.

b. Solving for the current flow through resistor R7 and one-half of the resistor R6 from:

$$I(R7 + R6/2) = \frac{V(R6 + R7)}{R6 + R7} = \frac{15 \ V}{2 \ k\Omega + 2 \ k\Omega} = 3.75 \ mA$$

c. Solving for the voltage drop across one-half (the remaining half) of resistor R6 from:

$$V(R6/2) = I(R7 + R6/2) \times R6/2 = 3.75 \ mA \times 2 \ k\Omega = 7.5 \ volts$$

NOTE: Since ground condition exists at the base of Q5 — wiper of variable resistor R6 — the voltage developed across the remaining half of the voltage variable resistor is the voltage out with respect to ground.

d. Using the direct method and solving for the voltage out with respect to ground:

$$Vo = \frac{V_{B5} \times R6/2}{R7 + R6/2} = \frac{15 \ V \times 2 \ k\Omega}{2 \ k\Omega + 2 \ k\Omega} = 7.5 \ volts$$

NOTE: If the wiper of the variable resistor R6 is varied to 3/4 position forming the 1 kΩ and 3 kΩ divider as illustrated in Figure 3-85(b), the voltage out would be:

167

$$V_0 = \frac{15 \text{ V} \times 1 \text{ k}\Omega}{2\text{k}\Omega + 3 \text{ k}\Omega} = 3 \text{ volts}$$

Note the 15 volts from the base of Q4 to the negative reference voltage still remains the same.

   e. Solving for the DC vltage drops around the circuit for mid-rotation of R6 conditions:

      1. $V_{B1} = V_0 + V_{BE1} = 7.5 \text{ V} + 0.6 \text{ V} = 8.1 \text{ volts}$

      2. $V_{B2} = V_{B1} + V_{BE2} = 8.1 \text{ V} + 0.6 \text{ V} = 8.7 \text{ volts}$

      3. $V_{C3} = V_{B2} = 8.7 \text{ V} + 0.2 \text{ V} = 8.9 \text{ volts}$

   f. Solving for the voltage drops across the circuit transistors:

      1. $V_{CE1} = V_{C1} - V_{E1} \approx 42.3 \text{ V} - 8.1 \text{ V} = 34.2 \text{ volts}$

      2. $V_{CE2} = V_{C2} - V_{E2} \approx 42.3 \text{ V} - 8.7 \text{ V} = 33.6 \text{ volts}$

      3. $V_{CE3} = V_{E3} - V_{C2} = 41.5 \text{ V} - 8.9 \text{ V} = 32.6 \text{ volts}$

      4. $V_{CE4} = V_{C4} - V_{E4} \approx 7.5 \text{ V} - (-0.6 \text{ V}) = 8.1 \text{ volts}$

      5. $V_{CE5} = V_{C5} - V_{E5} = 8.7 \text{ V} - (-0.6 \text{ V}) = 9.3 \text{ volts}$

   g. Solving for the ripple voltage:
      1. Across filter capacitor C1 from:

$$\Delta V_{C1} = \frac{T}{C_1}\left(\frac{V_{R3}}{R3} + \frac{V_{R4}}{R4} + \frac{V_{R8}}{R8} + \frac{V_0}{R6/2}\right) = \frac{1/120}{1000 \text{ } \mu\text{F}}\left(\frac{40.9 \text{ V}}{10 \text{ k}\Omega} + \frac{8.1 \text{ V}}{2 \text{ k}\Omega} + \frac{14.4 \text{ V}}{4 \text{ k}\Omega} + \frac{7.5 \text{ V}}{2 \text{ k}\Omega}\right) = 127 \text{ mV}$$

      2. Since R9 controls the I_L for C2, ΔV_{C2} remains the same at approximately 151 mV.

6. Top rotation of Resistor R6:   Varying the wiper of variable resistor R6 to full top condition places a ground directly at the output "positive" terminal and zero volts across the output terminals. Note, however, that the 15 volts difference between the base of Q5 and the negative reference voltage remains constant at 15 volts. This is illustrated in Figure 3-86.

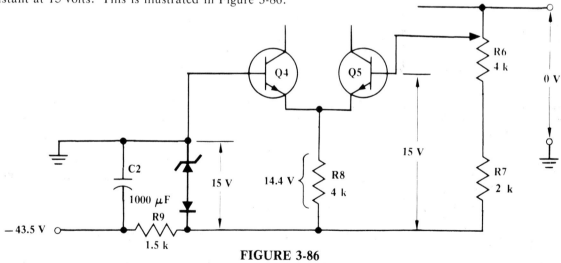

FIGURE 3-86

   a. Solving for the DC voltage drops around the circuit:

      1. $V_{B1} = V_0 + V_{BE1} = 0 \text{ V} + 0.6 \text{ V} = 0.6 \text{ volts}$

168

2. $V_{B2} = V_{B1} + V_{BE2} = 0.6 \text{ V} + 0.6 \text{ V} = 1.2$ volts

3. $V_{C3} = V_{B2} + V_{R2} = 1.2 \text{ V} + 0.2 \text{ V} = 1.4$ volts

b. Solving for the voltage drops across the circuit transistors:

1. $V_{CE1} = V_{C1} - V_{E1} \approx 42.3 \text{ V} - 0 \text{ V} = 42.3$ volts

2. $V_{CE2} = V_{C2} - V_{E2} \approx 42.3 \text{ V} - 0.6 \text{ V} = 41.7$ volts

3. $V_{CE3} = V_{E3} - V_{C3} = 41.5 \text{ V} - 1.4 \text{ V} = 40.1$ volts

4. $V_{CE4} = V_{C4} - V_{E4} \approx 0 \text{ V} - (-0.6 \text{ V}) = 0.6$ volts

5. $V_{CE5} = V_{C5} - V_{E5} = 1.2 \text{ V} - (-0.6 \text{ V}) = 1.8$ volts

c. The low voltage across Q4 at, and just above, a zero volt condition seemingly forces Q4 into a saturated condition. One solution to this problem is to reference the base of Q4 a few volts below ground (-1.4 volts is an example) and place an adjustable resistor in series with variable resistor R6 to provide a "zero" or ground condition when R6 is at full top rotation. This is shown in Figure 3-87. Therefore, with respect to ground, the base of Q4 and Q5 are at minus 1.4 volts and with the adjustable resistor set at approximately 190 ohms (for 7.5 mA of bleed current), the output voltage will be zero volts.

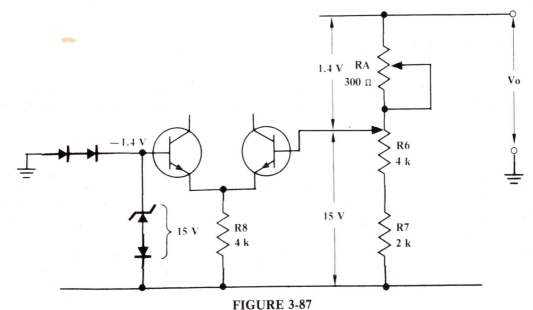

FIGURE 3-87

d. For a loaded condition of 100 ohms, the power dissipated across Q1 is greatest at full bottom rotation of variable resistor R6 because this condition provides maximum output voltage. However, the least V_{CE1} occurs at this maximum Vo condition.

1. For a 100 ohm load condition at bottom rotation:

$$P_{Q1} = V_{CE1} \times I_L \approx 12.3 \text{ V} \times 300 \text{ mA} = 3.69 \text{ W} \quad \text{where:} \quad I_L \approx V_o/R_L = 30 \text{ V}/100 \text{ }\Omega = 300 \text{ mA}$$

2. For a 100 ohm load condition at mid-rotation:

$$P_{Q1} = V_{CE1} \times I_L \approx 34.2 \text{ V} \times 7.5 \text{ mA} = 2.565 \text{ W} \quad \text{where:} \quad I_L \approx V_o/R_L = 7.5 \text{ V}/100 \text{ }\Omega = 75 \text{ mA}$$

3. For a 100 ohm load condition at (2 volt output) just before zero output:

$$P_{Q1} = V_{CE1} \times I_L \approx 41.3 \text{ V} \times 20 \text{ mA} = 826 \text{ mW} \quad \text{where:} \quad I_L = V_o/R_L = 2 \text{ V}/100 \text{ }\Omega = 20 \text{ mA}$$

e. **High Current-Voltage Conditions:** For increased load conditions approaching short circuit condi-

169

tions, the maximum power dissipated across Q1 occurs when V_{CE} is large. For instance, assuming 0.8 amps with a Vout of 2 volts:

$$P_{Q1} = V_{CE1} \times I_L = 40.5 \text{ V} \times 0.8 \text{ A} = 32.4 \text{ watts}$$

NOTE: A single series pass transistor will require a relatively large heat sink in order to dissipate 32.4 watts safely. However, if two such transistor are place in series or in parallel, the 32.4 watts can be shared.

EXPERIMENTAL OBJECTIVES

To investigate the voltage variable power supply having zero output voltage variability.

LIST OF MATERIALS

1. Transformer: 48 V C.T. or equivalent
2. Diodes: 1N4001 (twelve) or equivalent
3. Resistors: 1 Ω(three) 100 Ω(one) 330 Ω(one) 1.5 kΩ(one) 2.2 kΩ(two) 4.7 kΩ(one) 10 kΩ(one) 10 kΩ potentiometer (one)
4. Capacitors: 1000 µF (two) 100 µF (two)
5. Zener diode: 12 volt(one)
6. Transistors: 2N3055(one) 2N3053 or 2N6178 or equivalent(one) 2N3904(two) 2N3906(one)

EXPERIMENTAL PROCEDURE

1. Connect the circuit as shown in Figure 3-88 and, initially, set the R6 potentiometer to bottom rotation. Also, the unregulated supply, as shown in the sample problem, can be used if only a center tapped transformer is available, or separate secondary windings can be used instead.

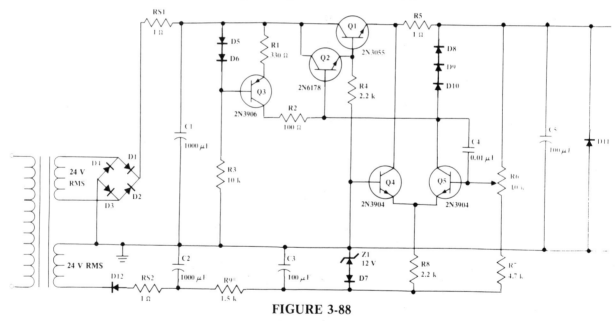

FIGURE 3-88

2. Across the secondary of the transformer driving the full wave bridge, perform either measurement:
   a. Measure the Vp-p with an oscilloscope.
   b. Measure the RMS voltage with a voltmeter.
   Across the secondary of the transformer driving the half wave rectifier, perform either measurement:
   a. Measure the Vp-p with an oscilloscope.
   b. Measure the RMS voltage with a voltmeter.
3. Calculate and measure:
   a. The Vp across filter capacitor C1.
   b. The Vp across filter capacitor C2.

4. Calculate the DC voltages across the filter capacitors C1 and C2 using the formula:

$$V DC = V_p - \Delta V_C/2 \approx V_p \quad \text{(for each filter capacitor)}$$

Measure the DC voltages across filter capacitors C1 and C2.

NOTE: ΔV_{C1} is calculated for full bottom rotation of R6. ΔV_{C2} calculations are solved simply from knowing I_{R9}, which must be large enough to insure zener and differential amplifier operation. Therefore, if V DC across C2 is lower than the value solved for in the sample problem, R9 will have to be changed.

5. For bottom rotation condition of variable resistor R6:
   a. Calculate the DC voltages, with respect to ground:
      1. At the base of Q4.

      2. At the emitters of Q4 and Q5 from: $V_{E4} = V_{E5} = V_{B4} - V_{BE4}$

      3. At the base of Q5 from: $V_{B5} = V_{E5} + V_{BE5}$

   b. Calculate the DC voltages across:

      1. The reference voltage from: $V_{ref} = V_{Z1} + V_{D7}$

      2. The resistor R8 from: $V_{R8} = V_{ref} - V_{BE4}$

      3. The resistor R7 from: $V_{R7} = V_{ref} = V_{R8} + V_{BE5}$

      4. The output terminals from: $Vo = \dfrac{V_{B5} \times R6}{R7}$

   c. Calculate the DC voltages, with respect to ground, at the:

      1. Emitter of Q1 from: $V_{E1} \approx Vo$

      2. Base of Q1 from: $V_{B1} = V_{E1} + V_{BE1}$

      3. Base of Q2 from: $V_{B2} = V_{B1} + V_{BE2}$

      4. Collector of Q3 from: $V_{C3} = V_{B2} + V_{R2}$

      5. Emitter of Q3 from: $V_{E3} = V_{C1} - V_{R1}$

      6. Base of Q3 from: $V_{B3} = V_{C1} - V_{(D5 + D6)} = V_{R1} + V_{BE3}$

   d. Measure the DC voltages calculated in the previous step.
6. Calculate the DC voltage drops across transistors Q1, Q2, Q3, Q4, and Q5.
   Measure the DC voltage drops across transistors Q1, Q2, Q3, Q4, and Q5.
7. Calculate the ripple voltage:
   a. Across filter capacitor C1:
      1. For a no load (bottom rotation) condition from: $\Delta V_{C1} = T/C \times I_L$

      where: $I_L \approx \dfrac{V_{R3}}{R3} + \dfrac{V_{R4}}{R4} + \dfrac{V_{R8}}{R8} + \dfrac{Vo}{R6}$

      2. For a load condition of 100 ohms from: $\Delta V_{C1} = T/C \times I_L$

      where: $I_L \approx \dfrac{V_{R3}}{R3} + \dfrac{V_{R4}}{R4} + \dfrac{V_{R8}}{R8} + \dfrac{Vo}{R6} + \dfrac{Vo}{RL}$

171

b. Filter capacitor C2:
   1. For a no load condition from: $\Delta V_{C1} = T/C \times I_{R9}$
   2. For a loaded condition of 100 ohms from: $\Delta VC1 = T/C \times I_{R9}$
c. Measure the ripple voltage across C1 and across C2 for no load and loaded conditions.
d. Measure the ripple voltage at the base of Q1, across the output, and across the zener diode Z1, for both loaded and unloaded conditions.

8. Set the variable R6 resistor to approximately mid-rotation.
   Calculate the Vo and then measure Vo, the voltage at the base of Q2, and the voltage drop across Q1, Q2, Q3, Q4, and Q5.
   b. Measure the ripple voltage at the output for no load and loaded conditions.

9. Set the variable R6 resistor to top-rotation condition.
   a. Calculate the Vo and then measure Vo and the voltage at the base of Q2.
   b. Measure the ripple voltage at the output for no load and for loaded conditions.

NOTE: Monitor the base of Q5 — to the negative reference voltage. This should stay constant throughout the varied voltage region. An exception could be in the very low output voltage region (<3 volts). This condition can be improved by placing the base of Q5 at a negative voltage condition of approximately minus 1.8 volts and placing a "zero adjust" variable resistor in series with resistor R6 — with one end connected at the positive output terminal. This will keep Q4 from a cutoff condition.

10. At full bottom rotation, test the short circuit condition.
    a. Calculated I_{SC} from $V_{R5}/R5$.
    b. Measure I_{SC}.
11.
    Measure and calculate the percent regulation at bottom rotation. Calculate from:

$$\%\text{reg} = \frac{V(NL) - V(L)}{V(NL)} \quad \text{where:} \quad RL = 100 \ \Omega$$

Insert the calculated and measure values, as indicated, into Table 3-15 and Table 3-16.

TABLE 3-15	Vp-p SEC.		V RMS SEC.		Vp		V DC	
	FWB	HW	FWB	HW	C1	C2	C1	C2
NO LOAD CALCULATED	///	///	///	///				
NO LOAD MEASURED								
100 Ω LOAD CALCULATED	///	///	///	///	///	///	///	///
100 Ω LOAD MEASURED	///	///	///	///	///	///		

	VDC BOTTOM ROTATION								
	V_{B4}	V_{E5}	V_{B5}	V_{E1}	V_{B1}	V_{B2}	V_{C3}	V_{E3}	V_{B3}
NO LOAD CALCULATED									
NO LOAD MEASURED									

	VDC BOTTOM ROTATION								
	$V_{(Ref)}$	V_{R6}	V_{R7}	V_{R8}	V_{CE1}	V_{CE2}	V_{CE3}	V_{CE4}	V_{CE5}
NO LOAD CALCULATED									
NO LOAD MEASURED									

TABLE 3-16	V_{CE} — MID-ROTATION					ΔV		BOTTOM ROTATION		
	Q1	Q2	Q3	Q4	Q5	C1	C2	B_{Q1}	Vo	V_{Z1}
NO LOAD CALCULATED	▨	▨	▨	▨	▨			▨	▨	▨
NO LOAD MEASURED										
100Ω LOAD CALCULATED	▨	▨	▨	▨	▨			▨	▨	▨
100 Ω LOAD MEASURED	▨	▨	▨	▨	▨					

	MID-ROTATION			TOP ROTATION			BOTTOM ROT.	
	Vo	B_{Q2}	ΔVo	Vo	B_{Q2}	ΔVo	I_{SC}	% Reg.
NO LOAD CALCULATED		▨	▨		▨	▨	▨	▨
MEASURED							▨	▨
100 Ω LOAD CALCULATED	▨	▨	▨	▨	▨	▨		
100 Ω LOAD MEASURED	▨	▨		▨	▨			▨

173

PART 3D — THE MONOLITHIC DRIVEN POWER SUPPLY

GENERAL DISCUSSION

There is no doubt that the trend in circuit design is moving towards the monolithic (complete circuit in a package) approach and away from the (all) discrete device approach used in the previous sections. The first move in that direction came when the high gain op-amp was used to replace the differential amplifier and driver for the series pass power transistor. This was soon followed with monlithic devices like the μA723, MC1456, LM309, and others, in which few external parts were required and excellent circuit performance was obtained.

THE OP-AMP DRIVEN ADJUSTABLE POWER SUPPLY

A power supply circuit that provides most of the features of the previous circuit, but which has fewer components, is the op-amp driven power supply circuit shown in Figure 3-89. The zener diode references the minimum voltage limit and also attenuates the ripple voltage present across filter capacitor C1. The op-amp is essentially a high gain differential amplifier. So, the DC voltage present at pin 3 also appears at pin 2 and, therefore, the ratio of R5 to (R4 +R5) determines the output voltage. Resistor R3 provides current limiting for the op-amp and resistor R2 is used to provide minimum DC offset at the output terminal pin 6. A filter capacitor is placed across the zener to provide futher attentuation of ripple and zener noise.

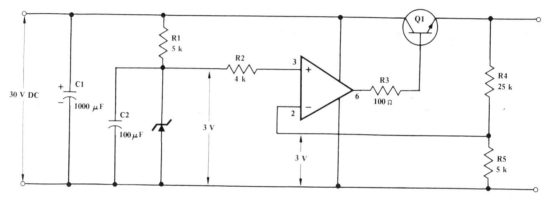

FIGURE 3-89

1. $V_{Z1} = 3$ volts

2. $V_{pin_3} = V_{pin_2} = V_{R5} = 3$ volts

3. $V_O = \dfrac{V_{R5}(R4 + R5)}{R5} = \dfrac{3V(5\ k\Omega + 25\ k\Omega)}{5\ k\Omega} = 18$ volts

4. $V_{B1} = V_O + V_{BE1} = 18\ V + 0.6\ V = 18.6$ volts

5. $V_{CE1} = V_{DC_{C1}} - V_O = 30\ V - 18\ V = 12$ volts

THE OP-AMP DRIVEN ADJUSTABLE POWER SUPPLY — CIRCUIT ANALYSIS

1. With reference to Figure 3-90, solving for the peak voltage across filter capacitor C1 from:

    $$V_{peak} = \sqrt{2} \times V\ RMS = 1.414 \times 20\ V\ RMS = 28.28\ V\ DC$$

2. Solving for the DC voltage developed across Filter capacitor C1:

    $$V\ DC = V_p - \Delta V_{C1}/2 = 28.28\ V - 180\ mV/2 = 28.19\ V \approx 28.2\ volts$$

3. Bottom Rotation — 1 kΩ Load

174

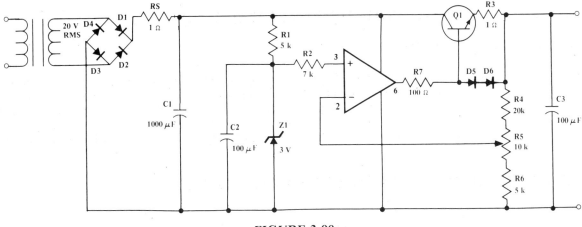

FIGURE 3-90

a. The voltage of zener diode Z1 is: $V_{Z1} = 3$ volts

b. Solving for the voltage drop across R1: $V_{R1} = VDC_{C1} - V_{Z1} = 21.2\ V - 3\ V = 18.2$ volts

c. Solving for the voltage at pin 3: $Vpin_3 = V_{Z1} - V_{R2} = 3$ volts

This is where no appreciable DC voltage is dropped across the R2 DC balancing resistor — where R2 is changed to 7 kΩ to approximate the mid-rotation DC balance condition.

d. Solving for the voltage at pin 2: $Vpin_2 \approx Vpin_3 = 3$ volts

NOTE: The inverting and non-inverting terminals of an op-amp are effectively the base inputs to a differential amplifier and the DC voltage should, therefore, follow each other in a working circuit.

e. Solving for the output voltage:

$$Vo = \frac{V_{R6}(R4 + R5 + R6)}{R6} = \frac{3\ V\,(20\ k\Omega + 10\ k\Omega + 5\ k\Omega)}{5\ k\Omega} = 21\ V$$

where: $V_{R6} = Vpin_2$

f. Solving for the voltage at the base of Q1:

$$V_{B1} = Vo + V_{BE1} = 21\ V + 0.6\ V = 21.6\ volts$$

g. Solving for the voltage at pin 6 of the op-amp:

$$Vpin_6 = V_{B1} + V_{R7} \approx 21.6\ V + 0.043\ V \approx 21.64\ volts$$

where: $I_{R7}R7 = 0.432\ mA \times 100\ \Omega = 0.043\ volts$

NOTE: Under 1 kΩ conditions, a total of 21.6 mA of "load" current flows and the base current for the Q1 transistor is 0.432 mA.

$$I_L = \frac{Vo}{R4 + R5 + R6} + \frac{Vo}{RL} = \frac{21\ V}{35\ k\Omega} + \frac{21\ V}{1\ k\Omega} = 0.6\ mA + 21\ mA = 21.6\ mA$$

$$I_{B1} = I_L/\beta_{Q1} = 21.6\ mA/50 = 0.432\ mA \quad where:\ \ \beta_{Q1} = 50$$

h. Solving for the ripple voltage across filter capacitor C1:

$$\Delta V_{C1} = T/C \times I_L = \frac{1/120}{1000\ \mu F} \times 21.6\ mA = 180\ mV$$

175

i. Solving for the voltage drop across VCE:

$$V_{CE1} = VDC_{C1} - V_{E1} \approx 28.2 \text{ V} - 21 \text{ V} = 7.2 \text{ volts}$$

NOTE: The voltage drop across the current limiting resistor R3 is insignifican (21.6 mV) and is ignored.

4. SHORT CIRCUIT CONDITIONS:

a. Solving for the short circuit current conditions:

$$I_{SC} = V_{R3}/R3 = 0.8 \text{ V}/100 \text{ }\Omega = 800 \text{ mA} \quad \text{where:} \quad D5 + D6 = 1.4 \text{ V} \quad \text{and} \quad V_{BE1} = 0.6 \text{ V}$$

b. Solving for the ripple voltage across C1:

$$\Delta V_{C1} \approx \frac{T}{C} \times I_{SC} = \frac{1/120}{1000 \text{ }\mu F} \times 800 \text{ mA} \approx 6.66 \text{ volts}$$

c. Solving for the DC voltage developed across the filter capacitor:

$$V \text{ DC} = Vp - \Delta V_{C1}/2 = 28.28 \text{ V} - 6.66 \text{ V}/2 = 24.95 \text{ volts}$$

d. Solving for the DC voltage developed across V_{CE1}:

$$V_{CE1} = VDC_{C1} - V_{E1} = 24.95 \text{ V} - 0.8 \text{ V} = 24.15$$

where: $V_{E1} = Vo + V_{R3} = 0 \text{ V} + 0.8 \text{ V} = 0.8 \text{ volts}$

e. Solving for the power dissipated by Q1:

$$P_{Q1} = V_{CE1} \times I_{SC} = 24.15 \text{ V} \times 800 \text{ mA} = 19.32 \text{ watts}$$

f. Solving for the base of Q1 current flow:

$$I_{BQ1} = I_{R7} = I_L/\beta_{Q1} = 800 \text{ mA}/50 = 16 \text{ mA}$$

NOTE: There is also the current flow through resistor R7 and out of the op-amp. Therefore, $V_{R7} = I_{R7} \times R7 = 16 \text{ mA} \times 100 \text{ }\Omega = 1.6$ volts, and 16 mA can shut down the op-amp and, hence, the circuit.

g. Solving for the voltage at Pin 6 of the op-amp:

$$Vpin_6 = V_{B1} + V_{R7} = 1.4 \text{ V} + 1.6 \text{ V} = 3 \text{ volts}$$

where: $V_{B1} = Vo + V_{BE1} + V_{R3} = 0.6 \text{ V} + 0.8 \text{ V} = 1.4 \text{ volts}$

5. Zener Diode Substitution: Since the intent of this variable supply is to keep the minimum voltage out low, a low voltage zener was selected. However, if further reduction is needed, diode substitution is an alternate method. The circuit of Figure 3-91 shows the technique — where the three series diodes provide a voltage reference of about 2.1 volts.

THE PLUS AND MINUS VARIABLE POWER SUPPLY

Plus and minus supplies can be easily achieved by connecting two variable supplies in series and referencing ground between the two. This method is used all the time in laboratory set-ups.

Another method is to utilize an NPN series pass device in the positive supply and a PNP series pass device in the negative supply. This technique is shown in Figure 3-92 and is, essentially, a duplication of the previously analized NPN series pass circuit of Figure 3-91 with the PNP series pass circuit providing a mirror image. For instance, with both potentiometers at full-bottom rotation, the supplies provide ±21 V DC, at mid-rotation they provide ±7.35 V DC, and at top rotation they provide ±4.9 V DC.

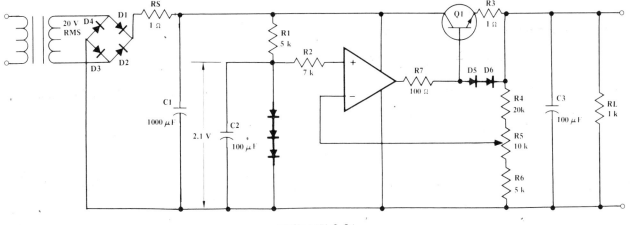

FIGURE 3-91

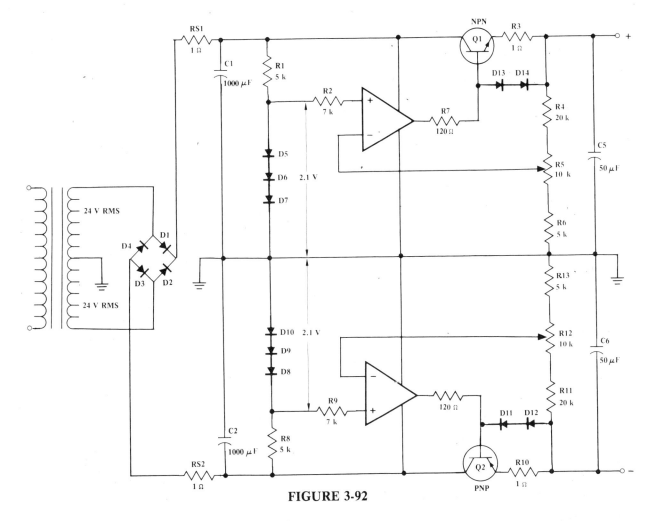

FIGURE 3-92

177

EXPERIMENTAL OBJECTIVES

To investigate the op-amp voltage variable power supply and the plus and minus variable op-amp driven power supply.

LIST OF MATERIALS

1. Transformer: 120 V : 24 V (two) or 120 V : 48 V CT
2. Diode: 1N4001 or equivalent (fourteen)
3. Resistors: 1 Ω (two) 2.7 Ω (two) 120 Ω (two) 1 kΩ (two) 3.3 kΩ (four) 6.8 kΩ (two)
    10 kΩ (two) 27 kΩ (two)
4. Capacitors: 50 μF (two) 100 μF (two) 1000 μF (two)
5. Transistors: 2N6178 (one) 2N6180 (one) or equivalents
6. Op-Amp: μA741 (one) or equivalent

EXPERIMENTAL PROCEDURE (PART 1):

1. Connect the circuit as shown in Figure 3-93. Use either a low voltage zener diode (approximately three volts) or the three diodes in series as the reference.

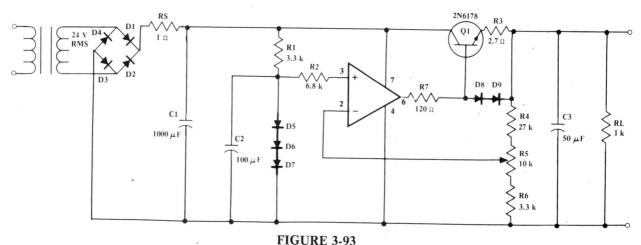

FIGURE 3-93

2. Across the secondary of the transformer perform either measurement:
    a. Measure the Vp-p with an oscilloscope.
    b. Measure the V RMS with a voltmeter.
3. Calculate and measure the Vp across the filter capacitor C1.
4. Calculate and measure the DC voltage across filter capacitor C1. Calculate from:

$$V\ DC = Vp - \Delta V_{C1}/2 \approx Vp$$

5. Bottom-Rotation Condition
    a. Calculate the DC voltages:
        1. Across the zener or diode reference — Pin 3 of the op-amp.
        2. Across resistor R6 — Pin 2 of the op-amp.
        3. Across the output terminals.
        4. At the base of Q1.
        5. At the output of the op-amp — Pin 6.
        6. Across the collector-emitter of Q1.
    b. Measure the DC voltages that were calculated in Step a.
6. Top-Rotation Condition
    a. Calculate the DC voltages:
        1. Across the zener or diode reference — Pin 3 of the op-amp.
        2. Across resistor R6 — Pin 2 of the Op-amp.
        3. Across the output terminals.
        4. At the base of Q1.

178

5. At the output of the op-amp — Pin 6.
6. Across the collector-emitter of Q1.
   b. Measure the DC voltages that were calculated in step a.
7. Calculate the ripple voltage across the filter capactior at bottom rotation.
   a. For the pre-loaded 1 kΩ "load" condition.
   b. For an additional load condition of 100 Ω connected in parallel with the 1 kΩ resistor.
8. Measure the ripple voltage for both the pre-loaded 1 kΩ condition and the 100 Ω load condition:
   a. Across the filter capacitor.
   b. Across the voltage reference — capacitor C2.
   c. Across the output terminals
9. Calculate and measure the percent regulation for bottom rotation using the pre-loaded condition and the V(NL) and the 100 Ω load condition as the V(L). Calculate from:

$$\% \text{ regulation} = \frac{V(NL) - V(L)}{V(NL)} \times 100$$

10. Test the short circuit condition at bottom rotation.
    a. Calculate I_{SC} from $V_{R3}/R4$.
    b. Measure the I_{SC}.
11. Insert the calculated and measured values, as indicated, into Table 3-17.

TABLE 3-17	SEC Vp-p	V RMS	Vp C1	V DC C1	V Pin 3	V Pin 2	Vo	V_{B1}
Bottom Rotation CALCULATED	//////	//////						
Bottom Rotation MEASURED								
Top Rotation CALCULATED	//////	//////	//////	//////				
Top Rotation MEASURED	//////	//////	//////	//////				
100 Ω Load Top Rotation CALCULATED	//////	//////	//////	//////	//////	//////	//////	//////
100 Ω Load Bottom Rotation MEASURED	//////	//////	//////	//////				

	V Pin 6	V_{CE}(Q1)	ΔV_{C1}	ΔV_{C2}	ΔVo	% Reg.	I_{SC}
Bottom Rotation CALCULATED				//////	//////	//////	
Bottom Rotation MEASURED						//////	
Top Rotation CALCULATED			//////	//////	//////	//////	//////
Top Rotation MEASURED			//////	//////	//////	//////	//////
100 Ω Load Bottom Rotation CALCULATED	//////	//////	//////	//////	//////		//////
100 Ω Load Bottom Rotation MEASURED						//////	//////

EXPERIMENTAL PROCEDURE (PART 2)

1. Connect the plus and minus variable power supply circuit as shown in Figure 3-94. Again, use either zeners or diode strings in providing the reference voltages. Adjust variable resistor R5 and R12 to bottom rotation.

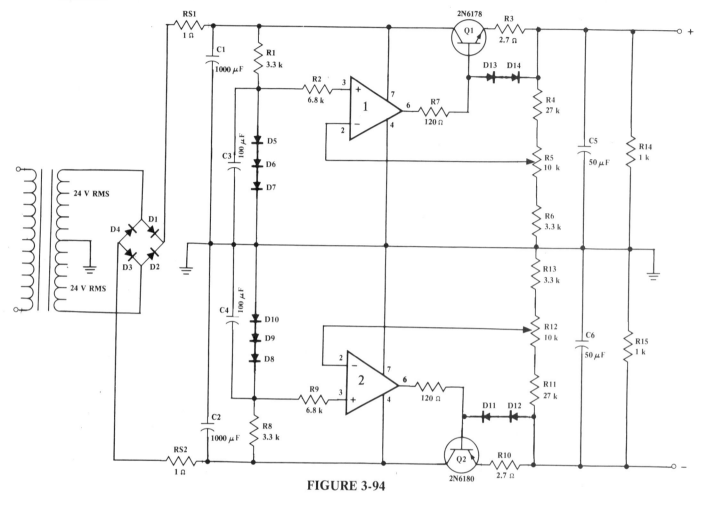

FIGURE 3-94

2. Across the secondary of the transformer, driving filter capacitor C1, perform either measurement:
   a. Measure the Vp-p with an oscilloscope.
   b. Measure the V RMS with a voltmeter.
3. Calculate and measure the Vp across filter capactiors C1 and C2.
4. Calculate and measure the DC voltages across filter capacitors C1 and C2. Calculate from:

$$V\,DC = Vp - \Delta V_C/2 \approx Vp \quad \text{(for each filter capacitor at both top and bottom rotation)}$$

5. The negative supply is a mirror image of the positive supply already calculated and measured. Therefore, it is necessary only to monitor the DC and ripple content to insure a similarly working circuit — that is, of course, if the plus and minus applied V DC is the same as that which was supplied to the single positive supply. Therefore, for both top and bottom rotation, taken with respect to ground:
   a. Measure the voltage across the reference diodes (Pin 3).
   b. Measure the voltage across resistor R13 — Pin 2 of op-amp #2.
   c. Measure the voltage across the output terminals.
   d. Measure the voltage at the base of Q2.
   e. Measure the voltage at the output of the #2 op-amp — Pin 6.
   f. Measure the voltage across the collector-emitter of Q2.
6. Monitor the ripple voltage for the 1 kΩ pre-loaded and 100 Ω loaded conditions:

180

a. Across the filter capacitor C2.
b. Across filter capactior C4.
c. Across the output terminals of the negative supply.
7. Measure the I_{SC} condition for both plus and minus supplies.
8. Calculate the percent regulation.
9. Insert the calculated and measured values, as indicated, into Table 3-18 and Table 3-19.

TABLE 3-18	SEC		Vp		V DC	
	Vp-p	V RMS	C1	C2	C1	C2
Bottom Rotation CALCULATED						
Bottom Rotation MEASURED						
Top Rotation CALCULATED						
Top Rotation MEASURED						

	NEGATIVE SUPPLY			POSITIVE SUPPLY			NEGATIVE SUPPLY		
	V Pin 3	V Pin 2	V Pin 6	V_{B1}	Vo	V_{CE1}	V_{B2}	Vo	V_{CE2}
Bottom Potation CALCULATED									
Bottom Rotation MEASURED									
Top Rotation CALCULTED									
Top Rotation MEASURED									

TABLE 3-19	POSTIVE SUPPLY			NEGATIVE SUPPLY			% REG. Supply		I_{SC} — B.R. Supply	
	ΔV_{C1}	ΔV_{C3}	Vo	ΔV_{C2}	ΔV_{C4}	ΔVo	+	—	+	—
PRE-LOADED CALCULATED										
PRE-LOADED MEASURED										
100 Ω Load CALCULATED										
100 Ω Load MEASURED										

NOTE: Insert the previously measured plus supply values into Table 3-18. This allows a comparison for the negative supply values.

PART 3E — MONOLITHIC REGULATORS

GENERAL DISCUSSION

Up until just a few years ago the decision whether to use monolithic regulators or discrete devices was still a toss-up. Monolithic devices were relatively expensive, susceptible to blowout during breadboarding, and required external transistors (normally series pass) and other components if current capabilities of much greater than 150 mA were required. Additionally, flexibility in design was limited to the device's availability.

Today, a considerable number of regulator devices are available and they are capable of fixed or variable output voltages in both positive and negative regulators. Additionally, some devices can be used in switching regulator circuits, have internal short circuit protection, and can deliver in excess of 1 ampere of output current with only a few additional external components.

EQUIVALENT CIRCUIT

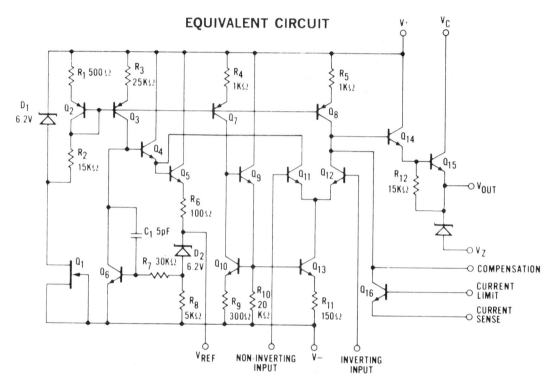

EQUIVALENT CIRCUIT

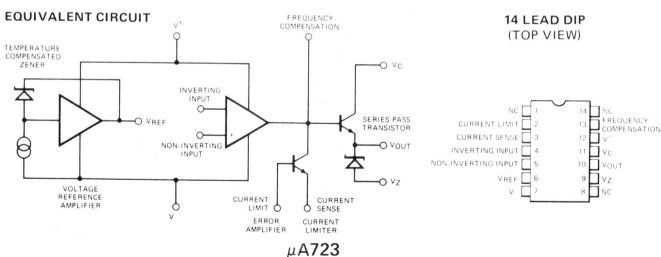

μA723

FIGURE 3-95   *Courtesy of Fairchild Semiconductor*

182

National Semiconductor, Signetics, Texas Instruments, Motorola, Fairchild, and Ratheon are some of the manufacturers of monolithic regulator devices, and data books and application notes on the regulator product line, for each of the above manufacturers, can be obtained from vendors, from the companies, or through technical book stores. The cost is minimal and some are free for the asking.

Most of the larger manufacturers produce the widely used regulators like the 723 and 309 devices, and the manufacturer can normally be identified by the letters preceding the device numbers. For instance, National uses LM723, Motorola MC723, Ratheon RM723, and Fairchild μA723. However, with regard to the 309 device, National uses LM309, Motorola MLM309, Ratheon LM309, and Signetics LM309. In this case, there is no clear indication as to whom the manufacturer might be. Too, signetics uses a variety of preceding letters including both LM and μA letters, which are normally identified with National and Fairchild, respectively.

THE μA723 MONOLITHIC REGULATOR — AN ANALYSIS

One of the first monolitic devices to appear on the market, the μA723 is capable of being used in positive or negative regulator circuits, as a current source, or as the "guts" of a switching circuit. It is available in both the 14 Pin DIP package and the 10 Pin TO-5 package, and it is capable of fixed or variable output voltages. The device has short circuit protection capability and a maximum current, without using external components, of 150 mA. Its only drawback is that it is highly susceptible to "blowout", so extreme care must be taken when breadboarding the device to make sure it has properly wired connections before the power is applied. The schematic diagram, the simplified functional schematic block diagram, and the device pin-out are shown in Figure 3-95.

The device has three major sections: the reference voltage section, the difference amplifier section, and the output section with short circuit protection capabilities. The reference voltage section consists of Q1 through Q5 and the zener output is approximately 7.2 volts. (7.2 V is the published reference voltage, but the measured values of the devices tested were nominally 7.18 volts.) The differential amplifier consists of transistors Q11 and Q12 and the zener output, when connected to the base of Q11, establishes the voltage at the base of Q12. The output stage consists of transistors Q14 and Q15 and these can be supplemented with external series pass transistors to increase the current carrying capability of the device. Q16 provides short circuit protection and an external resistor, placed between the base-emitter of Q16, provides the current limit. The functional schematic block diagram illustrates the building blocks of the device.

The μA723 can be connected in several different circuit applications and, in Figure 3-96, a representative voltage supply is shown where the current capabilities of the device have been extended to 600 mA. Note that the stepped-down zener reference (emitter of Q5) is tied to the base of Q11, establishing that voltage at the base of Q12. Hence a variety of circuit reference voltages can be obtained through voltage divider resistors, and maximum reference voltage can be obtained when Pins 5 and 6 are directly connected. When the short between Pins 5 and 6 is removed and replaced by the R16 and R17 resistors, the reference voltage can be varied down to about 1/7 of the zener voltage.

1. Pin 6 is the zener reference voltage and external resistors R16 and R17 step down the approximate 7.2 volt reference voltage to 3.24 volts. To solve for V_{R17}:

$$V_{R17} = \frac{V(\text{zener ref.}) \times R17}{R16 + R17} = \frac{7.2\ V \times 1.8\ k\Omega}{2.2\ k\Omega + 1.8\ k\Omega} = 3.24 \text{ volts}$$

2. The junction of R16 and R17 is connected to Pin 5, the base of Q11 of the differential amplifier. Hence, the voltage at the base of Q12 is also at 3.24 volts and will be the voltage out for the top-rotation condition. For bottom-rotation condition:

$$V_o = \frac{V_{B12}(R14 + R15)}{R15} = \frac{3.24\ V(10\ k\Omega + 1.8\ k\Omega)}{1.8\ k\Omega} = 21.24 \text{ volts, where: } V_{B12} = V(\text{Pin 4})$$

3. The short circuit current is limited to about 600 mA by the base-emitter junction of Q16 and the external 1 ohm resistor R13. This is obtained from:

$$V_{BE16}/R16 = 0.6\ V/1\ \Omega = 600 \text{ mA}$$

4. Capacitor C3, connected across the collector-base junction of transistor Q12, provides Miller effect frequency roll-off to minimize the oscillatory possibilities.

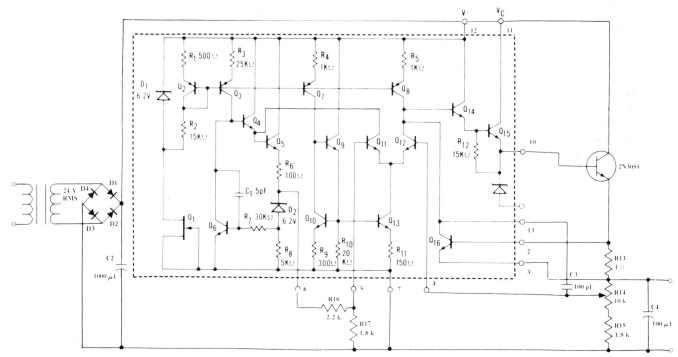

FIGURE 3-96

Another method used to connect a variable output regulator is to fix the resistors at the output to provide voltage multiplication and, then, vary the reference voltage. This kind of circuit is shown in Figure 3-97 where the functional schematic block diagram for the μA723 is used for circuit simplification.

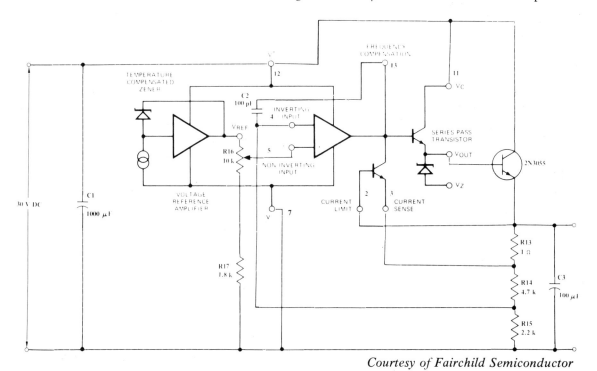

FIGURE 3-97

1. The reference voltage is varied between a high of approximately 7.2 volts (the zener voltage) to a low of about 1 volt. If the voltage at the base of Q11 is much lower than 1 volt, the constant current source Q13 saturates, the voltage at the base of Q12 does not follow the voltage at the base of Q11, and regulation

184

ceases. Therefore, if a 10 kΩ variable resistor is used, a 1.8 kΩ resistor will maintain the minimum voltage at the base of Q11 to approximately 1.1 volts, from:

$$V_{B11} = \frac{V(\text{zener ref.}) \times R16}{R16 + R17} = \frac{7.2\,V \times 1.8\,k\Omega}{10\,k\Omega + 1.8\,k\Omega} \approx 1.1 \text{ volts}$$

The manufacturers minimum voltage is suggested at 0.72 volts, so 1.1 volts is above the minimum limitation.

2. The wiper of variable resistor R16 is connected to the base of Q11. Hence, the voltage at the base of Q12 and the junction of resistors R14 and R15 is the same. Therefore, as the variable resistor R16 is varied between top and full-bottom rotation:

a.   For bottom rotation: $Vo = \dfrac{V_{B12}(R14 + R15)}{R15} = \dfrac{1.1\,V(4.7\,k\Omega + 2.2\,k\Omega)}{2.2\,k\Omega} = 3.45$ volts

b.   For top rotation: $Vo = \dfrac{V_{B12}(R14 + R15)}{R15} = \dfrac{7.2\,V(4.7\,k\Omega + 2.2\,k\Omega)}{2.2\,k\Omega} = 22.58$ volts

EXPERIMENTAL OBJECTIVES

To investigate the μA723 monolithic regulator as the control element in a variable (< 3 V to > 23 V) power supply.

LIST OF MATERIALS

1. Integrated Circuit: μA723 voltage regulator
2. Transformer: 24 V RMS or equivalent
3. Diodes: 1N4001 or equivalents
4. Transistor: 2N3055 or equivalent
5. Resistors: 1 Ω (one) 100 Ω (one) 1.2 kΩ (two) 2.2 kΩ (one) 10 kΩ potentiometer (one)
6. Capacitors: 100 pF (one) 100 μF (one) 1000 μF (one)

EXPERIMENTAL PROCEDURE

1. Connect the circuit as shown in Figure 3-98.

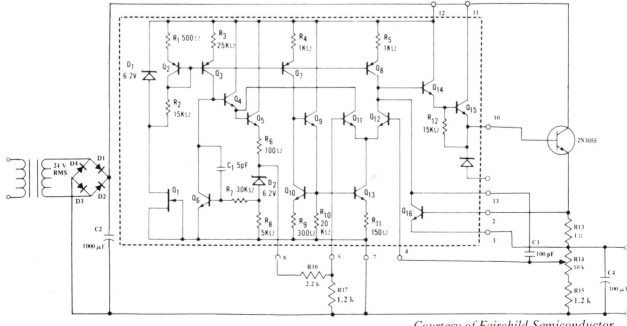

FIGURE 3-98　　　　　　　　　　　*Courtesy of Fairchild Semiconductor*

2. Measure the DC voltage:
    a. Across the C1 filter capacitor.
    b. Across the series resistors R16, R17 — Pin 6 with respect to ground. This is the zener reference of the device.
    c. Across R17 — Pin 5 with respect to ground. This is the circuit reference voltage and the minimum voltage out capability for the supply.
    d. At Pin 4 — wiper of variable resistor R14. This voltage should be approximately equal to the voltage of Pin 5, since the base of Q12 must follow the base of Q11 in the differential amplifier.
3. Use the measured zener voltage of approximately 7.2 volts and verify the measured values.

    a. $$V_{R17} = \frac{V(zener\ ref.) \times R17}{R16 + R17}$$

    b. $$V(Pin\ 4) = V(Pin\ 5) = V_{R17}$$

    c. $$Vo(top\ rotation) = V(Pin\ 4)$$

    $$Vo(bottom\ rotation) = \frac{V(Pin\ 4) \times (R14 + R15)}{R15}$$

4. Rotate variable resistor R14
    a. To bottom rotation and monitor Vo — across ouput capacitor C3.
    b. To top rotation and monitor Vo.
5. Connect a 100 Ω load across the output terminals.
    a. Vary variable resistor R14 to top and bottom rotation and monitor.
    b. Monitor the ripple voltage across the 100 Ω load resistor and across the C1 capacitor for top and bottom rotation.
6. Insert the calculated and measured values, as indicated, into Table 3-17.

| | | | | | | | 100 Ω LOAD | | | |
| | | | | | Vo | | Vo | | ΔV_{C1} | |
TABLE 3-17	$Vcap_1$	$V_{zener\ ref-pin\ 6}$	$V_{pin\ 5}$	$V_{pin\ 4}$	B.R.	T.R.	B.R.	T.R.	B.R.	T.R.
CALCULATED	///	///					///	///	///	///
MEASURED										

186

HIGHER POWER VOLTAGE REGULATORS

GENERAL DISCUSSION

The LM309 is primarily a fixed-voltage, positive regulator with current carrying capabilities up to 1 ampere. The fixed output voltage is 5 volts but voltage can be varied by simply adding a fixed resistor and a variable resistor. The LM320 and the LM340 are other examples of fixed-voltage regulators, but these monolithic devices come in a variety of output voltages. The LM340, a positive regulator, can be ordered with fixed output voltages of 5 V, 6 V, 8 V, 12 V, 15 V, 18 V, and 24 V. The LM320, a negative regulator, can be ordered with fixed voltages of -5 V, -5.2 V, -6 V, -8 V, -12 V, -15 V, -18 V, and -24 V. Both devices have current capabilities in excess of 1 ampere and both devices can be varied. Again, the LM designation denotes a National Semiconductor (and in some cases a Signetics) device label.

The μA7800 and μA7900 series of devices provide the same function as the LM340 and the LM320 devices. The μA7800 is equivalent to the LM340 and the μA7900 is equivalent to the LM320. In the series, μA7805 denotes a 5 volt regulator and μA7824 denotes a 24 volt regulator.

Like the LM320 and LM340, the μA7800 and μA7900 are primarily fixed voltage devices that can be modified to accommodate variability. The μA letters denote Fairchild Semiconductor as the primary source, but these devices can be second sourced under the same letters or by some other designation. All of the LM320, LM340, μA7800, and μA7900 devices are three terminal devices.

The μA78G and μA79G are positive and negative regulators, respectively, and they are designed as voltage variable devices. They have four terminals and the extra terminal is designated as a control terminal. Also, the LM117, a three terminal device, classifies as a variable voltage monolithic regulator because it has such low biasing current.

LM309 FIXED VOLTAGE REGULATOR

The LM309 can be connected at a fixed output voltage of 5 volts or as a variable supply. Figure 3-99 shows its extreme simplicity in usage. As with other devices, power handling capabilities are based on the heat sinking used. For instance, if the regulator is to be used at 5 volts for TTL applications, then the input voltage should not be much greater than 10 V DC and heat sinking should be provided. As a variable supply, the input voltage will necessarily have to be higher and, hence, current demand lower. Again, heat sinking should be used because the output voltage can be varied from a low of 5 volts to a high of approximately 23.15 V from: $Vo \approx Vpreset(R1 + R2)/R1 = 5 V(330 \Omega + 1 k\Omega)/330 \Omega$.   The input circuitry is typical of all the unregulated supplies studied previously.

The value of R1 is choses at 330 ohms (manufacturers recommend 300 ohms) to insure good regulation. And the reason 300 ohms must be used is covered more fully when the similarly constructed LM340 is discussed. Further LM309 device applications and typical performance characteristics can be found in manufacturer data sheets and application notes.

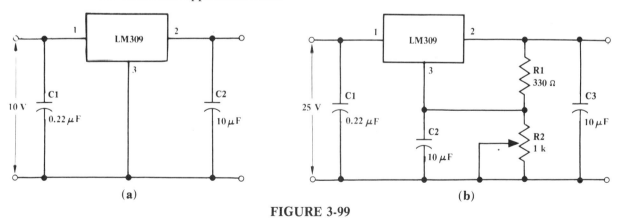

FIGURE 3-99

The LM320 and LM340 are connected similarly to the LM309 — only three connections are necessry. Typical circuit applications are illustrated in Figure 3-100(a), where both fixed and adjustable supplies are shown. Note that both devices end with a -18, which signifies the preset device voltage of 18 volts. The letters T and K designate the package used. Again, the input voltage is applied from a plus and minus, filtered, unregulated supply — similar to previously studied plus and minus applications.

187

In Figure 3-100 (b), the output voltages for both the LM320K and the LM340k are varied between a low of 5 volts, the preset device voltage, and a high of approximately 23.5 volts from:

$$Vo \approx (R1 + R2)/R1 = 5 \text{ V} (270 \text{ } \Omega + 1 \text{ k}\Omega)/270 \text{ } \Omega = 23.5 \text{ volts}$$

The value of R1 is critical to good regulation and the manufacturers recommend an R1 value of approximately 300 ohms or lower so that I_{R1} will be at least 10 times larger than the biasing current of the device $(I_{Pin\ 3})$.

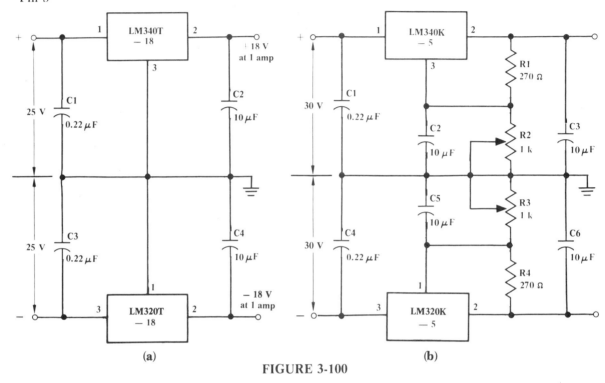

(a)   (b)

FIGURE 3-100

NOTE: With regard to the regulator circuits discussed, the LM340T-18 can be replaced by a μA7818V, the LM340K-5 by a μA7805K, the LM320T-18 by a μA7918V, and the LM320K-5 by a μA7805K. Essentially, each device second sources the other.

μA7800 ANALYSIS

The LM340 and μA7800 perform similar outboard functions but the internal circuitry differs. Since the internal circuitry of the μA7800 is easier to follow, it will be analized first. The schematic diagram of the device and the top view of the TO-220 and the higher power TO-3 packages are shown in Figure 3-101.

The μA7805, like all monolithic regulators, can be functionally separated into three sections — the reference voltage section, the differential amplifier section, and the output section. Q12-Q13 and Q5-Q6 provide the required differential amplifier, where the zener reference voltage at the base of Q12 is stepped down (in voltage) across the base-emitter diode junction of Q12, the R5 3.3 kΩ resistor, and the base-emitter diode juntion of Q13. Then, the voltage is stepped up across the base-emitter diode junction of Q5 and Q6 so that the voltage at the junction of resistors R19 and R20 is approximately equal to the zener reference voltage minus the voltage drop across R5. Therefore, the voltage at the junction of resistors R19 and R20, essentially, follows the voltage at the base of Q12 (minus the R5 voltage drop) and (for the μA7800 device) the voltage at the R19-R20 junction is 5 volts. Hence, if 5 volts is dropped across the 5 kΩ R19 resistor and 1 mA of current flows, then a 19 kΩ R20 resistor will provide 24 volts across R19 and R20 and the output voltage at Pin 2 will be 24 volts. Also, if R20 is made to equal zero ohms, 5 volts will appear across R19 and the Pin 2 output terminal. Obviously, varying the resistance value of R20 between 0-19 kΩ produces the remaining fixed voltage conditions. Additionally, Q17 and Q16 provide the output and driver transistors, respectively, and Q15 (along with resistor R11, provides short circuit protection.

Since the device has the required reference voltage (zener), differential amplifier, and output stage

188

EQUIVALENT CIRCUIT

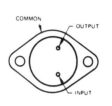

CONNECTION DIAGRAMS
TO-220 PLASTIC POWER PACKAGE
(TOP VIEW)

TO-3 PACKAGE
(TOP VIEW)

FIGURE 3-101   *Courtesy of Fairchild Semiconductor*

(along with short circuit protection) sections and is a complete regulator on one chip, it can be connected similarly to the LM340 device in the circuit of Figure 3-100. Because the reference voltages match (LM340-5 to μA7805), the circuits and output voltages will perform similarly.

LM340 ANALYSIS

Since the circuitry of the LM340 is difficult to analize, a reasonable question is: Why is it necessary? If the LM340K-5 data sheet provides sufficient data for the technologist to hook the device up and it performs to specifications, then internal analysis seems unnecessary. And this is a correct assumption to the point when the question is asked about the reasons for a manufacturer suggested 300 ohms for an outboard resistor. Then, the circuit has to be studied.

For instance, considering the circuit of Figure 3-102, the 2.6 kΩ resistor in series with R17 must be paralleled with a resistor ten time less in value. 300 ohms or 270 ohms does this. However, note that the

schematic and connection diagrams

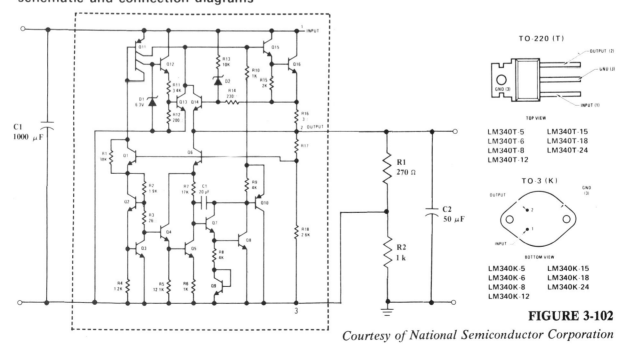

FIGURE 3-102

Courtesy of National Semiconductor Corporation

189

devices biasing (quiescent) current flows through R2 and the Vout is not just the ratio of $(R1 + R2)/R1$, because it must contain the effect of the voltage drop across R2.

For the LM340K-5 device, 5 volts appears at the output Pin 2. The nominal (typical) quiescent current of the device is 7 mA and 10 mA is given at maximum. So, if a worst case condition is assumed, where an R18 of 2.6 kΩ is considered, it becomes obvious that the lower the R1, and hence R2 resistor values, the closer the ratio of $(R1 + R2)/R1$ the output voltage becomes.

1. Therefore, assuming a quiescent current of 7 mA and a Vout of 5 volts and solving for Vout across R1 and R2:

    a. Solving for I_{R1} from:

$$I_{R1} = V_{R1}/R1 = 5\ V/270\ \Omega = 18.518\ mA$$

    b. $I_{R2} = I_{R1} + I_Q = 18.518\ mA + 7\ mA \approx 25.5\ mA$

    c. $V_{R2} = R2 \times I_{R2} = 1\ k\Omega \times 25.5\ mA = 25.5\ volts$

    d. $Vo = V_{R1} + V_{R2} = 5\ V + 25.5\ V = 30.5\ volts.$

2. Obviously, the quiescent current of the device plays a large role in the evential output voltage and, because the quiescent current changes with temperature (given in the data sheet as 1.3 mA), attempting to increase the voltage with outboard resistors is not the approach to take with fixed voltage devices. However, if a 30 volt output is required, use the LM340-24 device. Then, a small R2 resistor can be used and the effect of the quiescent current is minimized.

Essentially, the LM340 is a fixed voltage device and should be used in this manner. This is also true of the µA7800, even though it has slightly lower quiescent current and quiescent current change. Two devices that are not fixed voltage, and which have minimal quiescent current problems are the three terminal LM117 and the four terminal µA78G.

µA78G ANALYSIS

The µA78G device is essentially a µA7800 device except the R19 and R20 resistors are brought outboard. This rather clever, but simple, technique produced a fully adjustable monolithic regulator that requires only two outboard resistors and the normal filtering capacitors. A representative circuit is shown in Figure 3-103.

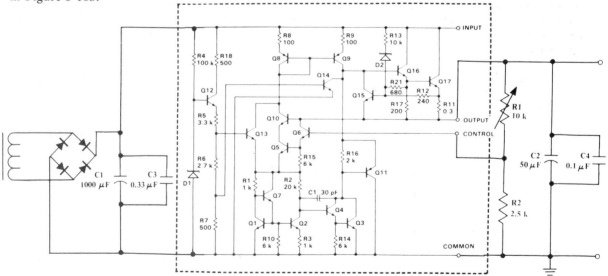

FIGURE 3-103   *Courtesy of Fairchild Semiconductor*

The zener reference voltage established at the base of Q12 again is stepped down across a base-emitter diode junction (Q12), across a 3.3 kΩ resistor (R5), and across the base-emitter diode junction of Q13. It is then stepped up across the base-emitter diode junctions of Q5 and Q6. However, instead of the voltage being developed across some fixed resistance within the device proper, it is brought out as a control lead to be connected to outboard components. The advantage of this connecton for variable purposes is obvious because, now, the outboard resistor can be in the greater than 1 kΩ category and the output voltage more closely follows the ratio of outboard resistors R1 and R2, since the base current of transistor Q6 is minimal.

Note that the circuitry of the μA78G is identical to the circuitry of the μA7800 except for the packaging — external versus internal control resistors. However, because the μA78G circuitry is similar to discrete differentail amplifier voltage adjustable supply circuity, the traditional method of solving for the output voltage is used: Vo = (R1 + R2)/R2. Then, if the nominal control terminal voltage of 5 volts is used, the output voltage can vary from about 5 volts to 25 volts.

1. $$Vo(max) = \frac{Vcontrol\ pin\,(R1\ +\ R2)}{R2} = \frac{5\ V\,(10\ k\Omega\ +\ 2.5\ k\Omega)}{2.5\ k\Omega} = 25\ volts$$

2. $$Vo(min) = \frac{Vcontrol\ pin\,(R1\ +\ R2)}{R2} = \frac{5\ V\ \times\ 2.5\ k\Omega}{2.5\ k\Omega} = 5\ volts$$

NOTE: The pin control current is typically less than 1 μA, and the quiescent current, while 2 mA, has little effect on the output voltage. However, this is also true of the fixed voltage devices as long as they are operated in the fixed voltage mode. Additionally, to insure stable operation over all operating conditions, the manufacturer recommends that compensations capacitors C3 and C4 be added.

LM317 ANALYSIS

The main drawback to using the fixed voltage LM340 type device as an outboard voltage other than at the designated fixed voltage was that outboard resistors had to be used. And the quiescent current of the device made the output voltages unpredictable because, first of all, it was high (nominally) at 7 mA and, secondly, it could drift with both temperature and loading changes.

The LM317 is a three terminal monlithic device that is packaged and connected much like the LM340, but it was developed primarily as an adjustable regulator. Therefore, the "quiescent" or adjustable terminal current (in this case) is extremely low. Instead of 7 mA it is about 100 μA. Therefore, the current flow through the outboard R2 resistor is minimal and when R1 is low in value, the output voltage is effectively controll by the ratio of (R1 + R2)/R2. Additionally, any change in the nominal 100 μA, because of loading or temperature changes, will have little effect, overall, on the output voltage.

Another advantage of the LM317 is that the reference voltage is low (≈ 1.2 V) and, therefore, the output voltage can be varied from a low of ≈ 1.2 volts to a high of greater than 30 volts. And if a plus and minus "bulk" unregulated supply is used, the output voltage can be varied down to a zero volt condition. A great many connections and further data on the device are given in the manufacturer data sheets and a representative connection is shown in Figure 3-104.

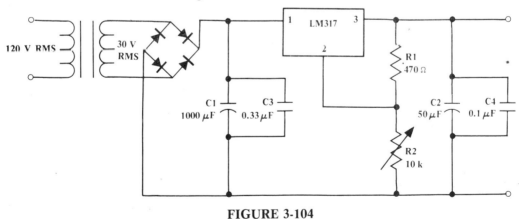

FIGURE 3-104

Note that the lower the value of R1 and R2, the less effect the 100 μA of quiescent current will have. Also, in the following calculations, the quiescent current is first included in the equation and then excluded in determining maximum output voltage. The maximum output voltage provides a worse case condition where the 100 μA flow through the 10 kΩ of the R2 resistor drops 1 volts. So, even with a low 100 μA, the quiescent current continues to play a role in the output voltage value.

191

1. Solving for maximum output voltage conditions with the 100 μA included in the calculations:

   a. $I_{R1} = \text{Vref}/R1 \approx 1.2 \text{ V}/470 \text{ }\Omega \approx 2.55 \text{ mA}$

   b. $I_{R2} = I_{R1} + I_{ADJ} = 2.55 + 0.1 \text{ mA} = 2.65 \text{ mA}$

   c. $V_{R2} = R2 \times I_{R2} = 10 \text{ k}\Omega \times 2.65 \text{ mA} = 26.5 \text{ volts}$

   d. $\text{Vo} = V + V_{R2} = 1.2 \text{ V} + 25.5 \text{ V} = 27.7 \text{ volts}.$

2. Solving for the maximum output voltage with the 100 μA excluded from the calculations:

$$\text{Vo} = \frac{\text{Vref}(R1 + R2)}{R1} = \frac{1.2 \text{ V}(470 \text{ }\Omega + 10 \text{ k}\Omega)}{470 \text{ }\Omega} = 26.7 \text{ volts}$$

DUAL TRACKING REGULATORS

Dual tracking regulators are designed to provide balanced positive and negative supply voltages that can be fixed or adjustable. For instance, the LM125 is a dual $\pm$ 15 volt tracking regulator that requires few external components even when the current is boosted by external power devices. Complete data on this device can be found in manufacturer data sheets.

Additionally, dual tracking regulators can be constructed by using back-to-back μA78G and μA79G devices or a combination of μA78G or μA79G with another power transistor. In the circuit of Figure 3-105, note that the monolithic portion is similar to the circuit of Figure 3-102 and the LM301 and PNP power transistor portion of the circuit is similar to previously discussed op-amp driven circuits.

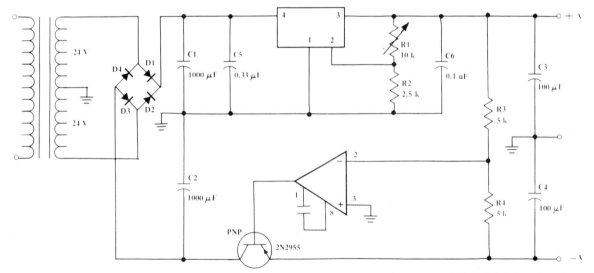

FIGURE 3-105   *Courtesy of Fairchild Semiconductor*

With reference to the circuit of Figure 3-105, when R1 is adjusted to zero ohms, 5 volts are developed across R2, a like voltage across R3, and this, in turn, causes a like negative voltage to be developed across resistor R4 for a $\pm$5 volt "tracking" supply. Increasing R1 to 10 kΩ allows 25 volts to be developed across R2, a negative 25 volts across R3, and a resultant $\pm$25 volts. Essentially, Pin 3 being grounded causes Pin 2 to be "virtually grounded" and, since R2 equals R3, the voltage drop across R3 equals the voltage drop across R2. Again, the recommended compensation capacitors are used to insure stability.

1. The transformer in the circuit of Figure 3-105 is 48 V CT. Therefore, the C1 and C2 filter capacitors each "see" a full wave center tapped circuit — filter capacitor C1 "sees" D1 and D2 and filter capacitor C2 "sees" D3 and D4. The DC voltage developed across both C1 and C2 filter capacitors is:

$$V \text{ DC (C1)} = V \text{ DC (C2)} = \sqrt{2} \times V \text{ RMS} = \sqrt{2} \times 24 \text{ V} \approx 33.94 \text{ volts}$$

2. Solving for the output voltage:

a. V(Pin 2) = V(ref) = 5 volts

b. Vo(min) = (5 V × 2.5 kΩ)/2.5 kΩ = 5 volts

c. Since V(Pin 3), with respect to ground, is 5 volts, then 5 volts are also developed across R3 with respect to ground. Also, since no appreciable current flows into the inverting terminal of the op-amp, then:

$$I_{R3} = V_{R3}/R3 = 5\ V/5\ k\Omega = 1\ mA$$

and $I_{R3} = I_{R4}$,

therefore: $V_{R4} = I_{R4} \times R4 = 1\ mA \times 5\ k\Omega = 5$ volts

NOTE: Again the junction of the R3 and R4 resistors is nearly at ground condition because Pin 2 of the op-amp follows the voltage level of Pin 3. Therefore, almost no current flows into the high impedance of the op-amp and $I_{E3} = I_{E4}$.

d. Increasing the value of R1 increases Vout and, hence, the $V_{R3} = V_{R4}$ tracking voltage output. The output voltage varies between ± 5 volts to ± 25 volts.

NOTE: All dual tracking power supplies operate the same way. They use the master-slave concept where one supply is varied and the other supply follows.

With regard to the circuit of Figure 3-105, almost any positive voltage regulator, whether monolithic or otherwise, can utilize the configuration successfully. For instance, an op-amp and a 2N3055 could have been used in place of the μA78G with equal results.

NEGATIVE REGULATORS

Both positive and negative regulators are used extensively. However, since there are more positive than negative monolithic devices available, positive regulators were emphasized and analized. Some examples of the negative regulators are the LM320, LM345, μA79G, and μA7900. Further information and applications for these devices can be found in manufacturer data sheets.

EXPERIMENTAL OBJECTIVES

To investigate the μA78G four terminal positive voltage regulator as the control element in both low power and higher power circuit configurations.

LIST OF MATERIALS

1. Integrated Circuit: μA78G (one)
2. Transistors: MJ2955 (one) or equivalent PNP power device and 2N3906 (one) or equivalent
3. Diodes: 1N4001 (four) or equivalents
4. Transformer: 24 volt
5. Resistors (one each): 1 Ω   4.7 Ω   100 Ω   470 Ω   2.2 kΩ   10 kΩ potentiometer
6. Capacitors (one each): 0.1 μF   0.33 μF   25 μF   1000 μF

EXPERIMENTAL PROCEDURE

1. Connect the circuit of Figure 3-106.
2. Measure the DC voltage:
   a. Across filter capacitor C1.
   b. Across resistor R2 — control voltage Pin 4 with respect to ground.
3. Use the measured reference voltage of Pin 4, across R2, and calculate the voltage out.

193

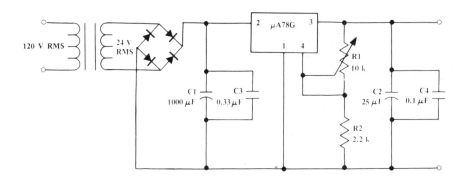

FIGURE 3-106

a. For bottom rotation ($V_o$max) from: $$Vout = \frac{Vpin\ 4 \times (R1 + R2)}{R2}$$

b. For top rotation ($V_o$min) from: $$Vout = \frac{Vpin4 \times R2}{R2} \approx Vpin\ 4$$

4. Measure the output voltage:
   a. For bottom, top, and mid-rotation of variable resistor R1.
   b. For bottom, top, and mid-rotation with a 470 ohm load resistor connected into the circuit.

NOTE: Attach the device to a heat sink or piece of aluminum to insure adequate heat sinking.

5. Monitor the ripple voltage for loaded conditions of 470 ohms at both bottom and top rotation of variable resistor R1.
   a. Across filter capacitor C1.
   b. Across the output.
6. Measure and calculate the percent regulation for bottom rotation. Calculate from:

$$\% \, reg = \frac{V(NL) - V(L)}{V(NL)} \times 100 \quad \text{where: } RL = 470 \, \Omega$$

7. Test the short circuit protection (momentarily) by placing a short across the device. Do this for maximum voltage out condition.
8. Insert the calculated and measured values, as indicated, into table 3-18.

TABLE 3-18	V DC		Vout			ΔV		% reg
	C1	Vpin₄	Vmax	Vmid	Vmin	C1	C2	
NO LOAD CALCULATED	/////	≈ 5 V		/////		/////	/////	
NO LOAD MEASURED								/////
470 Ω CALCULATED	/////	/////	/////	/////	/////		/////	
470 Ω MEASURED	/////	/////						/////

9. Connect the circuit of Figure 3-107, where a 2N2955 (or any other PNP power transistor) along with Q1 biasing resistor R3 is added. Additionally, transistor Q1 and resistor R4 are connected into the circuit to provide short circuit protection for the Q2 power transistor
10. Measure the output voltage and the voltage drop across the V_{CE} of Q2:
    a. For minimum, maximum, and mid-rotation of resistor R1.
    b. For minimum, maximum, and mid-rotation of resistor R1 with a 100 ohm load connected into the circuit.
11. Measure the ripple voltage for unloaded and then loaded conditions of 100 Ω.

194

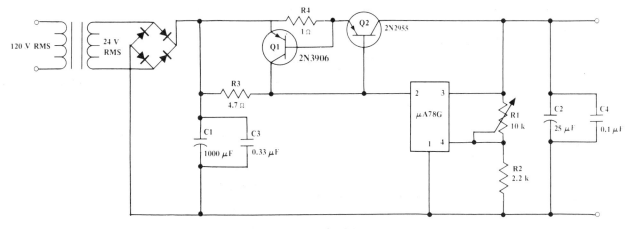

FIGURE 3-107

12. Calculate the percent reguattion for bottom rotation. Calculate from:

$$\% \text{ reg} = \frac{V(NL) - V(L)}{V(NL)} \times 100 \qquad \text{where:} \quad RL = 100 \ \Omega$$

13. Test the short circuit protection.

   a. Calculate I_{SC} from: $I_{SC} = V_{BE}(Q2)/R4$.

   b. Measure the I_{SC} condition either with an ammeter or by monitoring V_{R4} and knowing the R4 resistance.

14. Insert the calculated and measured values into Table 3-19.

TABLE 3-19	VDC_{C1}	Vmax	Vo Vmid	Vmin	V_{CE}(Q2) $V_o$max	$V_o$min	Vmid	ΔVo Vmax	% reg	I_{SC}
NO LOAD MEASURED									/////	/////
100 Ω CALCULATED	/////	/////	/////	/////	/////	/////	/////	/////		
100 Ω MEASURED									/////	

195

4 ‖ RESISTANCE-CAPACITANCE (RC) OSCILLATORS

GENERAL DISCUSSION

Oscillators are circuits which convert DC power to AC power at selected frequencies as determined by RC circuitry, LC circuitry, or frequency selective crystals. LC and crystal oscillators are primarily used for the sine wave generation of signals, but RC oscillators can be sine wave or non-sine wave, depending on the circuit design used.

The Wein-bridge, the twin-T, and the phase-shift are three standard RC oscillators used to produce a sine wave, the RC astable multivibrator is used to produce a square wave, and the UJT driven relaxation oscillator is used to produce a sawtooth wave. Additionally, a sine wave oscillator can be used to drive a Schmitt trigger in producing a square wave output at a more predictable frequency than the astable multivibrators. Other wave-shaping networks like integrators and differentiators can be used to provide a variety of shapes like the triangle wave, a reconstituted sine wave, or a shape similar to the unfiltered full-wave bridge output.

All sine wave oscillators can be explained by Barkhausen's criterion for oscillation which states that the overall gain (including the losses around the circuit) must be greater than unity and that the phase angle shift between input to the output and back to the input must be in phase. This in-phase condition requires either a 0° or a 360° phase shift.

Barkhausen's criterion for oscillation is illustrated in Figure 4-1, where the overall gain of 1 and the in-phase condition are analyzed. Assume that the amplifier has a gain of 100 and provides 1 Vp-p at the output for 10 mVp-p of input signal. Then, if 10 mVp-p is coupled back to the input, an overall gain of 1 is provided. Also, the phase shift must be maintained in phase by using a 0° phase shift or by using a 360° phase shift (two 180° phase shifts) to insure oscillatory conditions.

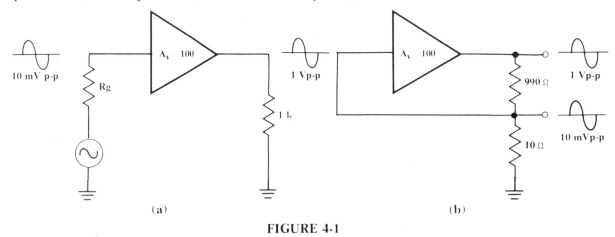

FIGURE 4-1

The gain of 100 and the stepdown of 100:1, by using the 990 Ω and 10 Ω resistors, is shown in Figure 4-1(b). The phase shift is shown in phase. Not shown are the frequency determining networks which could be LC, crystal, or RC networks.

PART 1 — RC SINE-WAVE OSCILLATORS

GENERAL DISCUSSION

The main advantage of RC oscillators is their ability to operate successfully at very low frequencies. Additionally, the RC component physical size does not differ too greatly when operated at frequencies of 1

196

MHz or 100 Hz. The LC components, on the other hand, need to be extremely large (especially the inductor) when frequencies of 100 Hz and below are used.

Three widely used RC oscillators are the Wein-bridge, the phase-shift, and the twin-T. Each of these oscillators can be varied over several decades simply by interchanging RC components and by providing a variable resistor or capacitor within the decade. The Wein-bridge, because it requires fewer variable components, is the main circuit used in laboratory sine-wave oscillators that operate below 1 MHz. Above 1 MHz, stray capacitance becomes too troublesome, and LC oscillators are then required. An exception to this is function generators that provide simulated sine waves to 5 MHz through use of wave-shaping techniques.

THE WEIN-BRIDGE OSCILLATOR

A Wein-bridge oscillator circuit is shown in Figure 4-2. The amplifier is operated Class A to make certain that a clean sine wave is produced. (This occurs when the overall gain of the circuit is slightly greater than unity.) The overall gain is provided by the Q1 and Q2 common emitter stages and the controlled loss is provided in the bridge by the voltage divider ratio of variable resistor R3 and resistor R4. The circuit obtains the 360° in-phase feedback by using the 180° phase shifts of Q1 and Q2. Additionally, the frequency of the ocillator is determined by the series-parallel networks (R_1C_1, R_2C_2) of the bridge — the R_1C_1 network determines the high pass and the R_2C_2 network determines the low pass in providing "band pass", zero phase shift conditions.

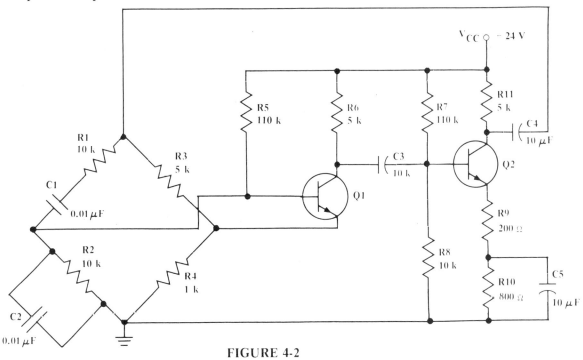

FIGURE 4-2

The Wein-bridge circuit of Figure 4-2 is deliberately capacitor coupled (and not direct coupled) to illustrate that the Q1 emitter stage is identical to the Q2 emitter stage and that resistors R2 and R4 are part of the bridge as well as being used as the biasing for Q1. For instance, resistor R2 serves the same functions as resistor R8 and, likewise, resistor R4 is the equivalent of resistors R9 + R10. Therefore, for the ease of both the DC and the AC analysis, the circuit is connected as an amplifier where resistors R2 = R12 = 10 kΩ and R4 = R13 = 1 kΩ, and these values are duplicated in the amplifier circuit of Figure 4-3 to illustrate their dual role in the oscillator circuit.

THE WEIN-BRIDGE OSCILLATOR — DIRECT CURRENT ANALYSIS

1. $V_{R5} = \dfrac{V_{CC} \times R5}{R5 + R12} = \dfrac{24\,V \times 110\,k\Omega}{110\,k\Omega + 10\,k\Omega} = 22\ volts$

197

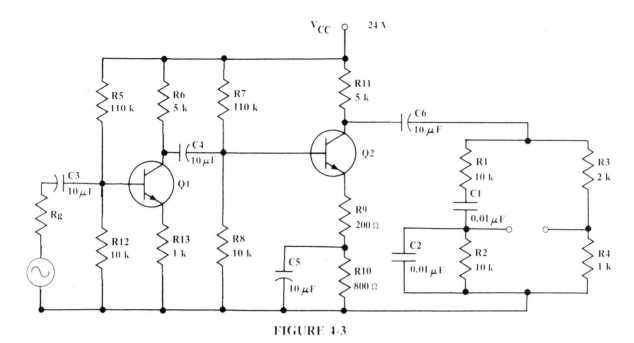

FIGURE 4-3

2. $V_{R12} = \dfrac{V_{CC} \times R12}{R5 + R12} = \dfrac{24 \text{ V} \times 10 \text{ k}\Omega}{110 \text{ k}\Omega + 10 \text{ k}\Omega} = 2 \text{ volts}$

3. $V_{R13} = V_{R12} - V_{BE1} = 2 \text{ V} - 0.6 \text{ V} = 1.4 \text{ volts}$

4. $I_{E1} = V_{R13}/R13 = 1.4 \text{ V}/1 \text{ k}\Omega = 1.4 \text{ mA}$

5. $V_{R6} \approx I_{E1} \times R6 = 1.4 \text{ mA} \times 5 \text{ k}\Omega = 7 \text{ volts}$

6. $V_{CE1} = V_{C1} - V_{E1} = 17 \text{ V} - 1.4 \text{ V} = 15.6 \text{ volts}$

    where: $V_{C1} = V_{CC} - V_{R6} = 24 \text{ V} - 7 \text{ V} = 17 \text{ V}$ and $V_{E1} = V_{R13} = 1.4 \text{ volts}$

7. $V_{R7} = \dfrac{V_{CC} \times R7}{R7 + R8} = \dfrac{24 \text{ V} \times 110 \text{ k}\Omega}{110 \text{ k}\Omega + 10 \text{ k}\Omega} = 22 \text{ volts}$

8. $V_{R8} = \dfrac{V_{CC} \times R8}{R7 + R8} = \dfrac{24 \text{ V} \times 10 \text{ k}\Omega}{110 \text{ k}\Omega + 10 \text{ k}\Omega} = 2 \text{ volts}$

9. $V_{E2} = V_{R8} - V_{BE2} = 2 \text{ V} - 0.6 \text{ V} = 1.4 \text{ volts}$

10. $I_{E2} = \dfrac{V_{(R9 + R10)}}{R9 + R10} = \dfrac{1.4 \text{ V}}{1 \text{ k}\Omega} = 1.4 \text{ mA}$

11. $V_{R11} \approx I_{E2} \times R11 = 1.4 \text{ mA} \times 5 \text{ k}\Omega = 7 \text{ volts}$

12. $V_{CE2} = V_{C2} - V_{E2} = 17 \text{ V} - 1.4 \text{ V} = 15.6 \text{ volts}$

    where: $V_{C2} = V_{CC} - V_{R11} = 24 \text{ V} - 7 \text{ V} = 17 \text{ V}$ and

                $V_{E2} = V_{(R9 + R10)} = 1.4 \text{ V}$

THE WEIN-BRIDGE OSCILLATOR — ALTERNATING CURRENT ANALYSIS

1. The center resonant frequency for the Wein-bridge oscillator is solved from:

$$f_o = \frac{1}{2\pi RC} \approx \frac{0.159}{10\ k\Omega \times 0.01\ \mu F} = \frac{0.159}{10^4 \times 10^{-8}} = \frac{0.159}{10^{-4}} = 1590\ Hz$$

where: $R = R1 = R2 = 10\ k\Omega$ and $C = C1 = C2 = 0.01\ \mu F$

2. Solving for the reactance of C1 and C2 at the center (oscillator) frequency where $C1 = C2 = C = 0.01\ \mu F$.

$$X_{C1} = X_{C2} = \frac{1}{2\pi fC} = \frac{0.159}{fC} = \frac{0.159}{(1.59 \times 10^3)(0.01\ \mu F)} = \frac{0.159}{1.59 \times 10^3 \times 10^{-8}} = \frac{0.159}{1.59 \times 10^{-5}} = 10\ k\Omega$$

NOTE: $f_o = 1/2\pi RC$ is a condition of $X_C = R$, where $X_C = 1/2\pi fC$ and $f = 1/2\pi RC$. And Since $R = X_C$, substitute R for X_C and $f = 1/2\pi RC$.

3. A balanced bridge condition exists when the series $R_1 + jX_{C_1}$ branch is twice the resistance value of the parallel $R2 \ /\!/ \ jX_{C_2}$ "branch", and where resistor R3 is twice the value of resistor R4.

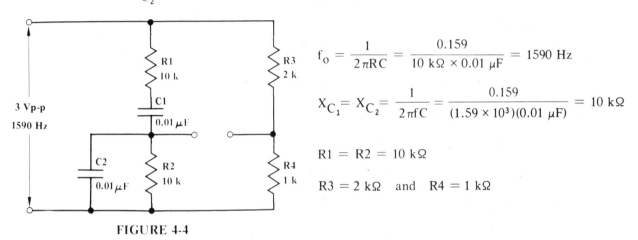

FIGURE 4-4

a. The series RC branch, where $R1 = 10\ k\Omega$ and $X_{C_1} = 10\ k\Omega$ is solved from:

$$R_1 - jX_{C_1} = 10\ k\Omega - j10\ k\Omega = \sqrt{(10\ k\Omega)^2 + (-j10\ k\Omega)^2}$$

$$= \sqrt{100\ k\Omega^2 + 100\ k\Omega^2} = \sqrt{200\ k\Omega^2} = 14.14\ k\Omega$$

and the angle associated with the $R = X_C = 10\ k\Omega$ values is:

$$\tan^{-1}\frac{X_C}{R} = \tan^{-1}\frac{10\ k\Omega}{10\ k\Omega} = \tan^{-1} 1 = 45°.$$

Therefore the resistance of $10\ k\Omega - j10\ k\Omega = 14.14\ k\Omega\angle -45°$, where the voltage laging the current (phase angle) associated with capacitors provides the minus sign ($\angle -45°$).

b. The parallel R2 and jX_{C_2} circuit is solved like resistors is parallel, but the calculations are complicated slightly by the phase angle requirements.

$$R2 \ /\!/ -jX_{C_2} = 10\ k\Omega \ /\!/ -j10\ k\Omega = \frac{10\ k\Omega \times -j10\ k\Omega}{10\ k\Omega - j10\ k\Omega} = \frac{-j\ 100\ k\Omega^2}{14.14\ k\Omega\angle -45°}$$

$$= \frac{100\ k\Omega^2\angle -90°}{14.14\ k\Omega\angle -45°} = \frac{100\ k\Omega^2\ \angle -90°\ \angle 45°}{14.14\ k\Omega} = 7.07\ k\Omega\angle -45°$$

199

NOTE: J notation is a convenient way of handling 90° angles, where $jX_C = X_C \angle 90°$ and $-jX_C = X_C \angle -90°$ and the values can be switched back and forth as in the solved parallel example.

c. Therefore, at the center frequency, the series $R_1 - jX_{C_1}$ circuit equals 14.14 kΩ $\angle -45°$ and the parallel of equal value components $(R2 \,/\!/ - jX_{C_2})$ equals 7.07 kΩ. Hence, if 3 Vp-p is applied to the bridge at the center frequency, 1 Vp-p would be developed across the parallel branch, or point A, with respect to ground. Additionally, since both networks have a $-45°$ phase angle, the net phase angle at point A is zero phase angle ($\angle 0°$).

Therefore, if R3 = 2 kΩ and R4 = 1 kΩ, then the 1 Vp-p will be developed across resistor R4 and, likewise, at point B, with respect to ground. Therefore, the voltage difference between point A and point B is zero volts at zero phase shift condition. This is illustrated in Figure 4-5, where the ratio of R3 to R4 and then the ratio of the series to the parallel RC networks are solved to emphasize the balanced zero phase shift conditions.

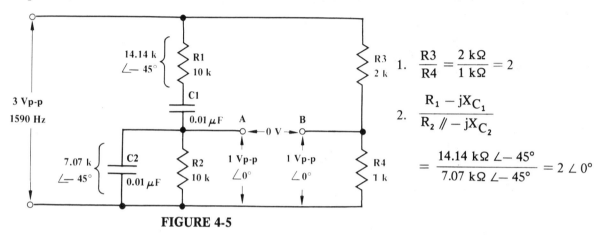

FIGURE 4-5

4. Solving for the voltage gain of the circuit to the collector of Q1 and then to Q2 at the operating frequency of 1590 Hz and at the balanced condtion:

a. $A_V(Q1) = \dfrac{R6 \,/\!/ \, \beta(ZinQ2)}{R13 + r_{e_1}} = \dfrac{5\,k\Omega \,/\!/\, 100(6.46\,k\Omega)}{1\,k\Omega + 18.6\,\Omega} = 2.77$

where: $r_{e_1} = r_{e_2} = 26\,mV/I_E = 26\,mV/1.4\,mA = 18.6\,\Omega$

Zin(Q2) = R7 $/\!/$ R8 $/\!/$ $\beta(R9 + r_{e_2})$ = 110 kΩ $/\!/$ 10 kΩ $/\!/$ 100(200 Ω + 18.6 Ω) = 6.458 Ω

b. $A_V(Q2) = \dfrac{R11 \,/\!/\, RL}{R9 + r_{e_2}} = \dfrac{5\,k\Omega \,/\!/\, 2.72\,k\Omega}{200\,\Omega + 18.6\,\Omega} = \dfrac{1762\,\Omega}{218.6\,\Omega} \approx 8.06$

where: RL = (R3 + R4) $/\!/$ ([R1 $-\,jX_{C_1}$] + [R2 $/\!/$ $-\,jX_{C_2}$])

= (2 kΩ + 1 kΩ) $/\!/$ (14.14 kΩ $\angle -45°$ + 7.07 kΩ $\angle -45°$)

= 3 kΩ $/\!/$ 21.21 kΩ $\angle -45°$ ≈ 2.72 kΩ $\angle 5°$

NOTE: The addition of 14.14 kΩ $\angle -45°$ and 7.07 kΩ $\angle -45°$ is solved by converting 14.14 kΩ $-45°$ back to rectangular form 10 kΩ $-$ j10 kΩ and 7.07 kΩ $\angle -45°$ back to rectanglular form 5 kΩ $-$ j5 kΩ, adding the reals and imaginaries: (10 kΩ $-$ j10 kΩ) + (5 kΩ $-$ j5 kΩ) = 15 kΩ $-$ j15 kΩ, and converting back to polar form: 15 kΩ $-$ j15 kΩ = 21.21 kΩ $\angle -45°$. The parallel combination of 3 kΩ and 21.21 kΩ $\angle -45°$ is similarly solved from (3 kΩ × 21.21 kΩ $\angle -45°$)/(3 kΩ + 21.21 kΩ $\angle -45°$) = (63.63 kΩ $\angle -45°$)/(3 kΩ + 15 kΩ $-$j15 kΩ) = (63.63 kΩ $\angle -45°$)/(23.43 kΩ $\angle -39.8°$) = 2.716 kΩ $\angle -5.2°$ ≈ 2.72 kΩ $\angle -5°$, where the 5° is ignored in the loading calculations.

c.   Therefore, the overall gain of the circuit is A_v(overall) = A_v(Q1) × A_v(Q2) = 2.77 × 8.08 = 22.4.

UNBALANCED BRIDGE CONDITIONS

Unbalanced bridge conditions are necessary for oscillatory conditions because, if a zero volt difference exists between points A and B, the gain of the amplifier circuit cannot serve its intended purpose. However, if an unbalanced condition is provided by lowering the value of R4, a difference voltage is obtained, where point A is more positive in signal level than point B, and while not obvious at this time, this is the direction that imbalance must occur if oscillatory conditions are to exist once the circuit is connected in oscillator form.

For instance, if R4 is lowered to a value of 900 Ω from 1 kΩ, the voltage at point A would remain at 1 Vp-p for 3 Vp-p of input voltage at the 1590 Hz center frequency, but the signal voltage across R3, and hence point B, would drop to 0.931 Vp-p and the difference voltage would increase from zero volts to 0.069 Vp-p. However, a V(diff) of 69 mVp-p is not enough to provide oscillatory conditions, if A_v = 22.4.

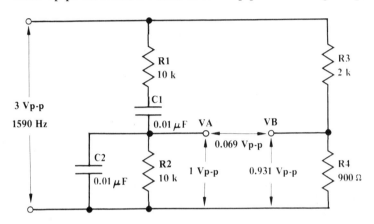

FIGURE 4-6

$$VA = \frac{Vp\text{-}p(R_2 \mathbin{/\mkern-5mu/} -jX_{C_2})}{(R_1 - jX_{C_1}) + (R_2 \mathbin{/\mkern-5mu/} -jX_{C_2})} = \frac{3\ Vp\text{-}p(7.07\ k\Omega \angle -45°)}{(14.14\ k\Omega \angle -45°) + (7.07\ k\Omega \angle -45°)}$$

$$= \frac{3\ Vp\text{-}p(7.07\ k\Omega \angle -45°)}{21.21\ k\Omega \angle -45°} = 1\ Vp\text{-}p$$

$$VB = \frac{Vp\text{-}p \times R4}{R3 + R4} = \frac{3\ Vp\text{-}p \times 900\ \Omega}{2\ k\Omega + 900\ \Omega} = 0.931\ Vp\text{-}p$$

$$V(\text{diff}) = VA - VB = 1\ Vp\text{-}p - 0.931\ Vp\text{-}p = 0.069\ Vp\text{-}p$$

NOTE: VA and VB can be given more generally in terms of Vop-p, where VA = Vop-p/3 and VB = Vop-p × R3/(R3 + R4), where Vop-p = 3 Vp-p.

5.   The gain of the amplifier required to amplify the difference voltages (VA − VB) so that 3 Vp-p can again be applied to the bridge circuit is solved by taking the inverse of the stepdown ratios or simply by dividing Vinp-p by the difference voltage. This will have to be the amplifier gain if R4 remains at 900 Ω.

a. $A_v = \dfrac{Vinp\text{-}p}{Vop\text{-}p} = \dfrac{3\ Vp\text{-}p}{0.069\ Vp\text{-}p} = 43.5$

$$A_v = \frac{1}{VA - VB} = \cfrac{1}{\dfrac{1}{3} - \dfrac{R4}{R3 + R4}} = \cfrac{1}{\dfrac{1}{3} - \dfrac{900\ \Omega}{2\ k\Omega + 900\ \Omega}} = \cfrac{1}{\dfrac{1}{3} - \dfrac{1}{3.222}} = 43.5$$

NOTE: VA is always ⅓ of Vop-p at the operating frequency, and VB is the ratio of the voltage divider equation R4/(R3 + R4).

6. Therefore, if R3 and R4 remained at 2 kΩ and 900 Ω respectively, the gain of the amplifier would have to be 43.5. However, if the gain is to remain at 22.4, then the stepdown factor of R3 and R4 will have to be increased where either R4 is increased or R3 is decreased.

a. Solving for the stepdown factor:

$$X = \frac{3A_v}{A_v - 3} = \frac{3 \times 22.4}{22.4 - 3} = \frac{67.2}{19.4} = 3.464$$

NOTE: The stepdown factor is the inverse of the stepdown ration of $1/X = 1/3.464 = 0.289$. Therefore, $1/3.464$ of the signal applied to the R3 — R4 voltage divider network is developed across the R4 resistor.

b. Solving for the value of R4 where R3 = 2 kΩ:

$$R4 = \frac{R3}{X - 1} = \frac{2\ k\Omega}{3.464 - 1} = \frac{2\ k\Omega}{2.464} = 812\ \Omega$$

c. Solving for the output of the amplifier where 100 mVp-p at 1590 Hz is applied to the input of the amplifier and the amplifier signal is then applied to the bridge. The signal voltage at VA is calculated from VA = ⅓Vop-p, the siganl voltage at VB is calculated at VB = (R4/[R3 + R4]) × Vop-p, and the difference voltage is solved from V(diff) = VA − VB. And if R4 is properly calculated, the voltage difference betwenn terminals A and B should be about 100 mVp-p, a minimum condition for oscillations to occur once the circuit is connected as an oscillator.

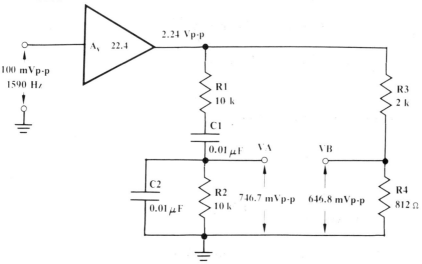

FIGURE 4-7

$$Vop\text{-}p = Vinp\text{-}p \times A_v = 100\ mVp\text{-}p \times 22.4 = 2.24\ Vp\text{-}p$$

$$VA = \frac{Vopp}{3} = \frac{2.24\ Vp\text{-}p}{3} = 7.467\ mVp\text{-}p$$

$$VB = \frac{Vop\text{-}p \times R4}{R3 + R4} = \frac{2.24\ Vp\text{-}p \times 812\ \Omega}{2\ k\Omega + 812\ \Omega} = 646.8\ mVp\text{-}p$$

$$V_{Diff} = VA - VB = 7.467\ mVp\text{-}p - 6.468\ mVp\text{-}p = 99.9\ mVp\text{-}p$$

202

7. Another technique that can be used to provide the approximate 100 mVp-p difference between points A and B is to increase the resistor value of R3 while maintaining R4 at 1 kΩ.

a. Solving for the stepdown factor:

$$X = \frac{3A_v}{A_v - 3} = \frac{3 \times 22.4}{22.4 - 3} = \frac{67.2}{19.4} = 3.464$$

NOTE: Again the signal voltage stepdown across resistor R4 is the inverse of the stepdown factor or $1/X = 1/3.64$.

b. Solving for the value of R3 where R4 = 1 kΩ:

$$R3 = R4(X - 1) = 1\,k\Omega(3.464 - 1) = 1\,k\Omega(2.464) = 2464\ \Omega$$

c. Solving for the V_{Diff} (VA − VB) for 100 mVp-p of input signal (including VA, VB, and Vop-p):

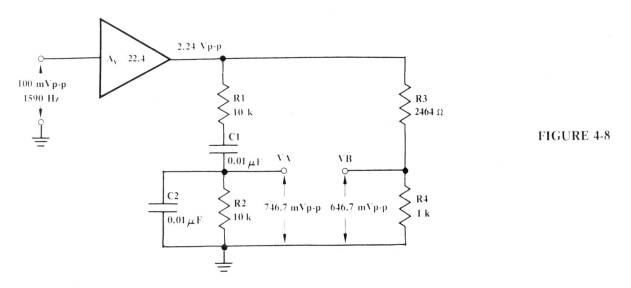

FIGURE 4-8

$$Vop\text{-}p = Vinp\text{-}p \times A_v = 100\ mVp\text{-}p \times 22.4 = 2.24\ Vp\text{-}p$$

$$VA = \frac{Vop\text{-}p}{3} = \frac{2.24\ Vp\text{-}p}{3} = 746.7\ mVp\text{-}p$$

$$VB = \frac{Vop\text{-}p \times R4}{R3 + R4} = \frac{2.24\ Vp\text{-}p \times 1\ k\Omega}{2464\ \Omega + 1000\ \Omega} = 646.7\ mVp\text{-}p$$

$$V_{Diff} = VA - VB = 746.7\ mVp\text{-}p - 646.7\ mVp\text{-}p = 100\ mVp\text{-}p$$

NOTE: The formula $X = 3A_v/(A_v - 3)$ is derived from $A_v(VA - VB) = 1$, where VA is always 1/3 and VB the unknown is 1/X. Therefore, $A_v(1/3 - 1/X) = 1$. Also, $R3 = R4(X - 1)$, which enables either R3 or R4 to be solved, is derived from $(1/X = (R4/[R3 + R4])$ equivalent ratios.

8. CONDITIONS FOR OSCILLATION: The circuit of Figure 4-9 is connected as a Wein-bridge oscillator. As previously stated, point VA must be higher in signal level than point VB, so point VA is connected to the base of Q1 and point VB is connected to the emitter.

A minimum condition of oscillation exists when either R3 or R4 are adjusted to provide a difference voltage at the base of Q1. And after being amplified by Q1 and Q2, the difference voltage is stepped down to point VA and VB to provide a signal voltage that is slightly higher in value than the original no difference voltage condition.

For instance, if R3 is adjusted to 2464 Ω and power is applied to the oscillator, all the capacitors in the circuit begin to charge, causing a disturbance signal at the input to be amplified, which presents some voltage to the bridge and some voltage difference between VA and VB. The base-emitter is therefore forward biased, because VA must be higher than VB (base voltage higher than emitter voltage for NPN) and oscillatory conditions begin. And to insure oscillatory conditions, R3 can be further increased in value to provide an even lower VB condition and a larger difference voltage.

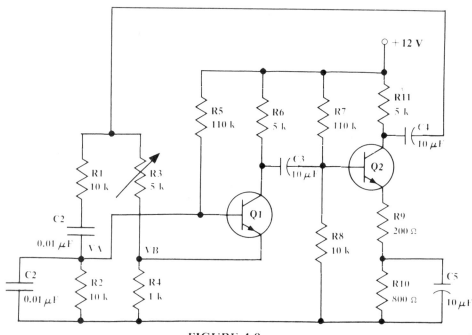

FIGURE 4-9

The cleanest sine wave output signal occurs when the overall gain of the oscillator, including the gains and losses, is slightly greater than unity. In the circuit of Figure 4-9, this will occur when R3 is adjusted to about 2500 Ω (2464 Ω), and where approximately zero phase shift (overall) occurs at the oscillatory frequency. Increasing either the gain of the amplifier or the difference voltage (VA − VB) much further can create distortion because frequencies other than the operating frequency can become amplified. In other words, when R4 is set to approximately 2464 Ω, 100 mV of signal difference exists between VA and VB at the frequency where 0° phase shift condition exists. At all other frequencies, 0° phase shift does not exist and less than 100 mV difference occurs between points A and B. therefore, if unity gain is set at the 0° phase shift condition, only that frequency will be amplified and all others, being less than unity, will not. Hence, the simple frequency sinewave condition exists at unity gain, and if the gain is increased much beyond unity, distortion will occur.

THE OP-AMP DRIVEN WEIN-BRIDGE OSCILLATOR

The operating principles of the discrete transistor oscillator are similar to those used with the op-amp oscillator. The op-amp is used as a block of gain and the amount of feedback is controlled either at the output of the op-amp in controlling the gain or in the bridge in controlling the balance. Both techniques are shown in Figure 4-10 and both are valid, but the circuit of Figure 4-10(b) is more critical, since resistor R3 serves in both the voltage gain and the voltage stepdown (balancing) functions. Again, terminal A must be at a higher voltage level than terminal B to provide positive oscillatory conditions.

THE OSCILLATOR IN AMPLIFIER FORM

In amplifier form, resistor R3 and R4 of the bridge circuit have to be duplicated. This is because R3 and R4 control the initial gain of the amplifier and they are also part of the bridge circuit. Also, note

204

that both op-amp driven circuits of Figure 4-10 use the non-inverting 0° phase-shift, feedback configuration as illustrated in Figure 4-11(b).

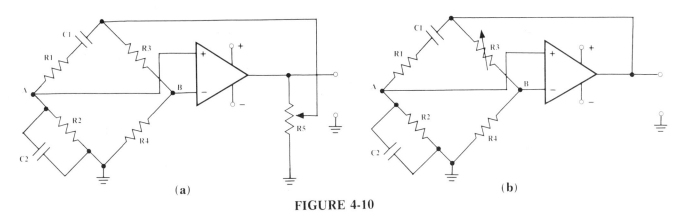

FIGURE 4-10

The oscillator circuit of Figure 4-11(a) is similar to the circuit connections of Figure 4-10. For instance, the circuit of Figure 4-10(a), when variable resistor R5 is at top rotation, is identical to the circuit of Figure 4-11(a), and the circuit of Figure 4-10(b), when R3 is a fixed value, is similar to the circuit of Figure 4-11(a). Therefore, connecting the basic circuit in amplifier form provides further simplification, where the gain at 3 and the voltage stepdown (at both A and B branches) provide a balance (0 volt difference) between A and B, as calculated.

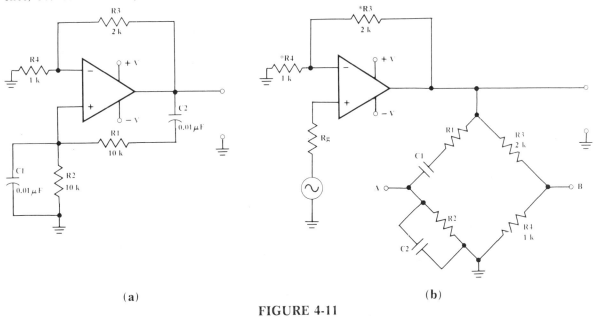

FIGURE 4-11

1. Most op-amps provide similar pin-out connections to those shown below. Therefore, the DC voltages that should be monitored for a working device having a ± 12 V DC power source are as follows:

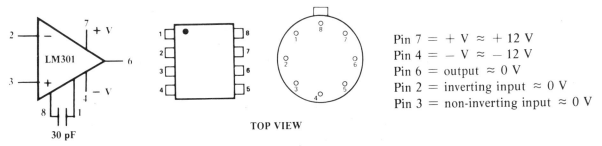

TOP VIEW

Pin 7 = + V ≈ + 12 V
Pin 4 = − V ≈ − 12 V
Pin 6 = output ≈ 0 V
Pin 2 = inverting input ≈ 0 V
Pin 3 = non-inverting input ≈ 0 V

NOTE: For the LM301 type devices, and external 30 pF of capacitance must be connected between pins 1 and 8.

2. Voltage gain and the peak-to-peak voltage output:

a. $A_V = \dfrac{R3 + R4}{R3} = \dfrac{1\ k\Omega + 2\ k\Omega}{1\ k\Omega} = 3$

b. $Vop\text{-}p = Vinp\text{-}p \times A_V = 1\ Vp\text{-}p \times 3 = 3\ Vp\text{-}p$

3. Voltage difference:

a. $VA = Vop\text{-}p \times \dfrac{Vop\text{-}p(R2 \mathbin{/\!/} jX_{C2})}{(R1 - jX_{C1}) + (R2 \mathbin{/\!/} jX_{C2})} = 3\ Vp\text{-}p \times 1/3 = 1\ Vp\text{-}p$

b. $VB = Vop\text{-}p \times \dfrac{R4}{R3 + R4} = 3\ Vp\text{-}p \times \dfrac{1\ k\Omega}{1\ k\Omega + 2\ k\Omega} = 1\ Vp\text{-}p$, and $VA - VB = 0\ Vp\text{-}p$

c. $VA - VB = 1\ Vp\text{-}p - 1\ Vp\text{-}p = 0\ Vp\text{-}p$

Once the circuit is connected in oscillator form, any voltage difference between A and B, where point A is higher than point B, will cause the circuit to oscillate, because the op-amp will amplify, greatly, any difference. Therefore, R3 must be increased slightly or R4 decreased slightly to insure oscillatory conditions.

Most laboratory grade Wein-bridge oscillators are used over many decades by switching in fixed capacitors and varying the frequency within that range, over a 10 : 1 frequency range, either by using ganged variable capacitors or ganged variable resistors. However, in order to do this and still maintain a constant output voltage level (and oscillatory conditions) automatic gain control circuitry must be used.

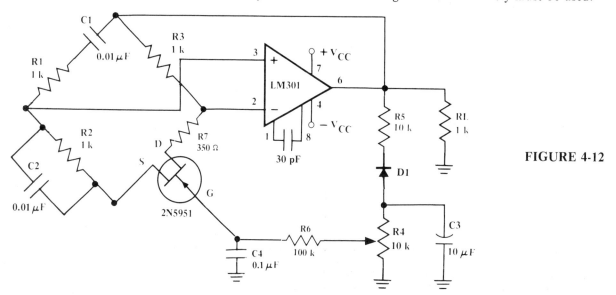

FIGURE 4-12

In the circuit of Figure 4-12, a FET replaces resistor R4 in the bridge and acts as a voltage controlled resistor. It has a low resistance when a small or no signal appears at the output and then it increases in resistance with increased output signal levels. Essentially, when little or no signal occurs at the output, the resistance of the drain-source is low and this creates a large voltage difference between terminals A and B which, in turn, causes a large voltage to appear at the output. However, the large voltage at the output is then developed across filter capacitor C3 and the increased negative DC voltage causes the device gate-source resistance to increase, lowering the voltage difference between terminals A and B which, in turn, lowers the amplitude at the output. This cause and effect (AGC) maintains the loop gain at slightly

206

greater than 1, which insures a clean non-distorted sine wave at the output.

A balanced bridge condition exists when the series resistance R7 and the source-to-drain resistance of the JFET equal 500 Ω. Therefore, if R7 equals 350 Ω, then the device should be about 150 Ω, and it can be adjusted by varying the R4 variable resistor. For the 2N5951, the drain-to-source voltage is lower (normally) than 0.1 volts, and the resistance of the JFET can be varied several hundred ohms.

EXPERIMENTAL OBJECTIVES

To investigate the Wein-bridge oscillator circuit.

LIST OF MATERIALS

1. Transistors: 2N3904(two) and 2N3819 or 2N5951 or equivalents
2. Op-amp: LM301 or equivalent
3. Resistors: 220 Ω(one)   1 kΩ(one)   1.2 kΩ(two)   2.2 kΩ(one)   3.3 kΩ(one)   4.7 kΩ(two)
      10 kΩ(four)   10 kΩ potentiometer(one)   100 kΩ(two)
4. Capacitors: 1000 pF(two)   30 pF(one)   10 μF(four)
5. Voltmeter
6. Oscilloscope
7. Counter

EXPERIMENTAL PROCEDURE (PART I):

1. Connect the Wein-bridge oscillator in amplifier form as shown in Figure 4-13.

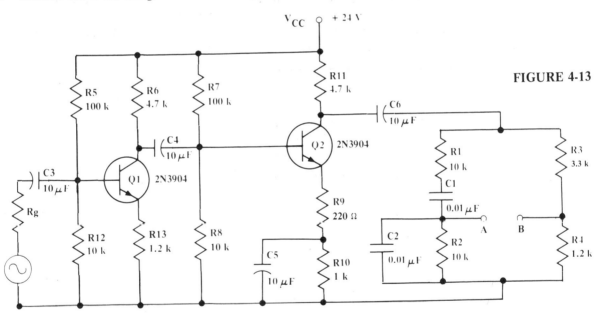

FIGURE 4-13

2. Calculate the DC voltage drops around the circuit.
3. Measure the DC voltage drops around the circuit.
4. Insert the calculated and measured values, as indicated, into Table 4-1.
5. Calculate:

   a.   The center frequency from: $f_o = \dfrac{1}{2\pi RC}$ where: R = 10 kΩ   and   C = 1000 pF

   b.   The voltage gain of the amplifier circuit (Q1 − Q2), including the loading effect of the bridge circuit.
   c.   The voltage output at the collector of Q2 for 100 mVp-p of input signal.
   d.   The voltage output at terminal A of the bridge and then terminal B of the bridge for 100 mVp-p of input signal.

207

TABLE 4-1	V_{R12}	V_{R13}	V_{R3}	V_{R4}	V_{R5}	V_{R6}	V_{R7}	V_{R8}	V_{R9}	V_{R10}	V_{R11}	V_{CE1}	V_{CE2}
CALCULATED													
MEASURED													

6. Apply 100 mVp-p of input voltage at the center frequency.
   a. Measure the Vop-p at the collector, at terminal A of the bridge, and at terminal B of the bridge.

NOTE: The voltage stepdown at terminal A of the bridge should be about 1/3 of the collector peak-to-peak voltage at center frequency. The peak-to-peak voltage at terminal B should be about R4/(R3 + R4) = 1.2 kΩ/(1.2 kΩ + 3.3 kΩ) of the collector peak-to-peak voltage.

   b. Calculate the difference voltage at the bridge from VAp-p − VBp-p.
   c. Monitor the phase shift from the input to Terminal A and then to terminal B.
7. Calculate the overall gain of the circuit from: $A_v(\text{overall}) = A_v(\text{gain}) \times A_v(\text{loss})$.

   where: $A_v(\text{loss}) = (1/VA) - (1/VB)$ and $A_v(\text{gain})$ is the amplifier gain from the input to the collector of Q2.

8. Insert the calculated and measured amplifier values, as indicated, into Table 4-2.

NOTE: The block gain can be adjusted with a variable potentiometer as shown in Figure 4-9, where R3 of the bridge is replaced by a potentiometer and adjusted.

9. Connect the Wein-bridge in oscillator form as shown in Figure 4-14 and monitor the oscillating frequency. Adjust variable resistor R11 for best linearity. Monitor the signal output off the collector of Q1.

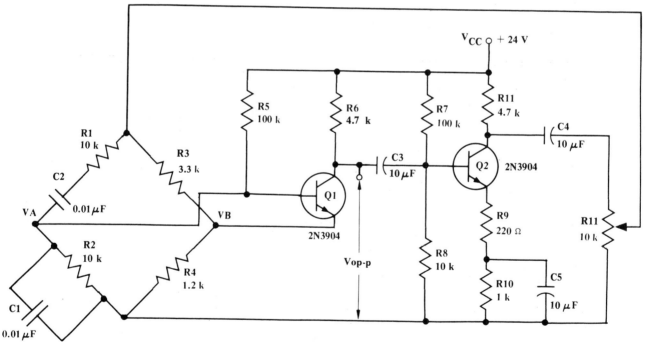

FIGURE 4-14

10. Insert the measured oscillator values, as indicated, into Table 4-2.

208

TABLE 4-2	f_o	A_v Collector	Vop-p	VA p-p	VB p-p	V(diff)	A_v Overall	OSCILLATOR	
								f_o	Vop-p
CALCULATED								/////////	/////////
MEASURED	/////////	/////////							

EXPERIMENTAL PROCEDURE (PART 2)

1. Connect the op-amp driven Wein-bridge oscillator circuit as shown in Figure 4-15.

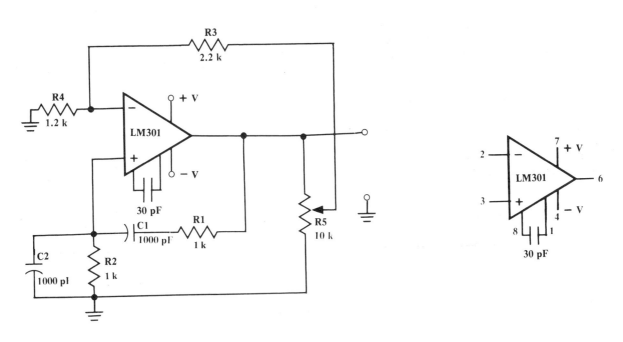

FIGURE 4-15

NOTE: The LM301 requires approximately 30 pF of capacitance between Pins 1 and 8.

2. Monitor the frequency of oscillations. Adjust R5 for linear sine wave conditions which occur just after unity gain is achieved.
3. Connect the oscillator in the slightly different circuit configuration shown in Figure 4-16.
4. Monitor the frequency of oscillation, adjusting variable resistor R3 for linear sine wave conditions.
5. Insert the measured values, as indicated, into Table 4-3.

NOTE: If the circuit does not oscillate, monitor the DC voltages of the circuit to make sure the op-amp is working. However, when DC voltage measurements are to be taken, adjust the variable potentiometer to a non-oscillating condition and then make the DC measurements.

6. Connect the op-amp Wein-bridge oscillator circuit with AGC (automatic gain control) as shown in Figure 4-17. (This circuit is the same as that of Figure 4-16 except that R4 have been replaced by a voltage controlled resistor (FET) and the DC rectifier feedback circuitry has been added.)
7. Monitor the DC voltages around the circuit.
8. Monitor the frequency of oscillation and adjust R3, while monitoring the output signal.
9. Insert the measured values, as indicated, into Table 4-3.

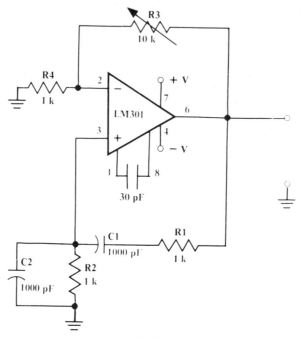

FIGURE 4-16

NOTE: LM301 requires approximately 30 pF of capacitance between Pins 1 and 8.

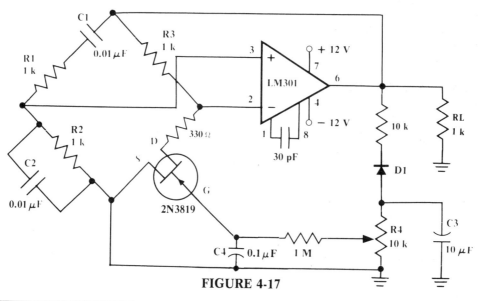

FIGURE 4-17

TABLE 4-3	Pin 2	Pin 3	Pin 6	Pin 7	Pin 4	fo	Vop-p
MEASURED Figure 5-15	≈ 0 V	≈ 0 V		+ 12 V	− 12 V		
MEASURED Figure 5-16	≈ 0 V	≈ 0 V		+ 12 V	− 12 V		
MEASURED Figure 5-17	≈ 0 V	≈ 0 V		+ 12 V	− 12 V		

PART 2 — THE PHASE-SHIFT OSCILLATOR

GENERAL DISCUSSION

The phase-shift oscillator uses an amplifier and a ladder type phase shift network in achieving 360° phase shift, where the amplifier provides 180° phase shift and the network the remaining 180° phase shift. The phase-shift network can be either phase lag, as illustrated in Figure 4-18(a), or phase lead, as illustrated in Figure 4-18(b), and the RC components can be all equal where R1 = R2 = R3 and C1 = C2 = C3, or they can be graded where, for example, R1 = 0.1 R2 = 0.01 R3 and C1 = 10 C2 = 100 C3. However, note that in both cases the time constants of all three branches are equal: $R_1C_1 = R_2C_2 = R_3C_3$.

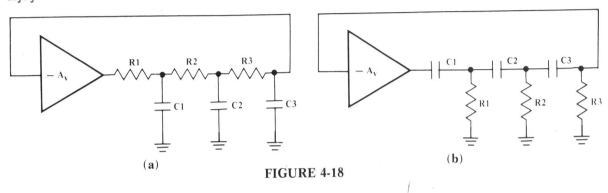

(a)   (b)

FIGURE 4-18

The phase-shift network usually has three branches, but it can have four. In the three branch network. each branch provides a 60° phase shift, and in four branches, each branch provides a 45° phase shift. Two branches are not possible because two branch capacitors, with any resistance, cannot achieve the required 180° phase shift. Also, four branches are rarely used because four ganged capacitors and resistors make it more difficult to provide a variable oscillator and the extra branch provides one more branch in which signal loss occurs, although the step-down per branch is not as great.

PHASE-SHIFT ANALYSIS OF GRADED RC BRANCHES

If each branch of the three branch phase shift is to provide 60°, where the effect of loading of each succeeding branch is not considered, the frequency of oscillation can be solved from $f_0 = 1/2\pi\sqrt{3}\,RC$, and the frequency of oscillation for the graded phase-shift oscillator of Figure 4-19 is 918.88 Hz. A FET input amplifier will be required to provide the 100 kΩ for Zin.

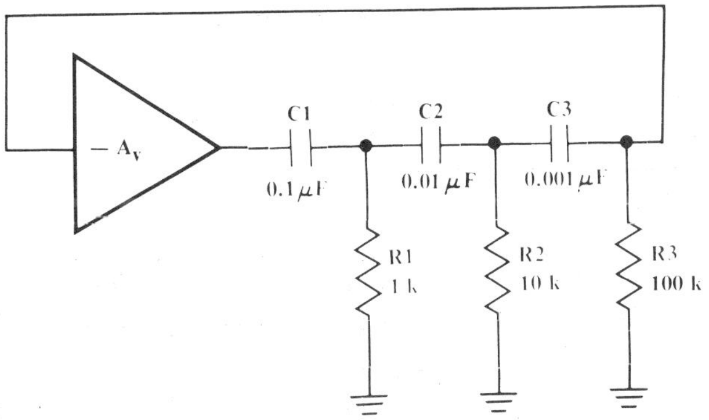

$$f_0 = \frac{1}{2\pi\sqrt{3}\,RC} = \frac{0.159}{1.73 \times 10^{-4}} = 918.88 \text{ Hz}$$

where: $R_1C_1 = 1\text{ k}\Omega \times 0.1\ \mu\text{F} = 10^{-4}$

$R_2C_2 = 10\text{ k}\Omega \times 0.01\ \mu\text{F} = 10^{-4}$

$R_3C_3 = 100\text{ k}\Omega \times 1000\text{ pF} = 10^{-4}$

FIGURE 4-19

The frequency of oscillation at $f_0 = 1/2\pi\sqrt{3}\,RC$ for the non-loading conditions of the graded phase-shift network at 60° can be better understood if a single branch is analized first at 45°, where $X_C = R$, and then, again, at 60°, where $X_C = \sqrt{3}\,R$.

211

a. 45° PHASE SHIFT

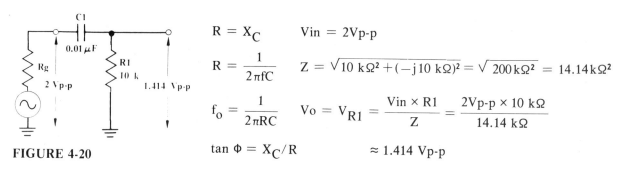

$R = X_C$ $\quad$ $V_{in} = 2V_{p-p}$

$R = \dfrac{1}{2\pi fC}$ $\quad$ $Z = \sqrt{10\ k\Omega^2 + (-j\,10\ k\Omega)^2} = \sqrt{200\ k\Omega^2} = 14.14\ k\Omega^2$

$f_o = \dfrac{1}{2\pi RC}$ $\quad$ $V_o = V_{R1} = \dfrac{V_{in} \times R1}{Z} = \dfrac{2V_{p-p} \times 10\ k\Omega}{14.14\ k\Omega}$

$\tan\Phi = X_C/R$ $\qquad\qquad$ $\approx 1.414\ V_{p-p}$

FIGURE 4-20

1. $f_o = \dfrac{1}{2\pi RC} = \dfrac{0.159}{10\ k\Omega \times 0.01\ \mu F} = \dfrac{0.159}{10^4 \times 10^{-8}} = 1590\ Hz$

2. $X_C = \dfrac{1}{2\pi fC} = \dfrac{0.159}{1.59 \times 10^3 \times 10^{-8}} = 10\ k\Omega$

3. $\tan^{-1}\dfrac{X_C}{R} = \tan^{-1}\dfrac{10\ k\Omega}{10\ k\Omega} = \tan^{-1} 1 = 45°$

NOTE: f_o can also be calculated from $f_o = 1/(2\pi RC \tan\Phi)$, where $\tan\Phi = X_C/R$ and $R\tan\Phi = X_C$. However, $\tan 45° = 1$ and the oscillating frequency is calculated from:

$$f_o = \frac{1}{2\pi RC \tan 45°} = \frac{1}{2\pi RC(1)} = \frac{1}{2\pi RC}\,.$$

b. 60° PHASE SHIFT

$R\tan\Phi = X_C$

$R\tan\Phi = \dfrac{1}{2\pi fC}$

$f_o = \dfrac{1}{2\pi RC \tan\Phi}$

FIGURE 4-21

1. $\tan\Phi = \tan 60° = 1.73 = \sqrt{3}$

2. $f_o = \dfrac{1}{2\pi RC \tan\Phi} = \dfrac{1}{2\pi RC \sqrt{3}} = \dfrac{0.159}{10\ k\Omega \times 0.01\ \mu F \times 1.73} = 918.9\ Hz$

3. $X_C = \dfrac{1}{2\pi fC} = \dfrac{0.159}{918.9\ Hz \times 0.01\ \mu F} = \dfrac{0.159}{918.9 \times 10^{-8}} = 17.3\ k\Omega$

4. $\tan^{-1}\dfrac{X_C}{R} = \tan^{-1}\dfrac{17.3\ k\Omega}{10\ k\Omega} = \tan^{-1} 1.73 = 60°$

NOTE: The frequency formula: $f_o = 1/(2\pi RC \tan\Phi)$ is valid for either phase lag or phase lead branches because both R and C are in the denominator.

5. The voltage for the graded (non-loading) three branch RC network can be solved by solving for the voltage loss across a single RC network and then stepping down the loss through the other two networks. The voltage loss in the single RC network is 1/2 and the loss through the three RC non-loaded networks is one eighth.

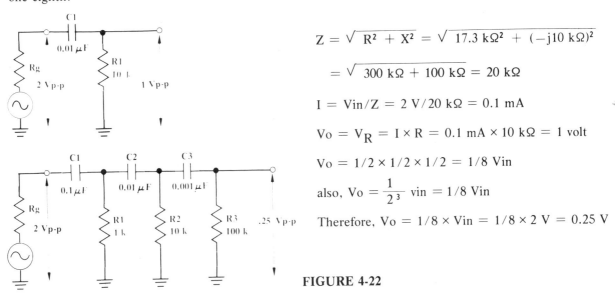

$$Z = \sqrt{R^2 + X^2} = \sqrt{17.3 \text{ k}\Omega^2 + (-j10 \text{ k}\Omega)^2}$$

$$= \sqrt{300 \text{ k}\Omega + 100 \text{ k}\Omega} = 20 \text{ k}\Omega$$

$$I = V_{in}/Z = 2 \text{ V}/20 \text{ k}\Omega = 0.1 \text{ mA}$$

$$V_o = V_R = I \times R = 0.1 \text{ mA} \times 10 \text{ k}\Omega = 1 \text{ volt}$$

$$V_o = 1/2 \times 1/2 \times 1/2 = 1/8 \text{ Vin}$$

also, $V_o = \dfrac{1}{2^3} \text{ vin} = 1/8 \text{ Vin}$

Therefore, $V_o = 1/8 \times V_{in} = 1/8 \times 2 \text{ V} = 0.25 \text{ V}$

FIGURE 4-22

NOTE: $X_{C_2} = 17.3 \text{ k}\Omega$ and R2 = 10 kΩ, but $X_{C_1} = 1.73 \text{ k}\Omega$ and R1 = 1 kΩ, and $X_{C_3} = 173 \text{ k}\Omega$ and R3 = 100 kΩ. In all instance, the ratio of X_C to R in each branch is $1.73 = \sqrt{3}$.

VARIABLE PHASE-SHIFT OSCILLATORS

Variable phase shift oscillators can use ganged capacitors or ganged resistors, where all three capacitors or all three resistors are varied at the same time. And the most readily available three-ganged capacitors or three-ganged resistors are those where the values of each component in the gang is equal in value. Therefore, R1 = R2 = R3 and C1 = C2 = C3 where, depending on the applications, phase leading or phase lagging networks will be used.

For instance, if a FET-bipolar driven phase-shift network, such as that in Figure 4-23(a), is used, then a phase-lead network is ideal. It minimizes components by using C3 as part of the phase-shift network and, again, as a coupling capacitor. On the other hand, if an op-amp is used to drive the phase-shift network, as shown in Figure 4-24(b), then a DC path is required between the output and input of the device (Pin 2 to Pin 6) and the resistive path of the phase-lag network (R1, R2, and R3) provides this.

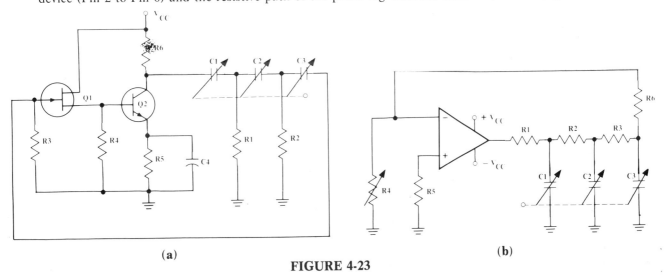

(a)   (b)

FIGURE 4-23

Again, the number of components is kept to a minimum because the components in the phase-shift network serve a dual role. However, one additional resistor must be used in the feedback path to isolate C3 from the essential ground condition of the inverting terminal and, also, this resistor will effect the frequency of oscillation.

AC ANALYSIS OF LIKE RC BRANCHES

If R1 = R2 = R3 and C1 = C2 = C3, then the loading effect of each branch on the other must be taken into account. The mathematical analysis for this can be accomplished by writing loop equations and solving by determining the frequency of oscillation and the voltage loss of the phase-shift circuit. Therefore, depending on whether the phase-shift is phase-leading or phase lagging or whether the driving impedance equals or does not equal the R's of the circuit, all will effect the frequency of oscillation of the circuit and, hence, overall loss of the phase-shift network.

For instance, if the RC of the output transistor equals the R's of the phase-lead, phase-shift network, then the frequency of oscillation is approximately: $f_0 = 1/(2\pi RC \sqrt{10})$. However, if the RC of the output transistor is small as compared to the R's of the circuit, then the frequency of oscillation is approximately: $f_0 = 1/(2\pi RC \sqrt{6})$, and the voltage gain required to obtain oscillatory conditions is slightly greater than 50, depending on the driving and loading impedances and the number of loop equations involved.

For the phase-lag network, where the driving impedance of the op-amp is low, the frequency of oscillation is approximately $f_0 = 1/(2\pi RC \sqrt{5})$. The loss across the resistors of the phase-lag network is relatively light and amplifier voltage gains of less than 5 are all that is required for oscillatory conditions.

THE HYBRID DRIVEN PHASE-SHIFT OSCILLATOR

The advantages of the FET-bipolar, amplifier-oscillator circuits of Figure 4-24 is that the FET provides a high input impedance, the bipolar provides voltage gain, and the combined devices provide a large current gain. The high input impedance of the FET insures that the RC branches of the phase-lead, phase-shift networks will be relatively equal. Also, components are minimized because R1 doubles as both the last resistor in the phase-shift network and as the gate resistor of the FET.

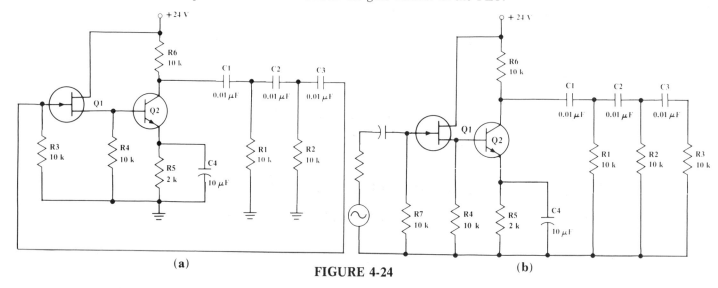

(a)   **FIGURE 4-24**   (b)

THE HYBRID DRIVEN PHASE-SHIFT OSCILLATOR — DC ANALYSIS

To begin the DC analysis, the oscillator is connected in amplifier form, as shown in Figure 4-24(b), where the value of R3 has been duplicated as R7 because it functions in both the amplifier and phase-shift networks.

1. Solving for the drain current of Q1:

$$I_D(Q1) = I_{DSS}(1 - V_{GS}/V_P)^2 = 6\,mA(1 - 3\,V/6\,V)^2 = 6\,ma \times 0.5^2 = 1.5\,mA$$

214

2. Solving for the voltage drop across resistor R4 from:

$$V_{GS} = V_{R4} = 3 \text{ volts, since } V_{R7} = 0 \text{ volts}$$

3. Solving for the voltage drop across resistor R5 from:

$$V_{R5} = V_{R4} - V_{BE_2} = 3 \text{ V} - 0.6 \text{ V} = 2.4 \text{ volts}$$

4. Solving for the emitter current of transistor Q2:

$$I_{E2} = V_{R5}/R5 = 2.4 \text{ V}/2 \text{ k}\Omega = 1.2 \text{ mA}$$

5. Solving for the voltage drop across resistor R6 from:

$$V_{R6} \approx I_C R6 = 1.2 \text{ mA} \times 10 \text{ k}\Omega = 12 \text{ V} \qquad \text{where: } I_C \approx I_E = 1.2 \text{ mA}$$

6. Solving for the single point voltages:

   a. $V_{C2} = V_{CC} - V_{R6} = 24 \text{ V} - 12 \text{ V} = 12 \text{ volts}$

   b. $V_{E2} = V_{R5} = 2.4 \text{ volts}$

   c. $V_{D1} = V_{CC} = 24 \text{ volts}$

   d. $V_{S1} = V_{R4} = 3 \text{ volts}$

7. Solving for the voltage drop across the Q1 transistor from:

$$V_{DS1} = V_{D1} - V_{S1} = 24 \text{ V} - 3 \text{ V} = 21 \text{ volts}$$

8. Solving for the voltage drop across the Q2 transistor from:

$$V_{CE2} = 12 \text{ V} - 2.4 \text{ V} = 9.6 \text{ volts}$$

NOTE: There are no DC voltage drops across R1, R2, and R3 because they are isolated by capacitors C1 and C2.

THE HYBRID DRIVEN PHASE-SHIFT OSCILLATOR — AC ANALYSIS

1. Solving for the frequency of oscillation where $R_C = R$ from:

   a. $f_r \approx \dfrac{1}{2\pi RC \sqrt{10}} = \dfrac{1}{6.28 \times 10 \text{ k}\Omega \times 0.01 \text{ } \mu\text{F} \times 3.16} = 503.3 \text{ Hz}$

   b. If $R_C \ll R$, for example where $R_C - 1 \text{ k}\Omega$ and $R = 10 \text{ k}\Omega$, then:

$$f_r \approx \frac{1}{2\pi RC \sqrt{6}} = \frac{1}{6.28 \times 10 \text{ k}\Omega \times 0.01 \text{ } \mu\text{F} \times 2.45} = 649.7 \text{ Hz}$$

NOTE: Writing loop equations for the phase equations, where both the reals and the imaginaries of the determinants are equated to zero, provides a frequency of oscillation formula for the imaginaries:

$$f_0 = \frac{1}{2\pi C \sqrt{4RR_C + 6R^2}}$$

and a (loss) required gain formula, which varies dramatically with the devices used, because of the variety of loops to be written. Therefore, when $R_C = R$:

$$f_0 = \frac{1}{2\pi C \sqrt{4RR_C + 6R^2}} = \frac{1}{2\pi C \sqrt{10R^2}} = \frac{1}{2\pi RC \sqrt{10}} ,$$

and when $R_C \ll R$:

$$f_0 = \frac{1}{2\pi RC \sqrt{6R^2}} = \frac{1}{2\pi RC \sqrt{6}}$$

where: $\sqrt{4RR_C}$ "drops" out in the calculations.

2. The voltage stepdown method used to solve for overall circuit loss across the phase-shift network is tedious, but it does provide better insight than the "plug and crank" determinant method.

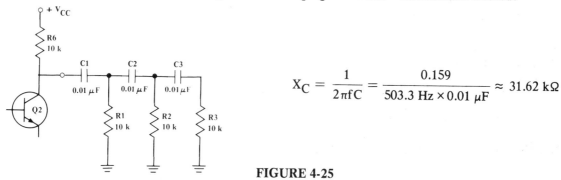

$$X_C = \frac{1}{2\pi fC} = \frac{0.159}{503.3 \text{ Hz} \times 0.01 \text{ } \mu F} \approx 31.62 \text{ k}\Omega$$

FIGURE 4-25

a. Solving for the loss of the third branch from:

$$\text{Loss(3rd BR)} = \frac{R3}{R3 - jX_{C3}} = \frac{10 \text{ k}\Omega}{10 \text{ k}\Omega - j31.6 \text{ k}\Omega} = \frac{10 \text{ k}\Omega}{33.14 \text{ k}\Omega \angle - 72.4°} = 0.3018 \angle 72.4°$$

b. Solving for the loss of the second branch from:

$$\text{Loss(2nd BR)} = \frac{R2 \,/\!/\, (R3 - jX_{C3})}{-jX_{C2} + (R2 \,/\!/\, [R3 - jX_{C3}])} = \frac{10 \text{ k}\Omega \,/\!/\, 33.14 \text{ k}\Omega \angle - 72.4°}{j31.6 \text{ k}\Omega + (10 \text{ k}\Omega \,/\!/\, 33.14 \text{ k}\Omega \angle - 72.4°)}$$

$$= \frac{8.86 \text{ k}\Omega \angle - 14.7°}{- j31.6 \text{ k}\Omega + (8.86 \text{ k}\Omega \angle - 14.7°)} = \frac{8.86 \text{ k}\Omega \angle - 14.7°}{- j31.6 \text{ k}\Omega + (8.57 \text{ k}\Omega - j2.26 \text{ k}\Omega)}$$

$$= \frac{8.86 \text{ k}\Omega \angle - 14.7°}{8.57 \text{ k}\Omega - j33.83 \text{ k}\Omega} = \frac{8.86 \angle - 14.7°}{34.85 \angle - 76.1°} = 0.254 \angle 61.4°$$

c. Solving for the loss of the first branch from:

$$\text{Loss(1st BR)} = \frac{R1 \,/\!/\, (-jX_{C2} + [R2 \,/\!/\, (R3 - jX_{C3})])}{-jX_{C1} + (R1 \,/\!/\, [-jX_{C2} + (R2 \,/\!/\, [R3 - jX_{C3}])])}$$

$$= \frac{10 \text{ k}\Omega \,/\!/\, 34.85 \text{ k}\Omega \angle - 76.1°}{-j31.6 \text{ k}\Omega + ((10 \text{ k}\Omega \,/\!/\, 34.85 \text{ k}\Omega \angle - 76.1°} = \frac{9.05 \text{ k}\Omega \angle - 14.6°}{-j31.6 \text{ k}\Omega + 9.05 \text{ k}\Omega \angle - 14.6°}$$

$$= \frac{9.05 \text{ k}\Omega \angle - 14.6°}{-j31.6 \text{ k}\Omega + (8.76 \text{ k}\Omega - j2.28 \text{ k}\Omega)} = \frac{9.05 \text{ k}\Omega \angle - 14.6°}{8.76 \text{ k}\Omega - j33.9 \text{ k}\Omega}$$

$$= \frac{9.05 \text{ k}\Omega \angle - 14.6°}{35.01 \text{ k}\Omega \angle - 75.5°} = 0.259 \angle 60.9°$$

d. Solving for the loading effect of the phase-shift network on the RC collector resistor:

$$RC \,/\!/\, (-jX_{C1} + [R1 \,/\!/\, (-jX_{C2} + [R2 \,/\!/\, (R3 = jX_{C3})])]) = 10 \text{ k}\Omega \,/\!/\, 35.01 \text{ k}\Omega \angle - 75.5°$$

216

$$= \frac{10 \text{ k}\Omega(35.01 \text{ k}\Omega \angle -75.5°)}{10 \text{ k}\Omega + 35.01 \text{ k}\Omega \angle -75.5°} = \frac{35.01 \text{ k}\Omega \angle -75.5°}{18.77 \text{ k}\Omega - j33.89 \text{ k}\Omega}$$

$$= \frac{35.01 \text{ k}\Omega \angle -75.5°}{38.74 \text{ k}\Omega \angle -61°} = 0.904 \angle -14.5°$$

e.   Therefore, the overall loss accrued across all three RC branches, and including the loading effect on the collector resistor is:

$$V(\text{Loss}) = 0.904 \times 0.259 \times 0.254 \times 0.3018 \approx 0.0189 = 1/52.9$$

Hence, the gain of the amplifier must be slightly greater than 53.

3.   The phase shift, as noted, has phase shifts of 60.9° for the first RC branch, 61.4° for the second branch, and 72.4° for the third branch. Also, included is the $\angle -14.5°$ phase shift due to loading. Therefore, the total phase shift is approximately:

$$\angle° \text{ Total} = 60.9° + 61.4° + 72.4° - 14.5° = 180.2°$$

THE OP-AMP PHASE-SHIFT OSCILLATOR

Using an op-amp as the amplifying device in the phase-shift oscillator requires that the phase-lag network be used so that the DC path can be accomplished by the resistor of the phase-shift network. In Figure 4-26, resistor R6 is used as an isolating resistor, where the higher the value of R6, the closer the calculated-measured frequency will be. A lower value R6 resistor limits the phase-shift effect of capacitor C3 and increases the operating (oscillating) frequency.

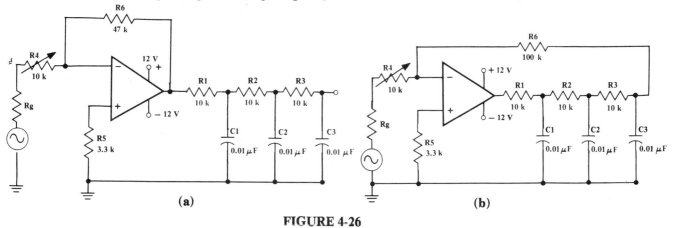

FIGURE 4-26

THE OP-AMP PHASE-SHIFT OSCILLATOR — DC ANALYSIS

The DC analysis for the op-amp is straight-forward. A DC balanced device has, theoretically, zero volts on Pins 2, 3, and 6; a positive 12 volts connected to Pin 7; and a negative 12 volts connected to Pin 4. The op-amp used in Figure 4-26(b) is no exception. However, when a LM301 type device is used, an approximte 30 pF capacitor is used between Pins 1 and 8 to make sure there is a sine wave and not a square wave output.

THE OP-AMP PHASE-SHIFT OSCILLATOR — AC ANALYSIS

The frequency of oscillation for the phase-lag network is solved from:

$$f_0 \approx \frac{1}{2\pi RC \sqrt{5}} = \frac{1}{6.18 \times 10 \text{ k}\Omega \times 0.01 \text{ }\mu\text{F} \times 2.24} = \frac{1}{14.05 \times 10^4 \times 10^{-4}} \approx 712 \text{ Hz}$$

Solving for the reactance of each of the 0.01 μF capacitances at the approximate frequency of oscillation:

217

$$X_C = \frac{1}{2\pi f C} = \frac{0.159}{712 \text{ Hz} \times 0.01 \ \mu F} = \frac{0.159}{7.12 \times 10^2 \times 10^{-8}} = \frac{0.159 \times 10^6}{7.12} = 22.35 \ k\Omega$$

The voltage loss of the phase-lag network can be solved similarly to the preceeding phase-lead network, where X_C is solved and then the voltage loss across each branch is found.

Since the output impedance of the op-amp is so much lower than the R's of the phase-shift network, it is not included in the calculations. For instance, writing the loop equations of the phase-lag network, where $R_C = R$, provides a frequency formula where: $f_0 = 1/(2\pi C \sqrt{4RR_C + 5R^2})$ and, when $R_C << R$, $\sqrt{4RR_C}$ "drops out" providing $f_0 = 1/(2\pi RC \sqrt{5})$. However, including the isolating resistor R6 will minimize the effect of C3 and cause the operating frequency to be increases. Therefore, the higher R6, the closer the calculated-measured values become, but the larger A_v has to be. However, A_v is easily achieved.

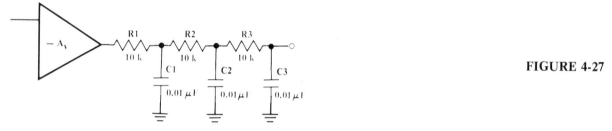

FIGURE 4-27

Solving for the total phase shift of the entire phase-lag, phase-shift network:

a. Solving for the loss across the third branch:

$$\text{Loss(3rd BR)} = \frac{-jX_{C3}}{R3 - jX_{C3}} = \frac{-j22.35 \ k\Omega}{10 \ k\Omega - j22.35 \ k\Omega} = \frac{22.35 \ k\Omega \ \angle -90°}{24.49 \ k\Omega \ \angle -65.9°} = 0.91 \ \angle -24.1°$$

b. Solving for the loss and phase shift across the second branch:

$$\text{Loss(2nd BR)} = \frac{-jX_{C2} \ /\!/ \ (R3 - jX_{C3})}{R2 + [X_{C2} \ /\!/ \ (R3 - jX_{C3})]} = \frac{-j22.35 \ k\Omega \ /\!/ \ 24.49 \ k\Omega \ \angle -65.9°}{10 \ k\Omega + (-j22.35 \ k\Omega \ /\!/ \ 24.49 \ k\Omega \ \angle -65.9°)}$$

$$= \frac{11.95 \ k\Omega \ \angle -78.5°}{10 \ k\Omega + 11.95 \ k\Omega \ \angle -78.5°} = \frac{11.95 \ k\Omega \ \angle -78.5°}{10 k\Omega + (2.38 \ k\Omega - j11.71 k\Omega)} = \frac{11.95 \ k\Omega \ \angle -78.5°}{17.04 \ k\Omega \ \angle -46.6°}$$

$$= 0.701 \ \angle -31.91°$$

c. Solving for the loss and phase shift across the first branch from:

$$\text{Loss(1st BR)} = \frac{-jX_{C1} \ /\!/ \ (R2 + [-jX_{C2} \ /\!/ \ (R3 - jX_{C3})])}{R1 + (-jX_{C1} \ /\!/ \ [R2 + (-jX_{C2} \ /\!/ \ [R3 - jX_{C3}])])}$$

$$= \frac{-j22.35 \ k\Omega \ /\!/ \ 17.04 \ k\Omega \ \angle -46.6°}{10 \ k\Omega + (-j22.35 \ k\Omega \ /\!/ \ 17.04 \ k\Omega \ \angle -46.6°)} = \frac{10.49 \ k\Omega \ \angle -66.6°}{10 \ k\Omega + 10.49 \ k\Omega \ \angle -66.6°}$$

$$= \frac{10.49 \ k\Omega \ \angle -66.6°}{10 \ k\Omega + (4.17 \ k\Omega - j9.63 \ k\Omega)} = \frac{10.49 \ k\Omega \ \angle -66.6°}{17.13 \ k\Omega \ \angle -34.2°} = 0.61 \ \angle -32.4°$$

NOTE: The overall loss of the phase-lag network is not as great as that of the phase-lead network because the loss in each branch is dropped across the resistor instead of the higher valued reactances.

d. Solving for the overall loss across the phase-lag, phase-shift network:

$$V(\text{Loss}) = 0.91 \times 0.7 \times 0.61 \approx 0.386$$

Therefore, the voltage gain need only be about 2.6 and, even with a large R6 value, A_v can remain less than 10.

1. The overall phase shift of the phase-lag network can be interpreted in terms of the phase lag where $90° - 32.4° = 57.6°$ for the first RC branch, $90° - 31.9° = 58.1°$ for the second RC branch, and $90° - 24.1° = 65.9°$ for the third RC branch.

2. Hence, the total phase shift of the entire phase-lag, phase-shift network is solved from:

$$\angle° \text{ Total } = 55.8° + 58.1° + 65.9° = 179.8°$$

NOTE: Phase-shifting of 60° per branch can be accomplished by noting a 60° leading across the resistor or −30° lagging across the capacitor, since the capacitor is at −90° to begin with and $60° - 90° = -30°$.

e. Therefore, by connecting the oscillator in oscillator form as shown in Figure 4-26(b), and adjusting R4 for a gain slightly greater than unity, will provide clean oscillatory conditions.

EXPERIMENTAL OBJECTIVES

To investigate the phase-shift oscillator using the phase-lead and phase-lag circuits.

LIST OF MATERIALS

1. Transistors: 2N3904 and 2N3819 or 2N5951 (or equivalents)
2. Op-amp: LM301A (or equivalent)
3. Resistors: 100 Ω (one)   2.2 kΩ (one)   3.3 kΩ (one)   10 kΩ (five)   10 kΩ pot (one)   47 kΩ (one)
4. Capacitors: 30 pF (one)   0.01 μF (three)   10 μF (two)

EXPERIMENTAL PROCEDURE (PART I)

1. Connect the circuit as shown in Figure 4-28.

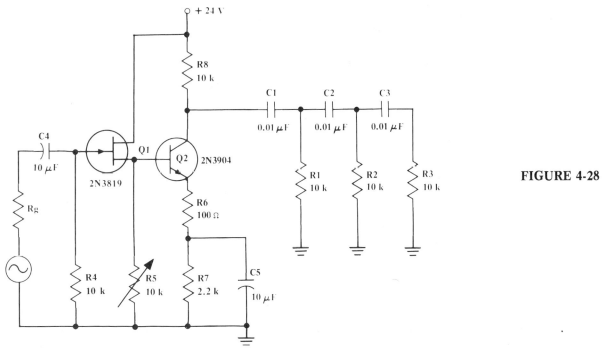

FIGURE 4-28

2. Monitor the collector of Q2 and adjust R5 until $V_{C2} \approx 16$ volts and calculate the voltage drops around the circuit. Or, set R5 at 10 kΩ and use the circuit analysis techniques acquired previously in this text to calulate the DC voltage drops around the circuit.
3. Measure the DC voltage drops around the circuit.
4. Insert the calculated and measured values, as indicated, into Table 4-4.

219

TABLE 4-4	V_{R4}	V_{R5}	V_{R6}	V_{R7}	V_{R8}	V_{DS1}	V_{CE2}	V_{C2}	V_{E2}	V_{S1}
CALCULATED										
MEASURED										

5. Calculated the approximate operating frequency from:

$$f_o = \frac{1}{2\pi RC\sqrt{10}} \quad \text{where:} \quad R = 10\ k\Omega \quad \text{and} \quad C = 0.01\ \mu F$$

6. Apply a 100 mVp-p signal to the input of the amplifier and monitor the signal level at the collector of Q2; at the junction of C1, C2, and R1; at the junction of C2, C3, and R2; and at the junction of C3 and R3 — all taken with respect to ground. Monitor the signal amplitude and ignore the distortion created by the successive RC branches.

7. Use a dual-trace scope (if available) and, while referencing the collector of Q2, monitor the phase shift across each of the branches, successively, so that (ideally) 60° phase shift, 120° phase shift, and then 180° phase shift can be monitored. Do this at the operating frequency.

NOTE: A single-trace scope can be used to monitor the phase shift if external triggering is used.

8. Verify, through calculations, the approximate gain of the amplifier (Figure 4-28) from: $A_v = R8/(r_e + R6)$, where the loading of the branch is ignored and Vin = 100 mVp-p.

NOTE: The calculations of voltage gains and losses of the phase shift network are tedious and are almost not worth the effort. Measurements can be just as valid in providing the necessary understanding of the circuit operation. Representative calculations are shown in the sample problem.

9. Connect the circuit in oscillator form as shown in Figure 4-29. Monitor the signal off the collector of Q2 and adjust the variable resistor R4 for best oscillatory conditions. Note that the cleanest sine wave occurs just after the circuit breaks into oscillations, where the overall gain of the circuit is slightly greater than unity.

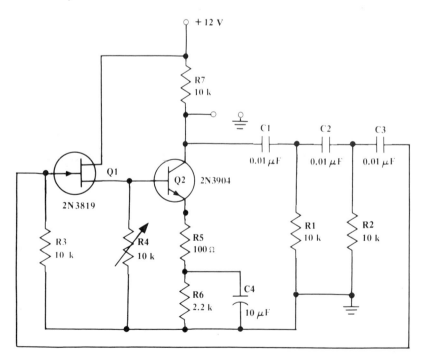

FIGURE 4-29

220

NOTE: It may be necessary to replace resistor R4 with a fixed 10 kΩ resistor and use a variable resistor in series with the C4 capacitor to control the gain of the circuit for the best output linearity of the sine wave. See Figure 4-30. Of course, more gain can be obtained by having the bypass circuit include the 100 ohm R5 resistor.

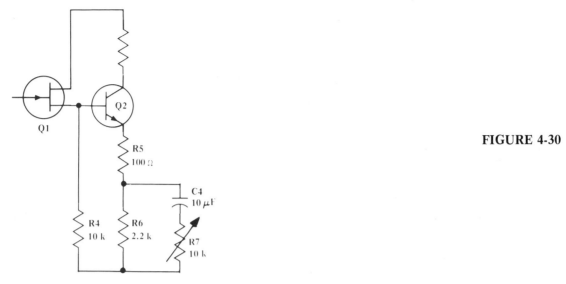

FIGURE 4-30

10.   Measure the frequency of oscillation and the amplitude of the sine wave.
11.   Insert the values, as indicated, into Table 4-5.

TABLE 4-5	100 mVp-p Input Signal				A_v Collector	PHASE SHIFT			OSCILLATION	
	$V_{C(Q2)}$	V_{R1}	V_{R2}	V_{R3}		R_1C_1	R_2C_2	R_3C_3	f_o	Vop-p
CALCULATED		////////	////////	////////		////////	////////	////////		
MEASURED					////////					

EXPERIMENTAL PROCEDURE (PART II)

1.   Connect the op-amp driven phase-shift, phase-lag circuit as shown in Figure 4-31.

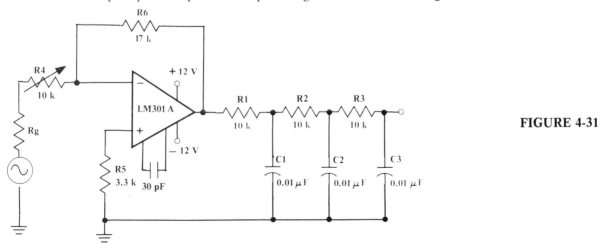

FIGURE 4-31

2.   Monitor the DC voltage drops around the circuit.
3.   Insert the measured values into Table 4-20.

TABLE 4-6	Pin 2	Pin 3	Pin 6	Pin 7	Pin 4	R4
CALCULATED	$\approx 0v$	$\approx 0\ V$		$+12\ V$	$-12\ V$	///////
MEASURED						

4. Calculated the operating frequency from: $f_o = \dfrac{1}{2\pi RC\sqrt{5}}$

5. Connect the circuit in oscillator form as shown in Figure 4-32 and monitor the output signal at Pin 6 of the op-amp. Adjust variable resistor R4 for best (cleanest) oscillatory conditions.

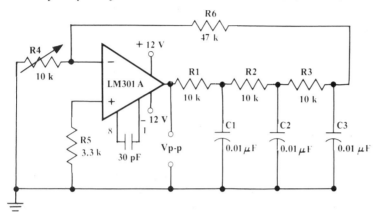

FIGURE 4-32

6. Measure both the oscillatory frequency and the amplitude of the wave at the output (Pin 6). For the LM301 type device a small capacitor of about 30 pF must be connected between Pins 1 and 8 to insure sine wave and not square wave output. Invrease or decrease the R6, 47 Ω, isolating resistor and note the frequency change, because at 47 kΩ it will effect C3 and cause the frequency to be higher than calculated.

7. Monitor the phase shift at Vo(Pin 6), V_{C1}, V_{C2}, and V_{C3}. Note the phase lag of about $-30°$ per branch and the apparent overall signal voltage phase shift of about 90°. Is something wrong? Then, why does the circuit oscillate?

8. Connect the circuit in amplifier form as shown in Figure 4-31 and continue the analysis, where the variable resistor R4 value is maintained at its oscillator connection value.

9. Apply 100 mVp-p of input signal at the operating frequency and monitor the voltage at the output of the op-amp (Pin 6) and then at the junction of R1, C1, and R2; at the junction of R2, R3, and C2; and at the junction of R3 and C3.

10. Use a dual-trace scope (if available) and, while referencing the output of the op-amp (Pin 6), monitor the phase shift across each of the RC branches.

NOTE: A single-trace scope can be used to monitor the phase shift if external triggering is used. Also, because the phase shift is monitored across the capacitors (C1, C2 and C3), $-30°$ of phase shift will be noted. However, it may be necessary to measure each branch separately because the overall phase shift might be difficult to observe.

11. Measure both the frequency of oscillation and the amplitude of the wave.
12. Insert the calculated and measured values, as indicated, into Table 4-7.

| TABLE 4-7 | SIGNAL VOLTAGE AT | | | | f_o | PHASE SHIFT | | | OSCILLATION | |
	Vo(Pin6)	V_{C1}	V_{C2}	V_{C3}		R_1C_1	R_2C_2	R_3C_3	f_o	Vop-p
CALCULATED	///////	///////	///////	///////		///////	///////	///////	///////	///////
MEASURED					///////					

222

PART 3 — THE TWIN T OSCILLATOR

GENERAL DISCUSSION

The twin T, or parallel T, oscillator is another form of null-type oscillator, where back-to-back high pass and low pass T circuits provide a high impedance and, theoretically, no current flow condition into the load at the operating frequency. The no-current condition occurs because the current flow out of the two back-to-back networks are equal but opposite, and the high impedance means that most of the input signal is dropped across the parallel T network and little or no signal is developed across the output load. Hence, there is a null, or dip, condition across the load at operating frequency.

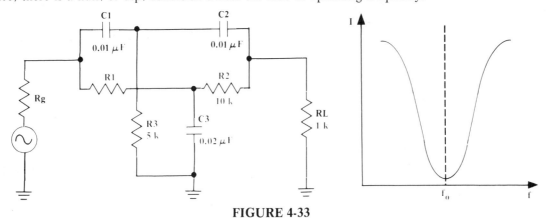

FIGURE 4-33

The frequency of oscillation for the twin T oscillator of Figure 4-33 is similar to that of the wein-bridge oscillator and is solved from: $f_o = 1/(2\pi RC)$ — where $R = R1 = R2$ and $C = C1 = C2$ and $C3 = 2C$. Therefore, if $R = R1 = R2 = 10\ k\Omega$ and $C = C1 = C2 = 0.01\ \mu F$, then $R3 = R/2 = 5\ k\Omega$ and $C3 = 2C = 0.02\ \mu F$. The frequency and reactances of X_{C1}, X_{C2}, and X_{C3} are solved from:

1. $f_o = \dfrac{1}{2\pi RC} = \dfrac{0.159}{10^4 \times 10^{-8}} = \dfrac{0.159}{10^{-4}} = 1590\ Hz$

2. $X_{C1} = X_{C2} = \dfrac{1}{2\pi f C} = \dfrac{0.159}{1.59\ kHz \times 0.01\ \mu F} = \dfrac{0.159}{1.59 \times 10^3 \times 10^{-8}} = 10\ k\Omega$

3. $X_{C3} = \dfrac{1}{2\pi f C} = \dfrac{0.159}{1.59\ kHz \times 0.02\ \mu F} = \dfrac{0.159}{1.59\ kHz \times 2 \times 10^{-8}} = 5\ k\Omega$

THE TWIN T AS AN AMPLIFIER

The twin T, like all oscillators, must have an overall gain of slightly greater than unity and it must provide an overall, in-phase 0° or 360° phase shift. At balanced conditions, the twin (or parallel) T circuit is considered to have an infinite impedance and it provides a no-phase-shift condition. Also, like the Wein-bridge, it requires a slight imbalance to provide a current flow into the load and, hence, a difference voltage that can be amplified (by the amplifier) to provide oscillations.

Unbalancing of the twin T network can occur by increasing or decreasing the input frequency or by increasing or decreasing the value of the R3 resistor. For instance, if the input frequency is increased or decreased about the operating frequency of a balanced circuit, leading and lagging 90° phase shifting will occur across the load as compared to the input of the bridge phase. Likewise, if the bridge is set at the operating frequency at balanced conditions, then increasing or decreasing the value of R3 will provide identical leading and lagging 90° phase shifting. If both varying of R3 and adjusting of frequency are used to provide phase leading, then 180° of phase leading can occur; and if both techniques are used to provide phase lagging, then 180° of phase lagging can occur. Two circuits that provide an additional 180° phase shift are shown in Figure 4-34, where both discrets and an inverting op-amp are used.

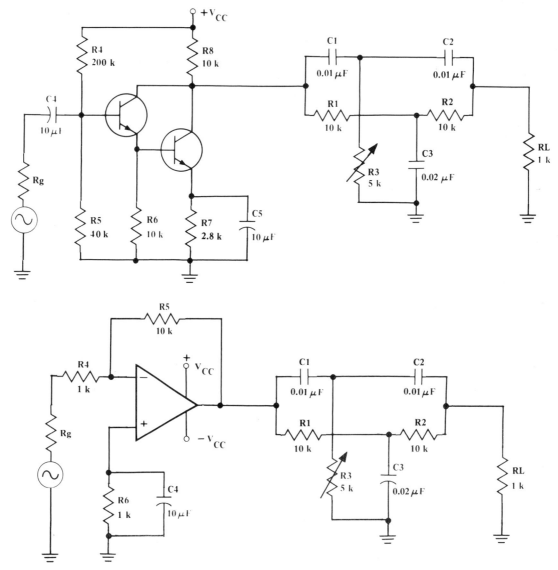

FIGURE 4-34

THE BALANCED TWIN T NETWORK

At the balanced condition, the operating frequency of the twin T circuit of Figure 4-35 is 1590 Hz, X_{C1} = X_{C2} = 10 kΩ, and X_{C3} = 5 kΩ. Therefore, the T networks have equal reactance values but they have opposite current flow which, when perfectly balanced, provides approximately no current flow into the load. Additionally, no phase shift should exist between the input and output of the twin T network. However, Point A will have a 90° phase lag and point B will have a 90° phase lead as compared to the input signal, and a 180° phase difference exists between point A and point B.

THE UNBALANCED TWIN T NETWORK

At a balanced condition, where the input frequency is at the operating frequency condition, the ouput signal across the load of the twin T network is at a minimum voltage level and it is in phase with the signal at the input. However, if the input frequency is maintained at the operating frequency and R3 is decreased below (R/2)5 kΩ, then a leading 90° phase shift will occur. Likewise, a lagging 90° phase shift in signal voltage will exist if R3 is increased above the (R/2)5 kΩ resistance. In addition, a voltage difference between point A and point B will exist, and the operating signal voltage level at the load will begin to come up out of the minimal (mull) condition.

224

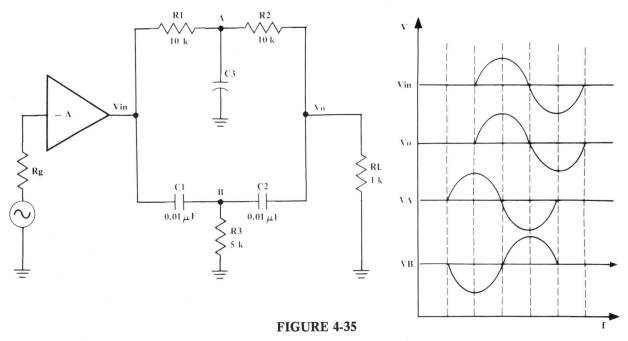

FIGURE 4-35

Too, by maintaining the balanced circuit condition and increasing the frequency will cause a leading 90° phase shift condition at the load and decreasing the frequency will cause a lagging 90° phase shift condition at the load as compared to the input of the network — similar to the R3 adjustment technique.

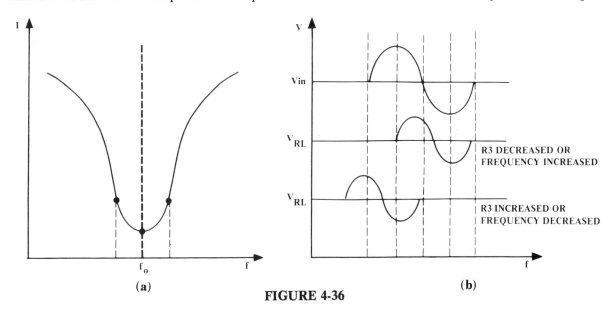

FIGURE 4-36

Therefore, if R3 is decreased, for instance, and the input frequency is adjusted to slightly above the operating frequency, then 180° phase shifting will be noted, where the output signal across the load leads the input signal by approximately 180°. Likewise, increasing resistor R3 above 5 kΩ and decreasing the frequency below the operating frequency will provide 180° of lagging phase shift across the load as compared to the input of the twin T network. Therefore, the twin T network is capable of three useful phase-shift conditions, 180°, 0°, and − 180°.

THE TWIN T CIRCUIT CONNECTED IN OSCILLATOR FORM

The twin T network is connected in oscillator form as shown in Figure 4-37. With reference to the circuit of Figure 4-37(b), adjusting variable resistor R3 to below 5 kΩ causes both conditions of 90° phase phase shift to occur, because lowering the resistance below 5 kΩ provides a 90° phase shift and, when

225

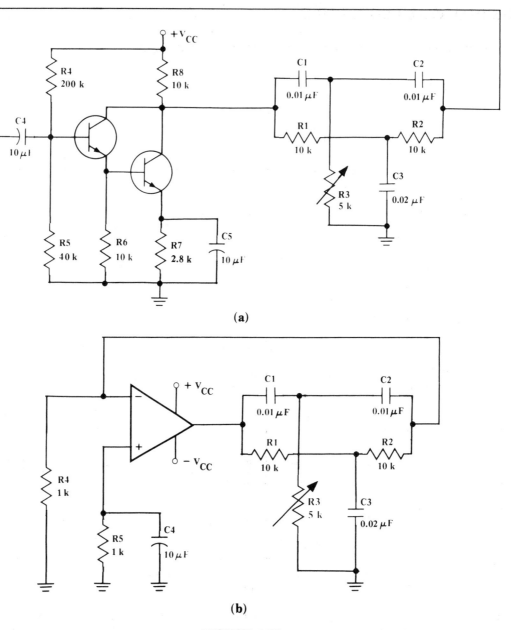

(a)

(b)

FIGURE 4-37

resistance is lowered, the frequency is also increased above the (null) balanced circuit frequency and an additional 90° occurs. Therefore, the 90° from the lowered resistance plus the 90° from the increased frequency (since the oscillator frequency is free to shift with component change, unlike the fixed frequency of the generator driven amplifier) provides 180° phase shift — the required phase shifting for oscillatory conditions. Figure 4-37(a) shows another representative twin T oscillator circuit.

EXPERIMENTAL OBJECTIVES

To investigate the twin T oscillator circuit.

LIST OF MATERIALS

1. Op-amp: LM301 (or equivalent)
2. Resistors: 1 kΩ (three)   10 kΩ (three)   10 kΩ pot (one)
3. Capacitors: 0.01 μF (two)   0.02 μF (one)   10 μF (one)

226

EXPERIMENTAL PROCEDURE

1. Connect the twin T oscillator in amplifier form as shown in Figure 4-38.

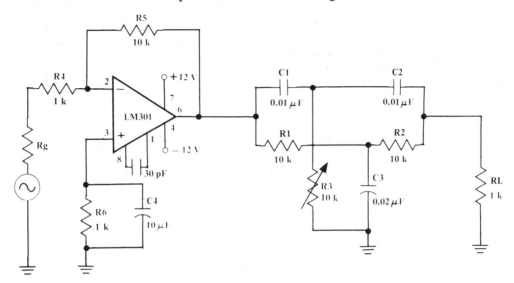

FIGURE 4-38

2. Monitor the DC voltage drops aroung the circuit.
3. Insert the measured voltage drops into Table 4-8.

TABLE 4-8	PIN 2	PIN 3	PIN 6	$+V_{CC}$	$-V_{CC}$	f_o	AT BALANCE Vop-p	R3
CALCULATED	≈ 0 V	≈ 0 V	///////	+ 12 V	− 12 V		///////	/////
MEASURED								

4. Calculate the center frequency from: $f_o = 1/(2\pi RC)$ where: R = 10 kΩ and C = 0.01 μF
5. Apply 100 mVp-p of input signal at the center frequency and monitor the voltage at the load. Adjust R3 for "best" balanced conditions, which should provide the lowest voltage condition at the load of the operating frequency. Measure f_o, Vop-p, the R3 resistor, and insert the values into Table 4.8.

NOTE: The voltage across RL at a balanced condition is obtained by adjusting both the input frequency and R3. The net result will be the smallest obtainable signal at the load that could be distorted.

6. Use a dual trace scope and monitor one signal at the input of the network (output of the device) and monitor the other at the load. Begin with a balanced condition at the operating frequency.
    a. At balanced conditions, monitor the in-phase signal condition at the load — with reference to the input signal to the network (output Pin 6 of the op-amp).
    b. Decrease resistor R3. Note the increase in amplitude and 90° phase lead conditions of the signal at the load as compared to the input signal to the network. Adjust the frequency above the operating frequency and note the additional phase shifting. A total of about 180° phase shift across the network should occur. Since the inverting op-amp provides 180° phase shift, the 360° phase shifting across the entire network at approximately the operating frequency is obtained.
7. Insert the previously calculated and measured values, as indicated, into Table 4-9

NOTE: 180° lagging phase shift can also be obtained but it is not required since it is not vital to the circuit operation. Also, 90° phase shifting can be obtained by adjusting the frequency alone when a balanced circuit is maintained. Experiment!!!

TABLE 4-9	f_o	R3 for 90°	R3 and f for 180°			f_o	Vopp
			f_o	R3			
CALCULATED		//////	//////	//////	//////	//////	//////
MEASURED	//////						

8. Connect the circuit in oscillator form as shown in Figure 4-39 and adjust R3 for an oscillatory condition.

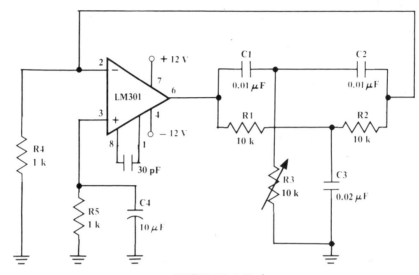

FIGURE 4-39

9. Measure the frequency and peak-to-peak voltage swing of oscillation at the output of the op-amp (Pin 6) and insert the f_o and Vop-p into Table 4-9.

228

PART 4 — THE ASTABLE NON-SINEWAVE OSCILLATOR

GENERAL DISCUSSION

The definition of astable is quasi-stable or non-stable and both the square-wave, free-running oscillator and the sawtooth-wave, relaxation oscillator fall into this category. Both are astable oscillators.

THE ASTABLE MULTIVIBRATOR

Before monolithic IC's came into common usage, multivibrators that used discrete devices were the rule. Today, discrete multivibrator circuits are still used but generally only in those areas where IC's cannot be used — such as higher power or higher frequency conditions.

The astable multivibrator circuit of Figure 4-40 is a free-running, square-wave oscillator that relies on the RC discharge of capacitor C1 through RB2, and then the discharge or capacitor C2 through RB1, in establishing the oscillating frequency. The square-wave output, at either of the Q1 or Q2 collectors, is obtained by the devices being switched on and off, between saturation and cutoff, at the oscillating frequency. There is a 180° phase difference between the Q1 and Q2 outputs.

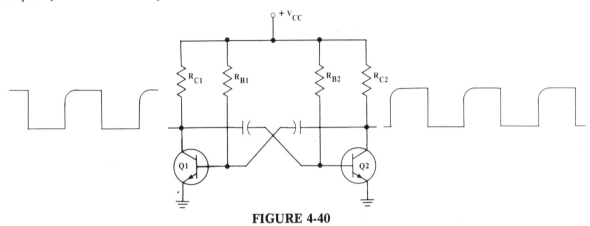

FIGURE 4-40

THE TRANSISTOR AS A SWITCH

An ideal switch allows current to flow when the switch is closed which causes all of the voltage to be developed across an in-series resistor as shown in Figure 4-41 (a), and it allows no current to flow when the switch is open and therefore no voltage is dropped across an in-series resistor as shown in Figure 4-41 (b).

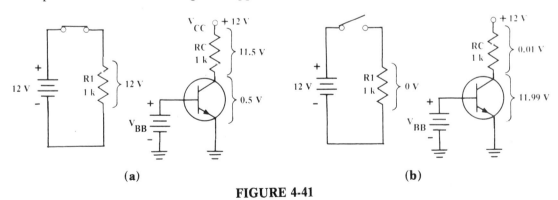

(a)   (b)

FIGURE 4-41

When a transistor is used as a switch, it operates, but at less than ideal conditions. This is because the collector-emitter of the device in the "switch on" or saturated condition will always have some voltage developed across it (although small), and the device in the "switch off" or cutoff condition will always have some reverse collector saturation current (although small) flowing through the load resistor. This is shown in Figure 4-41 where nominal $V_{CE}(SAT)$ values are about 0.5 V and nominal I_{CO} values are about 10 μA.

229

THE TRANSISTOR IN SATURATION

Turning on, or saturating, the transistor is accomplished by forward biasing the base-emitter diode junction so that the collector current flow becomes limited by the collector resistor. Any further attempt to increase the collector current, by increasing the base current for instance, provides no further collector current. Therefore, the collector current is at a maximum-saturated $I_C(SAT)$ condition. This is shown in the following examples, where the beta of the transistor is 100, the collector resistor is 1 kΩ, the $V_{CE}(SAT)$ is 0.2 volts, and the base resistor is varied.

1. In this first example, the base resistor value is selected at 190 kΩ so that the device is operated in the middle of the active region.

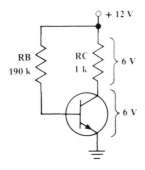

$$V_{RB} = V_{CC} - V_{BE} = 12 \text{ V} - 0.6 \text{ V} = 11.4 \text{ V}$$

$$I_B = V_{RB} / R_B = 11.4 \text{ V}/190 \text{ k}\Omega = 60 \text{ }\mu\text{A}$$

$$I_C = \beta I_B = 100 \times 60 \text{ }\mu\text{A} = 6 \text{ mA}$$

$$V_{RC} = I_C R_C = 6 \text{ mA} \times 1 \text{ k}\Omega = 6 \text{ V}$$

$$V_{CE} = V_{CC} - V_{RC} = 12 \text{ V} - 6 \text{ V} = 6 \text{ V}$$

2. In this second example, the base resistor value is lowered to 96.61 kΩ so that the base current, and hence collector current, is increased and the device is operated just at a saturated condition of 0.2 volts. In saturation, the resistance of the collector-emitter drops to about 17 ohms and the DC input resistance, at the base, to about 5.2 kΩ.

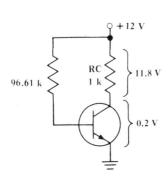

$$V_{RB} = V_{CC} - V_{BE} = 12 \text{ V} - 0.6 \text{ V} = 11.4 \text{ V}$$

$$I_B = V_{RB}/R_B = 11.4 \text{ V}/96.61 \text{ k}\Omega = 118 \text{ }\mu\text{A}$$

$$I_C = \beta I_B = 100 \times 118 \text{ }\mu\text{A} = 11.8 \text{ mA}$$

$$V_{RC} = I_C R_C = 11.8 \text{ mA} \times 1 \text{ k}\Omega = 11.8 \text{ V}$$

$$V_{CE} = V_{CC} - V_{RC} = 12 \text{ V} - 11.8 \text{ V} = 0.2 \text{ V}$$

$$R(SAT) = V_{CE}(SAT)/I_C = 0.2 \text{ V}/11.8 \text{ mA} = 17 \text{ }\Omega$$

$$R_{BE}(SAT) = V_{BE}(SAT)/I_B = 0.6 \text{ V}/118 \text{ }\mu\text{A} \approx 5.2 \text{ k}\Omega$$

3. Further reducing the base resistance to 50 kΩ increases the base current, but it cannot increase the collector current further. So, although the collector current would normally increase if a large V_{CC} were used, it cannot, and it is limited by both the voltage drop across RC and the value of the RC resistor. Also, the R(SAT) remains at about 17 ohms but the DC input resistance decreases to about 2.63 kΩ.

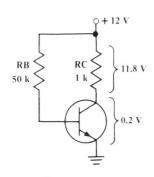

$$V_{RB} = V_{CC} - V_{BE} = 12 \text{ V} - 0.6 \text{ V} = 11.4 \text{ V}$$

$$I_B = V_{RB}/R_B = 11.4 \text{ V} /50 \text{ k}\Omega = 228 \text{ }\mu\text{A}$$

NOTE: $I_C = \beta I_B = 100 \times 22.8 \text{ }\mu\text{A} = 22.8 \text{ mA}$ is only possible if V_{CC} is high. However, for $V_{CC} = 12$ volts, the maximum collector current $I_{C(SAT)} = 11.8 \text{ mA}$. Therefore:

$$R(SAT) = V_{CE}(SAT)/I_C = 0.2 \text{ V}/11.8 \text{ mA} \approx 17 \text{ }\Omega$$

$$R_B(SAT) = V_{BE}(SAT)/I_B = 0.6 \text{ V}/228 \text{ }\mu\text{A} = 2.63 \text{ k}\Omega$$

4. Therefore, I_C(SAT) is solved from:

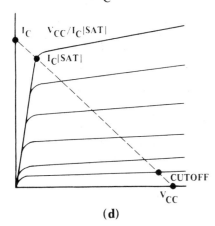

$$I_C(SAT) = \frac{V_{CC} - V_{CE}(SAT)}{R_C}$$

$$= \frac{12\text{ V} - 0.2\text{ V}}{1\text{ k}\Omega} = \frac{11.8\text{ V}}{1\text{ k}\Omega} = 11.8\text{ mA}$$

also,

$$I_C(SAT) = V_{RC}/RC = 11.8\text{ V}/1\text{ k}\Omega = 11.8\text{ mA}$$

(d)

FIGURE 4-42

THE TRANSISTOR IN CUT-OFF

Reverse biasing the base-emitter junction cuts off the flow of base current and causes the collector current to almost cease flowing. Approximately zero volts will be dropped across the RC collector resistor. The cutoff collector current flow is the reverse collector saturation current ($I_{CO} \approx 10\ \mu A$) and, since this minimal current flow cause about a zero volt drop across the collector resistor, then approximately 12 volts are dropped across the device. This is shown in the examples in Figure 4-43.

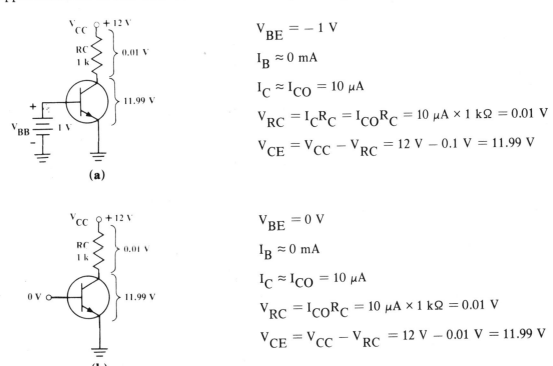

$V_{BE} = -1\text{ V}$

$I_B \approx 0\text{ mA}$

$I_C \approx I_{CO} = 10\ \mu A$

$V_{RC} = I_C R_C = I_{CO} R_C = 10\ \mu A \times 1\text{ k}\Omega = 0.01\text{ V}$

$V_{CE} = V_{CC} - V_{RC} = 12\text{ V} - 0.1\text{ V} = 11.99\text{ V}$

(a)

$V_{BE} = 0\text{ V}$

$I_B \approx 0\text{ mA}$

$I_C \approx I_{CO} = 10\ \mu A$

$V_{RC} = I_{CO} R_C = 10\ \mu A \times 1\text{ k}\Omega = 0.01\text{ V}$

$V_{CE} = V_{CC} - V_{RC} = 12\text{ V} - 0.01\text{ V} = 11.99\text{ V}$

(b)

FIGURE 4-43

NOTE: Reverse biasing the base-emitter diode junction creates less leakage current than when the base is allowed to float. The increase in leakage current is caused by the addition of ICBO, an added leakage current attributable to open base conditions.

DIRECT CURRENT ANALYSIS

The DC analysis of the astable multivibrator, like any oscillator, must be obtained when the circuit is

231

not oscillatory. Disconnecting capacitor C1 or C2, or both, will accomplish this.

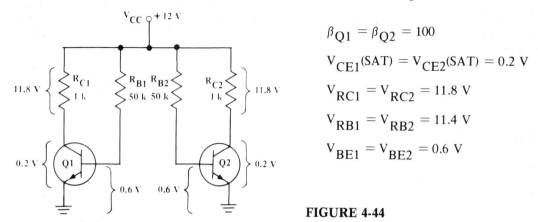

$$\beta_{Q1} = \beta_{Q2} = 100$$

$$V_{CE1}(SAT) = V_{CE2}(SAT) = 0.2 \text{ V}$$

$$V_{RC1} = V_{RC2} = 11.8 \text{ V}$$

$$V_{RB1} = V_{RB2} = 11.4 \text{ V}$$

$$V_{BE1} = V_{BE2} = 0.6 \text{ V}$$

FIGURE 4-44

The previous analysis shows that both the Q1 and Q2 stage are in saturation, where $V_{CE1} = V_{CE2} = 0.2$ V and $V_{RC1} = V_{RC2} = 11.8$ V. Therefore, the collectors of both Q1 and Q2 are at 0.2 volts. However, if a negative voltage of minus 1 volt, for example, were connected to the base of Q1, the collector voltage would increase approximately 12 volts and the device would be in cutoff. This would also be true for the indentically biased Q2 stage.

ALTERNATINC CURRENT ANALYSIS

Connecting the C1 and C2 coupling capacitors into the circuit, as shown in Figure 4-45, and turning on the power will cause an oscillatory condition to begin and, once in oscillation, the devices alternate between cutoff and saturated condions, where if Q1 is in saturation then Q2 is in cutoff and vice versa. Therefore, the square waves at the collector output of Q1 and Q2 are 180° out of phase.

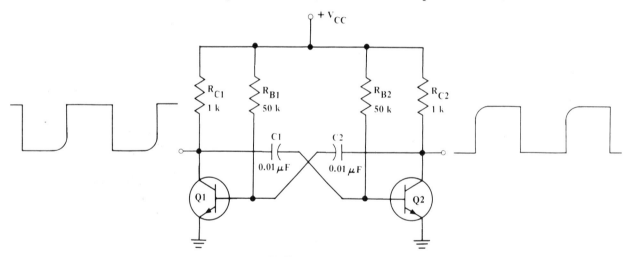

FIGURE 4-45

CIRCUIT OPERATION

The moment the power is applied to the circuit of Figure 4-46, both Q1 and Q2 go into saturation (momentarily) because of the DC biasing. At the same time, both the C1 and C2 coupling capacitors attempt to charge to V_{CC} through the relatively low resistance of the saturated devices and the collector resistors. However, only one device will actually remain in saturation, forcing the other device into cutoff.

For instance, assume that Q1 conducts first and goes into saturation. Then, the collector of Q1, at an approximate DC ground condition, ties the positive plate of the C1 coupling capacitor to an effective ground condition which makes the negative plate of the C1 coupling capacitor at a negative voltage with respect to ground. This condition forces transistor Q2 into cutoff and capacitor C1 then discharges through the base resistor of Q2 towards the positive V_{CC}. However, as the capacitor discharges from a

232

minus voltage condition (assume -10 V), it holds transistor Q2 off and would continue to charge towards $+12$ volts if the "charging" path were not intercepted by Q2 conducting and then going into saturation, at about a zero volt condition. At this same point in time, Q2 is in saturation and coupling capacitor C2 then discharges through the base resistor of Q1, forcing Q1 into cutoff. The rate of oscillation is determined by the cutoff time of the devices.

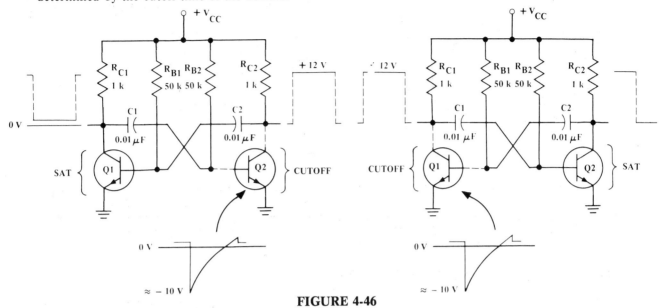

FIGURE 4-46

The charge time of capacitors C1 and C2 is usually much faster than the discharge time even if the charging and discharging begin at the same time. The charging path for the C1 capacitor, for instance, begins as Q1 comes out of saturation into cutoff and the voltage at the collector of Q1 rises from

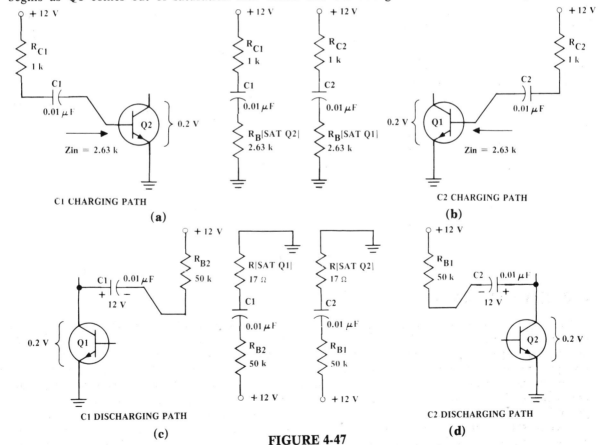

FIGURE 4-47

233

$V_{CE}(SAT)$ towards V_{CC}. At this time, Q2 is already in saturation (a condition which forced Q1 into cutoff) and capacitor C1 charges to V_{CC} (+12 V) through collector resistor R_{C1} and the saturated base resistance of transistor Q2. The charging of coupling capacitor C2 is identical to that of capacitor C1, but in this case, transistor Q2 is in cutoff, transistor Q1 is in saturation, and the capacitor charges through collector resistor R_{C2} in series with the base resistance of the saturated Q1 transistor.

The discharge path of C1 through R_{B2} +R(SAT Q1) is shown in Figure 4-47(c) along with the charging path of capacitor C1 through R_{C1} +R_B(SAT Q2) in Figure 4-47(a). In Figure 4-47(d), C2 is shown discharging through R_{B1} +R(SAT Q2), and in Figure 4-47(b), C2 is shown charging through R2 + R_B(SAT Q2). The devices in cutoff have minimal loading effect. Also, note that R_B(SAT), the imput impedance of the saturated device, is 2.63 kΩ and R(SAT), the saturated collector-emitter resistrance, is 17 ohms (as previously calculated), and the previous calculations apply for both devices.

For instance, if R1 = R2, R_{B1} = R_{B2}, R_B(SAT Q1) = R_B(SAT Q2), and R(SAT Q1) = R(SAT Q2), then the discharge time of C1 and C2 is 500 μsec and the charge time of Q1 and Q2 is about 36 μsec.

1. CHARGING TIME

    a. t(charge C1) = C1[R_{C_1} + R_B(SAT Q2)] = 0.01 μF(1 kΩ + 2.63 kΩ) = 36 μsec

    b. t(charge C2) = C2[R_{C_2} + R_B(SAT Q1)] = 0.01 μF(1 kΩ + 2.63 kΩ = 36 μsec

2. DISCHARGING TIME

    a. t(discharge C1) = C1[R_{B2} + R(SAT Q1)] = 0.01 μF(50 kΩ + 17 Ω) = 500 μsec

    b. t(discharge C2) = C2[R_{B1} + R(SAT Q2)] = 0.01 μF(50 kΩ + 17 Ω) = 500 μsec

NEGATIVE VOLTAGE CONCEPTS

If the positive side of a DC power source is tied to ground, then the negative side is negative with respect to ground. Since this is so, then a positively charged capacitor, when connected to ground in a similar manner, will also provide a negative voltage, with respect to ground, at the negative plate if the positive side is connected to ground. But unlike a power source, the capacitor will not retain the charged voltage and immediately begins to discharge through the associated circuit resistance.

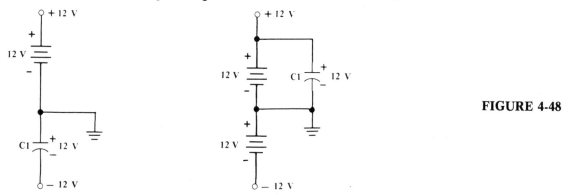

FIGURE 4-48

However, with reference to Figure 4-46, capacitors C1 and C2 will not charge or discharge from a minus 12 volt condition because, once the circuit is in oscillation, voltage (DC) averaging around the circuit will not allow it to happen. And capacitors C1 and C2 will only (charge) discharge about two thirds of V_{CC}. For a V_{CC} of 12 volts, the negative-going "sawtooth" waveshape will be about 8 volts. (This will be measured in the laboratory experiment.

FREQUENCY OF OSCILLATION

The frequency of oscillation for the symmetrical astable multivibrator is determined by the discharge paths of capacitor C1 through base resistor R_{B2} and capacitor C2 through base resistor R_{B1}. It is solved from:

$$f_o = \frac{1}{2t} = \frac{1}{2(\ell n2)RC} \quad \text{where:} \quad R = R_{B1} = R_{B2} \quad \text{and} \quad C = C1 = C2$$

234

Therefore, if $R_{B1} = R_{B2} = 50$ kΩ, $C1 = C2 = 0.01$ μF, and R(SAT) is too small to be considered, then:

$$f_O = \frac{1}{2(\ell n2)RC} = \frac{1}{2(0.693) \times 50\text{ k}\Omega \times 0.01\text{ μF}} = \frac{1}{6.931 \times 10^{-4}} = 1442.7\text{ Hz}$$

If the frequency of oscillation is determined by each discharge path separately, then it can be solved from:

$$f_O = \frac{1}{(\ell n)RC + (\ell n)RC} = \frac{1}{0.693RC + 0.693RC}$$

$$= \frac{1}{0.693 \times 50\text{ k}\Omega \times 0.01\text{ μF} + 0.6.93 \times 50\text{ k}\Omega \times 0.01\text{ μF}} = 1442.7\text{ Hz}$$

GENERATED WAVESHAPES

The voltages at the collectors of the transistors, in a symmetrical astable multivibrator, switch "on" and "off", from a saturated to a cutoff condition, at the oscillating frequency. Therefore, the waveshape, at either transistor collector, is a 0.2 volt to 12 volt square wave (as shown in Figure 4-49). And the voltage at the base of either transistor switches from about 0.6 volts at saturation to an approximate (theoretical) minus 12 volts. Then, it increases back toward 0.6 volts as the coupling capacitor charges toward the positive 12 volts of VCC.

Therefore, the coupling capacitor charge line is (theoretically) from minus 12 volts to 12 volts because, while the coupling capacitor discharges into the base resistor and becomes less and less negative as it approaches zero volt condition, it would continue to charge, nevertheless, towards V_{CC} (12 V) if it were not for the previously cutoff transistor conducting at about 0.6 volts, causing the charge line to be interrupted at that point. The blip, or overshoot, when the device is "turned on", is illustrated in Figure 4-49.

Therefore, the slope of the charge line is determined (theoretically) by the capacitor charging to about twice V_{CC} or from minus 12 volts to 12 volts, and it is mathematically given as $\ell n2 = 0.693$, from $\ell n(1 + [VBp\text{-}p/V_{CC}]) = \ell n(1 + [12\text{ V}/12\text{ V}]) = \ell n2$. However, in working circuits, the amplitude of the negative "sawtooth" wave may reach only two thirds of V_{CC} because of the voltage averaging, once the circuit is in oscillation. For instance, if $VBp\text{-}p \approx 8$ V, as shown in figure 4-49, then $\ell n(1 + [VBp\text{-}p/V_{CC}]) = \ell n(1 + [8\text{ V}/12\text{ V}]) \approx \ell n1.67 = 0.51$, and the frequency would then increase from 1442.7 Hz to 1957.6 Hz.

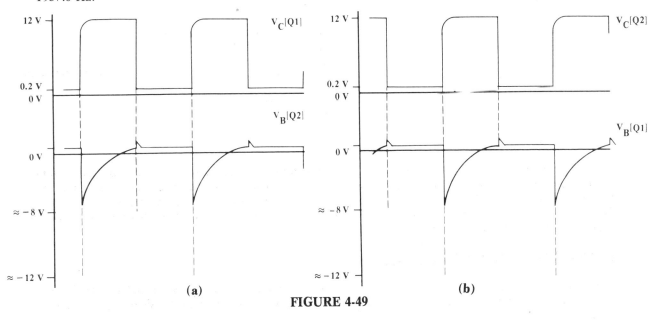

FIGURE 4-49

NOTE: The $\ell n2$ assumes that the sawtooth waveshape amplitude, at the base of either device, is 12 volts. And, while the $\ln 2$ is only approximate, it is used in literature, because only measured VBp-p can provide closer calculations.

DUTY CYCLE

The duty cycle of the symmetrical astable multivibrator is 50 percent. In other words, the output device is on for 50 percent of the time and off for 50 percent of the time. If C1 is charged so that it is 3 times larger than C2, then Q1 is on for 25 percent of the time and off for 75 percent of the time for a 25 percent duty cycle. However, Q2 then has a 75 percent duty cycle. See Figure 4-50.

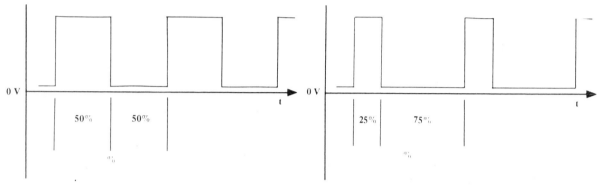

FIGURE 4-50

For instance, if C2 = 0.01 μF and C1 = 0.03 μF, then the frequency of oscillation is solved from:

$$f_o = \frac{1}{(\ln 2)R_{B1}C1 + (\ln 2)R_{B2}C2} = \frac{1}{0.693 \times 50 \text{ k}\Omega \times 0.03 \text{ }\mu\text{F} + 0.693 \times 50 \text{ k}\Omega \times 0.01 \text{ }\mu\text{F}}$$

$$= \frac{1}{3.466 \times 10^{-4} + 10.397 \times 10^{-4}} = \frac{1}{13.86 \times 10^{-4}} = 721.3 \text{ Hz}$$

Since Q2 is held in cutoff 3 times longer than transistor Q1, if the collector output of Q2 were monitored, the voltage would be plus 12 volts for three fourths of the time and Q1 would have a 75 percent duty cycle. Also, as shown in Figure 4-50, the collector output of Q1 would have a 25 percent duty cycle.

NOTE: The frequency of oscillation for the symmetrical astable multivibrator is 1442.7 Hz, and for the 75 percent duty cycle (or 25 percent duty cycle, depending on the collector output) the frequency is 721.3 Hz. Hence, any change in duty cycle will cause the oscillating frequency to change.

IDEAL SQUARE WAVE

An ideal square wave has a "sharp" rise and fall time. However, with reference to the output waveshape of Figure 4-49, notice the rounding corners during rise time. This rounding is caused by the transistor attempting to go from a saturated to a cutoff condition (0.2 V to 12 V), but it must wait for the coupling capacitor to be fully charged before the V_{CC} (12 V) can be realized. For instance is Q2 saturates, then Q1 is forced into cutoff, and the collector of Q1 rises toward plus 12 volts. However, as the collector of Q1 rises toward plus 12 volts, capacitor C1 begins to charge and, until it is fully charged, the voltage at the collector of Q1 cannot rise to plus 12 volts. The rounding-off is a consequence of this. However, if another path is provided for the C1 coupling capacitor to charge, then a sharp rise time can be obtained. An additional resistor and a diode added to the circuit, as shown in Figure 4-51, can provide this sharp rise time. In this circuit, at cutoff, transistor Q2 rises immediately to V_{CC}, turning diode D1 off, and allowing capacitor C2 to charge through resistor R1.

RISE TIME

An ideal square wave has a flat top and 90 degree sides (slopes). However, as long as some capacitance exists in a circuit, rise time (or fall time for that matter) cannot be perfect. The rise time is described in

236

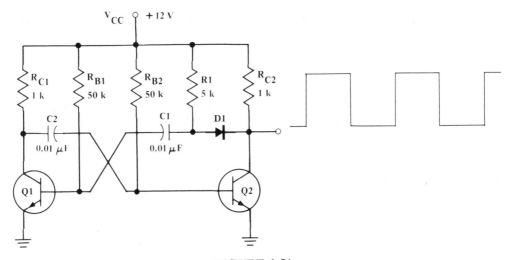

FIGURE 4-51

the literature as the time it takes for the wave to rise from 10 percent to 90 percent of the applied amplitude of the wave. For instance, if the square wave is 1 volts and the time measured on the scope between the voltage levels of 0.1 volt to 0.9 volts is 2 μsec, the rise time is 2 μsec. For a square wave amplitude where $V_{CC} = 12$ V, $V_{CE}(SAT) = 0.2$ V, and the amplitude of the wave is 11.8 volts, 10 percent of the amplitude is 1.18 volts and 90 percent is 10.62 volts. Therefore, with reference to ground, where $V_{CE}(SAT) = 0.2$ volts, the 10 percent point occurs at 1.38 volts and the 90 percent point occurs at 10.82, as shown in Figure 4-52.

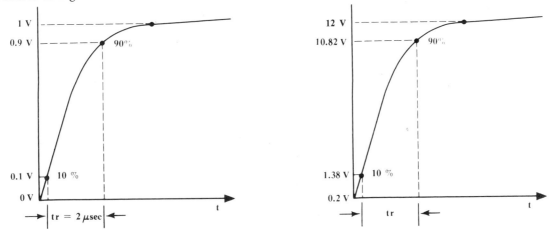

FIGURE 4-52

237

EXPERIMENTAL OBJECTIVES

To investigate the astable square wave oscillator.

LIST OF MATERIALS

1. Transistors: 2N3904 (two)   or equivalents
2. Diode: 1N4001 (one)
3. Resistors: 1 kΩ (two)   10 kΩ (one)   47 kΩ (two)
4. Capacitors: 0.01 μF (two)   0.05 μF (one)

EXPERIMENTAL PROCEDURES

1. Connect the circuit as shown in Figure 4-53.

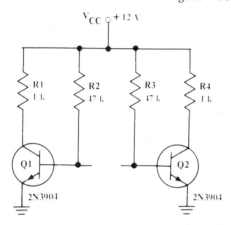

FIGURE 4-53

2. Calculate the DC voltage drops around the circuit of the non-oscillating multivibrator:

   a. Calculate the voltage drop across resistor R2 from: $V_{R2} = V_{CC} - V_{BE1}$

   b. Calculate the voltage drop across resistor R3 from: $V_{R3} = V_{CC} - V_{BE2}$

   c. Calculate the voltage drop across resistor R1 from: $V_{R1} = V_{CC} - V_{CE}(\text{SAT Q2})$

   d. Calculate the voltage drop across resistor R4 from: $V_{R4} = V_{CC} - V_{CE}(\text{SAT Q2})$

   e. Calculate the $I_B(\text{SAT})$ of Q1 from $V_{R2}/R2$.

      Calculate the $I_B(\text{SAT})$ of Q2 from $V_{R3}/R3$.

   f. Calculate the $I_C(\text{SAT})$ of Q1 from $V_{R1}/R1$.

      Calculate the $I_C(\text{SAT})$ of Q2 from $V_{R4}/R4$.

3. Measure the DC voltage drops around the circuit.
4. Measure the $I_B(\text{SAT})$ and $I_C(\text{SAT})$ of Q1 and Q2.
5. Insert the calculated and measured values into Table 4-10.

TABLE 4-10	V_{R1}	V_{R2}	V_{R3}	V_{R4}	$V_{CE}(\text{SAT})$ Q1	Q2	$V_B(\text{SAT})$ Q1	Q2	$I_C(\text{SAT})$ Q1	Q2	$I_B(\text{SAT})$ Q1	Q2
CALCULATED					░	░	░	░				
MEASURED												

6. Connect the symmetrical astable multivibrator of Figure 4-54.

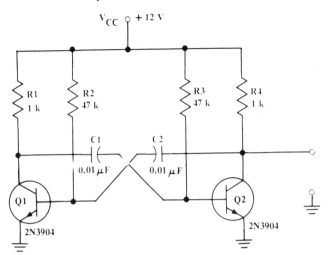

FIGURE 4-54

7. Calculate the oscillating frequency from:

$$fo = \frac{1}{2(RC\ln 2)} \approx 1/1.39RC \qquad \text{where:} \quad R = 47 \text{ k}\Omega \quad \text{and} \quad C = 0.01 \text{ }\mu F$$

8. Calculate the amplitude of the square-wave output at the collector of the Q1 and Q2 transistors from:

   a. $V\text{op-p}(Q1) \approx V_{CC} - V(\text{SAT Q1})$

   b. $V\text{op-p}(Q2) \approx V_{CC} - V(\text{SAT Q2})$

9. Calculate the amplitude of the sawtooth wave at the base of the Q1 or Q2 transistors from:

$$V_{B1} = V_{B2} \approx 0 \text{ V} - V_{CC}$$

NOTE: The theoreticl amplitude of the sawtooth wave is about equal to V_{CC} but, in a working circuit, it can be as small as 2/3 of V_{CC}. Also, once VBp-p is measured, f_0 can be recalculated to provide closer to measured results from: $f_0 = 1/RC \ln(1 + [VBp\text{-}p/V_{CC}])$.

10. Monitor and measure the square wave at the collector of Q1 and then Q2. If a dual-trace oscilloscope is available, monitor both collectors simultaneously and note the phase inversion. Also, note the rounding of the rise time and the sharpness of the fall time. Measure the rise time.

11. Monitor and measure the sawtooth wave at the base of Q1 and Q2. If a dual-trace oscilloscope is available, monitor both bases simultaneously and note the phase inversion. Also, note the blip and measure its amplitude with respect to ground potential.

12. If a dual trace oscilloscope is available, monitor the base of Q1 and the collector of Q2 and then Q1. Note how Q1 is being cut off by the negative-going "sawtooth" pulse.

13. Connect the additional resistor and diode to the collector of the Q2 transistor as shown in Figure 4-55, but where C1 = C2 = 0.01 μF as in Figure 4-54. Monitor the wave shape at the collector of Q2 and note the improvement of the rise time. Measure the rise time of both Q1 and Q2.

14. Monitor the fall time. Is there improvement or no improvement? Comment.

15. Insert the calculated and measured values into Table 4-11. Sketch the waveshapes at the base of Q1 and at the collector of Q1 where the improved rise-time circuitry was used. Include the appropriate on and off times in microseconds.

TABLE 4-11	f_0	Vop-p Q1	Q2	Vp-p(base) Q1	Q2	RISE TIME Q1	Q2	DUTY CYCLE Q1	Q2	Diode R.T. Q1	Q2
CALCULATED						//////	//////			//////	//////
MEASURED											

239

16. Connect the circuit as shown in Figure 4-55, where the circuitry for improving rise time is used.

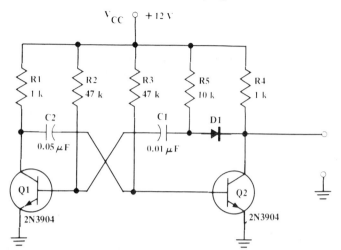

FIGURE 4-55

17. Calculate the frequency of oscillation from:

$$fo = \frac{1}{R3C2 \ln 2 + R2C1 \ln 2}$$

18. Calculate the duty cycle of the wave shapes at the collectors of Q1 and Q2.
19. Measure the duty cycle and amplitudes of the wave shapes at the collectors of Q1 and Q2. Record the frequency.
20. Measure the sawtooth-wave ampitudes at the base of Q1 and Q2.
21. Sketch the waveshapes at the collectors of Q1 and Q2. Include the appropriate on and off times in microseconds. Measure the rise time of the waveshapes.
22. Insert the calculated and measured values, as indicated, into Table 4-12.

TABLE 4-12	f_o	Vop-p		DUTY CYCLE		RISE TIME		Vp-p(base)	
		Q1	Q2	Q1	Q2	Q1	Q2	Q1	Q2
CALCULATED		////	////			////	////	////	////
MEASURED									

240

PART 5 — THE OP-AMP DRIVEN ASTABLE MULTIVIBRATOR

GENERAL DISCUSSION

The op-amp driven astable multivibrator uses an op-amp, a capacitor, and three resistors to provide a square wave output that is only limited in amplitude by the positive and negative saturation voltages of the op-amp. With reference to the circuits of Figure 4-56, by changing the ratio of R2 to R3, the frequency can be changed. For instance, when R2 = R3 = 10 kΩ the voltage divider ratio of R2 to R3 allows C1 to charge to 1/2 Vop-p. However, if R2 is reduced to 5 kΩ C1 can charge to 2/3 Vop-p, and the lowered charging time decreases the operating frequency. Likewise, if R3 is decreased the charging time is decreased and the frequency increased. Also, the frequency can be changed by changing either R1 or C1. However, both circuits of Figure 4-56 maintain a 50 percent duty cycle.

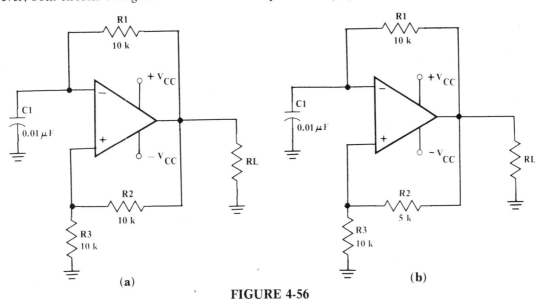

FIGURE 4-56

DEVICE OPERATION

If the device were operated in a negative feedback condition only, as shown in Figure 4-57(a), where resistors R1 and R2 are connected to the inverting input terminal of the op-amp and balancing resistor R3 is connected to the non-inverting terminal of the op-amp, then zero volts would exist at pins 2, 3, and also at pin 6 if perfect balancing is obtained.

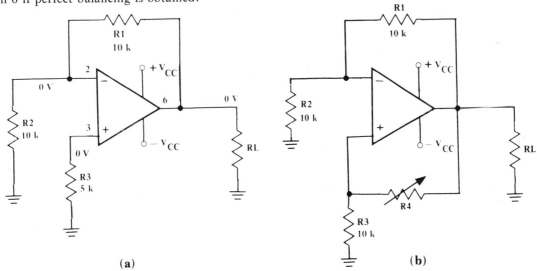

FIGURE 4-57

241

However, it the difference voltage at the input terminals has an off-set voltage, then a positive or negative DC voltage will exist on the output terminals. If the difference voltage is large, then the output will be forced into saturation either towards the positive or negative supply voltage. Too, remember that the difference voltage does not have to be too large because of the high (DC) open loop gain of the op-amp.

For instance, if the inverting terminal is made more negative than the non-inverting terminal, then a positive voltage (approaching $+V_{CC}$) will exist at the output terminals. If the inverting terminal is made more positive than the non-inverting terminal, then a negative voltage (approaching $-V_{CC}$) will exist at the output terminals. A method to provide an off-set voltage is to connect a feedback resistor from the output terminal to the non-inverting terminal, as shown in Figure 4-57(b). Then, when power is switched on and the output goes to a positive voltage condition, the voltage at the inverting terminal must be negative with respect to the non-inverting terminal. Adjusting the value of any resistor in the "bridge" can assure this. Likewise, a negative voltage at the output will require a positive voltage at the inverting terminal, with reference to the non-inverting terminal. Remember that the non-inverting terminal provides the op-amp gain reference.

CIRCUIT OPERATION OF THE ASTABLE CIRCUIT

The moment power is turned on to the circuit shown in Figure 4-58, the voltage across capacitor C1 should be low and the output of the device should switch to a positive (approaching $+V_{CC}$) voltage condition. If it is assumed that this happens, then the C1 capacitor will charge towards a positive voltage condition and will continue to charge until the voltage becomes slightly larger than the voltage across the R3 resistor because, at this point in time, the inverting terminal is more positive than the non-inverting terminal and the output voltage switches to a negative (approaching $-V_{CC}$) voltage condition.

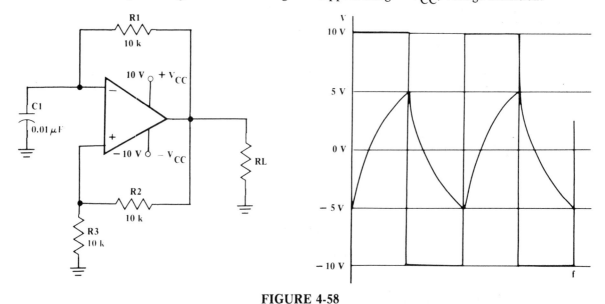

FIGURE 4-58

Therefore, the C1 capacitor must now discharge from a positve ($V_{CC}/2$) voltage condition toward an approximate minus V_{CC} voltage condition. However, once the capacitor charging towards approximately minus V_{CC} becomes slightly more negative than the non-inverting terminal ($-V_{CC}/2$), then the output switches back to a positive voltage and another cycle begins.

For instance, if the output of the device switches between ± 10 volts, then the voltage at the inverting terminal is limited to about ± 5 volts. Therefore, when capacitor C1 charges slightly above 5 volts, the output switches negative because of the inverting amplification of the op-amp. When capacitor C1 charges slightly below minus 5 volts, the output switches to positive. The capacitor's charging and discharging wave and the square wave switching of the output is shown in Figure 4-58.

FREQUENCY OF OSCILLATION

The frequency of oscillation for the multivibrator of Figure 4-58 is: $f_0 = \dfrac{1}{2\left[R_1 C_1 \ln\left(\dfrac{R2 + 2R3}{R2}\right)\right]}$

242

It is derived from the fact that the charging path of C1 is limited by the voltage divider circuit of R2 and R3. For instance, if R1 = 10 kΩ, C1 = 0.01 μF, and R2 = R3 = 10 kΩ, then the frequency of oscillation is:

$$f_o = \frac{1}{2\left[R_1 C_1 \ln\left(\frac{R2 + 2R3}{R2}\right)\right]} = \frac{1}{2\left[10\ k\Omega \times 0.01\ \mu F \times \ln\left(\frac{10\ k\Omega + 2[10\ k\Omega]}{10\ k\Omega}\right)\right]}$$

$$= \frac{1}{2[10^4 \times 10^{-8} \ln(30\ k\Omega/10\ k\Omega)]} \approx \frac{10^4}{2 \times 1.1} = 4.55\ kHz$$

VARYING THE FREQUENCY

With reference to the circuit of Figure 4-59, if resistor R2 is changed to 5 kΩ the voltage at the junction of R2 and R3 is at either 6.6 volts or minus 6.6 volts (because of the R2 and R3 voltage divider network), and capacitor C1 must charge and discharge to those voltage levels. Therefore, increased charging time and, then, discharging time means decreased frequency. Since the charge-discharge times are the same (although larger in amplitude), the duty cycle remains at 50 percent.

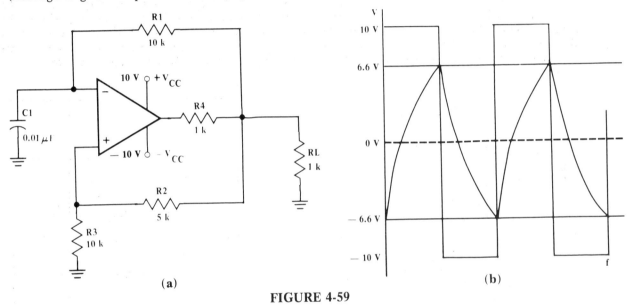

FIGURE 4-59

The frequency of oscillation for the op-amp multivibrator of Figure 4-59 is:

$$f_o = \frac{1}{2\left[R_1 C_1 \ln\left(\frac{R2 + 2R3}{R2}\right)\right]} = \frac{1}{2\left[10\ k\Omega \times 0.01\ \mu F \times \ln\left(\frac{5\ k\Omega + 2 \times 10\ k\Omega}{5\ k\Omega}\right)\right]}$$

$$= \frac{1}{2(10^4 \times 10^{-8} \times \ln \times 25\ k\Omega/5\ k\Omega)]} = \frac{10^4}{2 \times 1.61} \approx 3.11\ kHz$$

However, if R3 is changed to 5 kΩ and R2 remains at 10 kΩ, the voltage at the junction of R2 and R3 would drop to 3.3 volts and minus 3.3 volts. The charge and discharge time would decrease, causing the frequency to increase. The frequency of oscillation would be:

$$f_o = \frac{1}{2\left[R_1 C_1 \ln\left(\frac{R2 + 2R3}{R2}\right)\right]} = \frac{1}{2\left[10\ k\Omega \times 0.01\ \mu F \times \ln\left(\frac{10\ k\Omega + 2 \times 5\ k\Omega}{10\ k\Omega}\right)\right]}$$

$$= \frac{1}{2(10^4 \times 10^{-8} \times \ln \times 20\ k\Omega/10\ k\Omega)} = \frac{10^{-4}}{2 \times 0.693} = 7213.5\ Hz$$

243

The circuit and the amplitude of the output and the charging-discharging waves, for this latter example, are shown in Figure 4-60.

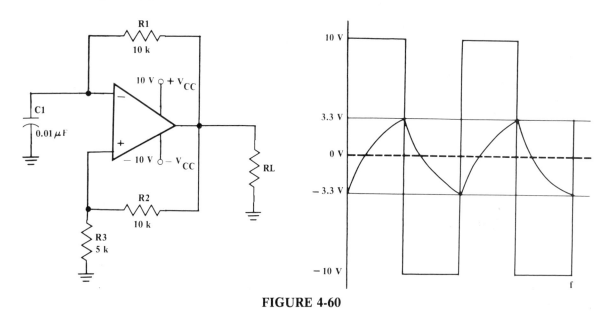

FIGURE 4-60

DUTY CYCLE

The charge-discharge time of the three previously discussed circuits is 50 percent because the magnitudes of the charge and discharge waves are equal. Only the frequency was changed by varying either R2 or R3 or by varying C1 or R1. However, duty cycle can be changed if the charge or discharge magnitudes are made unequal — where the power supply votages of plus volts and minus volts are either increased or decreased.

EXPERIMENTAL OBJECTIVES

To investigate the op-amp driven astable multivibrator.

LIST OF MATERIALS

1. Op-amp: LM301 or equivalent
2. Resistors: 1 kΩ (one)   2.2 kΩ (one)   4.7 kΩ (three)
3   Capacitor: 0.01 μF (one)

EXPERIMENTAL PROCEDURE

1. Connect the op-amp driven astable multivibrator as shown in **Figure 4-61**.

2. Calculate the approximate frequency of oscillation from: $f_o = \dfrac{1}{2\left[R_1 C_1 \, \ell n \left(\dfrac{R2 + 2R3}{R2} \right)\right]}$

3. Calculate the duty cycle from: Duty Cycle ≈ 50%, if ± power supplies are equal.

4. Calculate the amplitude of the square wave at the output (Pin 6) of the device from: $\approx \pm V_{CC}$.

5. Calculate the amplitude of the charging-discharging waveshape across C1 from: $\dfrac{2V_{CC} \times R3}{R2 + R3}$

6. Monitor the square wave at the output of the device.
   a. Measure the amplitude of the square wave.

244

b. Measure the frequency of oscillation.
c. Measure the duty cycle.

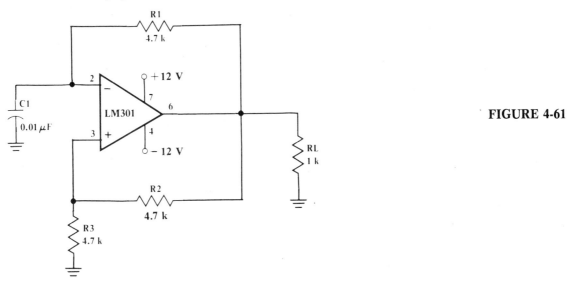

FIGURE 4-61

NOTE: The 30 pF capacitor is not used between pins 1 and 8 because it would effect the high frequency rolloff and minimize rise time.

7. Monitor the charging-discharging waveshape across capacitor C1 — between Pin 2 and ground.
   a. Measure the amplitude.
   b. Measure the charge-discharge ratio across the C1 capacitor.
8. Sketch the waveshapes, comparing the magnitude and timing. Use a dual-trace oscilloscope, if available, to monitor both waves simultaneously. Measure at the output and across capacitor C1.
9. Insert the calculated and measured values, as indicated, into Table 4-13.

TABLE 4-13	DEVICE (Pin 6)			V_{C1}p-p	ACROSS C1 CHARGE-DISCHARGE RATIO
	f_O	Duty Cycle	Vop-p		
CALCULATED					
MEASURED					

10. Connect a circuit where R2 = 2.2 kΩ and R3 = 4.7 kΩ, as shown in Figure 4-62.

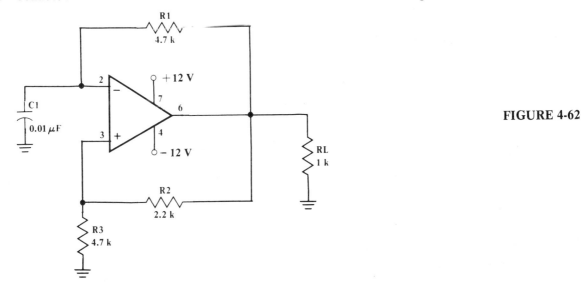

FIGURE 4-62

245

11. Calculate the approximate frequency of oscillation from

$$f_o = \frac{1}{2R_1C_1\ln([R2 + 2R3]/R2)} \qquad \text{where:} \quad R2 = 2.2\ k\Omega \quad \text{and} \quad R3 = 4.7\ k\Omega$$

12. Calculate the duty cycle at the output from: Duty Cycle ≈ 50%, if ± power supplies are equal.

13. Monitor the waveshape at the output of the device.
    a. Measure the amplitude.
    b. Measure the frequency of oscillation.
    c. Measure the duty cycle.
14. Monitor the amplitude of the charging-discharging wave.
    a. Measure the amplitude at the junction of R2 and R3 and, also, across C1.
    b. Measure the charge-discharge ratio across the C1 capacitor. Measure the duty cycle across R3.
15. Sketch the waveshape, comparing the magnitude and time.
16. Insert the calculated and measured values into Table 4-14.

TABLE 4-14	DEVICE (Pin 6)			VR3(Pin 3)			CAPACITOR C1	
	f_o	Vop-p	Duty Cycle	Vop-p	Duty Cycle	$V_{C1}p\text{-}p$		Charge-Discharge Ratio
CALCULATED		/////		/////	/////	/////	/////	/////
MEASURED								

17. Connect a circuit where R2 = 4.7 kΩ and R3 = 2.2 kΩ, as shown in Figure 4-63.

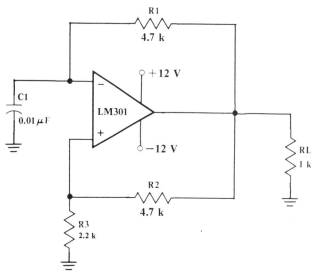

FIGURE 4-63

18. Calculate the approximate frequency of oscillation from:

$$f_o = \frac{1}{2R_1C_1\ln([R2 + R3]/R2)} \qquad \text{where:} \quad R3 = 2.2\ k\Omega \quad \text{and} \quad R2 = 4.7\ k\Omega$$

19. Calculate the duty cycle at the output from: Duty Cycle ≈ 50%, if ± power supplies are equal.

20. Monitor the waveshape at the output of the device.
    a. Measure the amplitude.
    b. Measure the frequency of oscillation.
    c. Measure the duty cycle.
21. Monitor the amplitude of the charging-discharging wave.
    a. Measure the amplitude at the junction of R2 and R3 and, also, across C1.

b. Measure the charge-discharge ratio across the C1 capacitor, and the duty cycle across R3.
22. Sketch the waveshape, caparing the magnitudes and time.
23. Insert the calculated and measured values, as indicated, into Table 4-15.

TABLE 4-15	f_o	DEVICE (Pin 6)		V_{R3} (Pin 3)			C1 Capacitor	
		Vop-p	Duty Cycle	Vop-p	Duty Cycle	V_{C1}p-p	Charge-discharge Ratio	
CALCULATED		/////		/////	/////	/////	/////////	
MEASURED								

PART 6 — THE 555 TIMER IN THE ASTABLE MULTIVIBRATOR MODE

GENERAL DISCUSSION

A 555 timer is a versatile IC that requires only a single-ended power supply and few components to provide a square-wave output capable of fixed or variable duty cycle.

Essentially, the "guts" of the 555 timer are comprised of two high-gain comparators, a resistor string that provides the reference voltages, and an output stage that can be switched from approximately zero volts to V_{CC} volts. In addition, there is a discharge path, by way of a switchable transistor that makes possible single power supply capabilities and, also, there is reset circuitry which, in the astable mode, is disabled by tying the terminal high to V_{CC}. An equivalent circuit is shown in Figure 4-64.

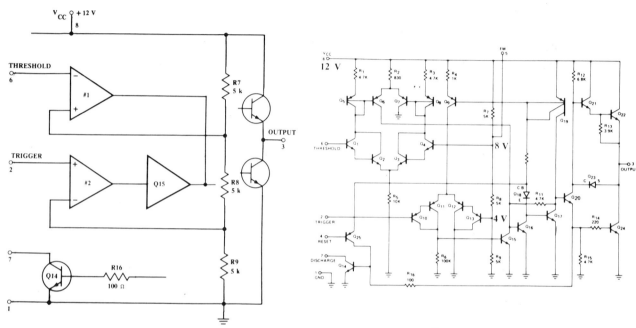

FIGURE 4-64   *Courtesy of Signetics Corporation*

With reference to the circuit of Figure 4-64, if the V_{CC} voltage is 12 volts, then the voltage at the base of Q4 (because of the three 5 kΩ resistors) is at 2/3 the applied V_{CC} or 8 volts, and the voltage at the base of Q13 is at 1/3 of the applied V_{CC} or 4 volts. Therefore, if more than 8 volts are applied to the threshold (Pin 6), Q1 is turned on, the upper comparator #1 switches states, and the output is switched to a zero voltage condition. If less than 4 volts is applied to the trigger (Pin 2), Q10 is turned on, the lower comparator #2 switches states, and the output is switched to an approximate V_{CC} voltage condition. Note that both comparators tie into Q15 and are maintained 180° removed from each other by the collector-base connections. Also, when the output is switched high to V_{CC}, Transistor Q14, which is connected to Pin 7, is switched into cutoff. When the output is switched to approximately ground condition, transistor Q14 is switched into saturation.

CIRCUIT OPERATION

When the 555 timer is connected as an astable multivibrator, as shown in Figure 4-65(a), it will produce a square wave output that switches between approximately zero volts to 12 volts at a frequency of about 4.8 kHz. The duty cycle will be about 66 percent and the amplitude of the charging and discharging voltage across capacitor C1 is about 4 volts, as shown in Figure 4-65(b).

When the power is applied to the circuit of Figure 4-65, capacitor C1 charges toward V_{CC} through resistors R1 and R2 and, because (initially) the voltage across C1 is low, the voltage at the output (Pin 3) is switched to approximately V_{CC} (+ 12 V). However, when the capacitor charges to slightly greater than 8 volts, comparator #1 is activated and the voltage at the output (Pin 3) is switched to about zero volts. This, in turn, forces the discharge transistor (Q14) to be switched into saturation and capacitor C1 to

248

discharge through resistor R2 toward ground condition. However, when the discharging voltage across capacitor C1 discharges to slightly lower than 4 volts, comparator #2 is activated, switching the output (Pin 3) high. This forces transistor Q14 into cutoff and, again, capacitor C1 begins to charge toward V_{CC}, beginning a new cycle. The charging and discharging paths are indicated in Figure 4-65(a).

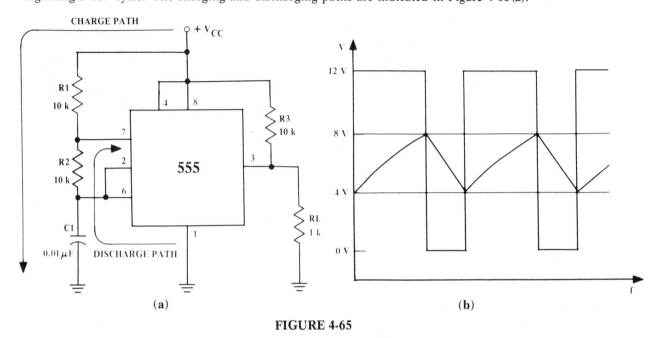

FIGURE 4-65

NOTE: Both the trigger and threshold connections (Pins 2 and 6) are connected to the charging and discharging C1 capacitor and Pin 4 (the reset circuit) is disabled by connecting it to V_{CC}. Additionally, pull-up resistor R3 has been added (although it is not totally necessary to circuit operation) to minimize the effect the output circuit capacitance has on rise time.

CIRCUIT ANALYSIS

1. The frequency of oscillation is solved from:

$$f_o = \frac{1}{(\ell n2 \times R_2C_1) + ([\ell n2][R1 + R2]C1)}$$

$$= \frac{1}{(\ell n2 \times 10 \text{ k}\Omega \times 0.01 \ \mu\text{F}) + ([\ell n2][10 \text{ k}\Omega + 10 \text{ k}\Omega]0.01 \ \mu\text{F})}$$

$$= \frac{1}{(0.693 \times 10^{-4}(\ + (0.693 \times 2 \times 10^{-4})} = \frac{1}{2.08 \times 10^{-4}} = 4.8 \text{ kHz}$$

NOTE: The capacitor charges through resistors (R1 + R2) but discharges through resistor R2 only.

2. The duty cycle is solved from:

$$\text{Duty Cycle} = \frac{(R1 + R2)100}{R2 + (R1 + R2)} = \frac{(10 \text{ k}\Omega + 10 \text{ k}\Omega)100}{10 \text{ k}\Omega + 20 \text{ k}\Omega} = 66.7\%$$

NOTE: Since the charge time is twice the discharge time, the output terminal voltage is at VCC or 2/3 of the total on time for a duty cycle of 66.7%.

249

DUTY CYCLE FLEXIBILITY

The main drawback to the circuit of Figure 4-65 is that the duty cycle can never go below 50% because the charge time is through both resistors and the discharge time is through resistor R2 only. However, connecting a diode across R2 so that it is forward biased during charging conditions, as shown in Figure 4-66, allows capacitor C1 to charge through the R1 resistor and not through R2, since R2 is bypassed by the almost nonexistent resistance of the forward biased D1 diode. During discharge time, diode D1 is reverse biased and capacitor C1, as usual, discharges through R2 to ground. Therefore, if R1 = R2 = 10 kΩ, the duty cycle is about 50% and the frequency of oscillation is solved like any symmetrical astable multivibrator.

$$f_o = \frac{1}{(\ln 2 \times R_1 C_1) + (\ln 2 \times R_2 C_1)} = \frac{1}{2(\ln 2 \times RC)} = \frac{1}{2(0.693 \times 10 \text{ k}\Omega \times 0.01 \text{ }\mu\text{F})} \approx 7.2 \text{ kHz}$$

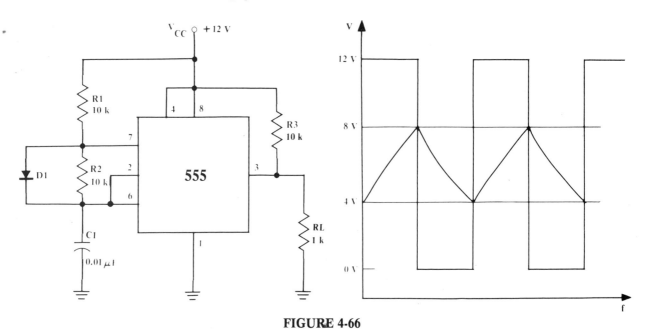

FIGURE 4-66

Further duty cycle flexibility can be achieved by varying the resistor values of either R1 or R2. For instance, if R1 were lowered to 5 kΩ, the duty cycle would go to 33.3%, and if R2 were lowered to 5 kΩ, the duty cycle would go to 66%. In both cases, the frequency would change with the duty cycle variability, although (in this case) they would both be the same. Therefore, if R1 were replaced by a 100 kΩ variable resistor, where R2 is still in parallel with a forward biased diode, the duty cycle can be varied from approximately less than 10% to greater than 90%. However, a constant frequency is not maintained. Further, if a constant duty cycle is to be maintained, where the frequency is varied, then a dual-ganged potentiometer is used, where variable resistances for R1 and R2 provide a symmetrical 50% duty cycle waveshape.

EXPERIMENTAL OBJECTIVES

To investigate the 555 timer in the astable multivibrator mode.

LIST OF MATERIALS

1. Integrated Circuit: 555 Timer
2. Diode: 1N4001
3. Resistors: 1 kΩ (two)   4.7 kΩ (two)   10 kΩ (one)
4. Capacitors: 0.01 μF   0.1 μF   (one each)

EXPERIMENTAL PROCEDURES

1. Connect the 555 timer as an astable multivibrator as shown in Figure 4-67.

250

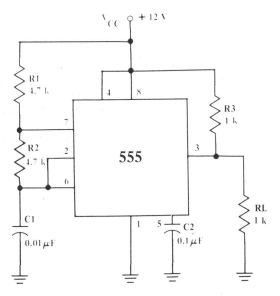

FIGURE 4-67

2. Calculate the approximate frequency of oscillation from:

$$f_o = \frac{1}{(\ln 2 \times R_2 C_1) + ([\ln 2][R1 + R2]C1)} \qquad \text{where:} \quad R1 = R2 = 4.7 \text{ k}\Omega$$

3. Calculate the duty cycle from: Duty Cycle $= \dfrac{R1 + R2}{R1 + 2R2} \times 100$

4. Calculate the square-wave amplitude at the output terminal at approximately V_{CC}.

5. Calculate the approximate charging-discharging waveshapes across capacitor C1 from:

$$2/3\ V_{CC} - 1/3\ V_{CC}$$

6. Monitor the output square wave at the Pin 3 output terminal.
    a. Measure the amplitude of the square wave.
    b. Measure the frequency of oscillation where $f_o = 1/t$ and both the on and off time are included.
    c. Measure the duty cycle.
7. Monitor the charging-discharging waveshapes across capacitor C1, between Pin 6 and Pin 1 (ground).
    a. Measure the amplitude.
    Measure the charge discharge ratio.
    c. Compare the on-off duty cycle to the charge-discharge ratio. Use a dual-trace oscilloscope.
    d. Sketch the waveshape, comparing the amplitude and timing.
8. Insert the measured and calculated values, as indicated, into Table 4-16.

TABLE 4-16	f_o	Vo Duty Cycle	Vop-p	V_{C1}p-p	V_{C1} Charge-Discharge Ratio
CALCULATED				/////	/////
MEASURED					

9. Connect the 555 astable multivibrator circuit where a forward biased diode is connected directly across the R1 resistor, as shown in Figure 4-68.

10. Calculate the approximate frequency of oscillation from: $f_o = \dfrac{1}{(\ln 2 \times R_1 C_1) + (\ln 2 \times R_2 C_1)}$

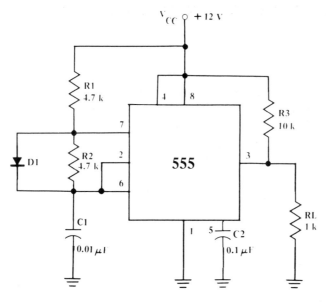

FIGURE 4-68

11. Calculate the duty cycle from: Duty Cycle $\approx \dfrac{R1 \times 100}{R1 + R2}$

12. Monitor the output square wave at the output terminal (Pin 3).
   a. Measure the amplitude of the square wave.
   b. Measure the frequency of oscillation.
   c. Measure the duty cycle.
13. Monitor the charging-discharging slope across capacitor C1.
   a. Measure the amplitude.
   b. Measure the charge-discharge ratio.
   c. Compare the on-off duty cycle to the charge-discharge ratio. Monitor on a dual-trace oscilloscope.
14. Sketch the waveshape, comparing the amplitude and timing.
15. Insert the calculated and measured values, as indicated, into Table 4-17.

TABLE 4-17	f_o	V_o Duty Cycle	Vop-p	V_{C1}p-p	V_{C1} Charge-Discharge Ratio
CALCULATED			/////////	/////////	/////////////////////
MEASURED					

16. Monitor the output or the charging-discharging waveshapes, or both, using a dual-trace oscilloscope. Momentarily, connect a 10 kΩ resistor in parallel with the R1 resistor and then the R2 resistor. Note the duty cycle change.
17. Futher investigate the capabilities of the 555 timer as a variable frequency, constant duty cycle, astable multivibrator. Both circuits are discussed under **DUTY CYCLE FLEXIBILITY** in the theory section of this exercise.

NOTE: The addition of capacitor C2, on Pin 5, minimizes jitter to the output square wave, since Pin 5 is the FM input and any signal input (even noise or hum) can cause the output signal to be frequency modulated.

PART 7 — THE UJT DRIVEN RELAXATION OSCILLATOR

GENERAL DISCUSSION

A UJT driven relaxation oscillator will produce a sawtooth output waveshape that can be improved with a constant current source. The symbol and the equivalent circuit for the UJT are shown in Figure 4-69 (a). The static characteristic curve is shown in Figure 4-69(b). Note that a peak point, a valley point, and a saturation point are indicated. also, note that the area between Vp and V_V is a negative resistance region.

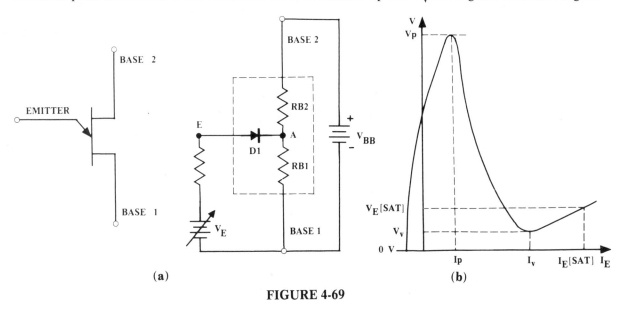

(a) **(b)**

FIGURE 4-69

DEVICE OPERATION

With reference to the circuit of Figure 4-69, if a voltage is applied across Base 1 and Base 2, where V_E is kept relatively low, a voltage at point A, as determined by the internal voltage divider resistance of RB1 and RB2 occurs. For instance, if V_{BB} is 12 volts and RB1 = RB2 = 2 kΩ, then VA = 6 volts. The junction at VA also describes the intrinsic standoff ratio which, essentially, defines the ratio of RB1 to (RB1 + RB2) and, in this case, it is 0.5.

Now, if the voltage V_E is increased to 6.7 volts so that diode D1 conducts, the voltage at the emitter of the unijunction transistor is at Vp and emitter current flows through the RB1 resistance to ground. Also, at this point in time, the resistance of RB1 begins to decrease, and continues to decrease, until it reaches saturation. Then the resistance of RB1 stops decreasing, which coincides with the valley emitter current flow (I_V) and minimum emitter voltage (V_V). Note that the series resistance must be large enough so that the voltage at the emitter will not be clamped by the power source.

THE BASIC UJT RELAXATION OSCILLATOR

The basic relaxation oscillator configuration, along with the typical waveforms across capacitor C1 and the B1 and B2 terminals are shown in Figure 4-70.

CIRCUIT OPERATION

The moment power applied to the circuit of Figure 4-70, capacitor C1 begins to charge towards V_{BB} through resistor R3. At that same time, a reference voltage is established at point A, because of the voltage divider of (RB1 + R1) and (RB2 + R2). Therefore, when the charging voltage across the capacitor equals the sum of the voltage at point A plus the diode voltage of approximately 0.7 volts, the diode becomes forward biased and the capacitor discharges through the low resistance of RB1 and R1. The capacitor continues to discharge until the voltage across C1 becomes so low that the diode no longer remains forward biased. Then, the discharge path is opened and the C1 capacitor begins to charge toward V_{BB} through resistor R3 and another cycle begins. Both point A and the equivalent circuit of diode D1 are shown in Figure 4-69.

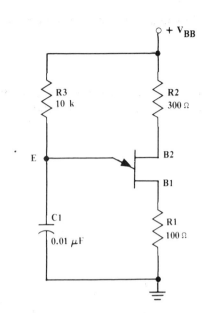 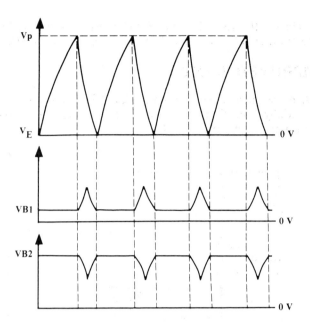

FIGURE 4-70

FREQUENCY OF OSCILLATION

The frequency of oscillation for the basic UJT relaxation oscillator is solved from:

$$f_o = \frac{1}{R_3 C_1 (\ell \eta 1/[1-\eta])} = \frac{1}{R_3 C_1 (\ell \eta 1/[1-0.5])} = \frac{1}{10 \text{ k}\Omega \times 0.01 \text{ }\mu\text{F} \times \ell n 2}$$

$$= \frac{1}{10^4 \times 10^{-8} \times 0.693} \approx 14.4 \text{ kHz}$$

2. The intrinsic standoff ration (η) determines the voltage at the junction of RB1 and RB2 and is solved like any voltage divider is solved. For instance, if RB1 = RB2 = 2 kΩ, then:

$$\eta = \frac{RB1}{RB1 + RB2} = \frac{2 \text{ k}\Omega}{2 \text{ k}\Omega + 2 \text{ k}\Omega} = 0.5$$

NOTE: The intrinsic ratio of most UJT's is generally between 0.4 and 0.8. The instrinsic ratio for the 2N2646 UJT is given as being between 0.56 and 0.75 with 0.65 being nominal.

CONDITIONS FOR OSCILLATION

In order for the UJT circuit to oscillate, the operating emitter current of the device must fall within the extreme current conditions of Ip and I_v. Therefore, R3 must be chosen accordingly, even though the choice is not critical.

For instance, the 2N2646 has a typical Ip of 0.4 μA and a typical I_v of 6 mA and the operating emitter currents must stay within theses extreme limitations.

1. Therefore, if the peak voltage is 6.7 volts, the R3 resistor value is 10 kΩ, and V_{BB} = 12 volts, Ip is:

$$Ip = \frac{V_{BB} - Vp}{R3} = \frac{12 \text{ V} - 6.7 \text{ V}}{10 \text{ k}\Omega} = 530 \text{ }\mu\text{A}$$

2. And solving for I_v, using the same R3 resistor, when V_v is at about 2 volts:

$$I_v = \frac{V_{BB} - V_v}{R3} = \frac{12 \text{ V} - 2 \text{ V}}{10 \text{ k}\Omega} = 1 \text{ mA}$$

3. Hence, the Ip, using the 10 kΩ resistor, is well within the devices limits and the calculations would indicate that R3 can be much larger, approaching 1 MΩ.

IMPROVED RAMP

A more linear ramp can be achieved if a constant current source, as shown in Figure 4-71, is added. Essentially, capacitor C1 charges into the high impedance of the collector of Q1 at a constant current condition as determined by Q1.

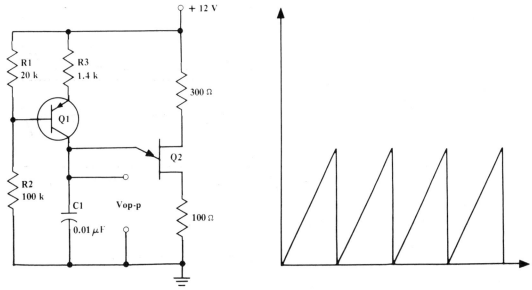

FIGURE 4-71

1. $V_{R1} = (VCC \times R1)/(R1 + R2) = 2$ volts   3. $V_{R3} = V_{R1} - V_{BE1} = 1.4$ volts

2. $V_{R2} = (V_{CC} \times R2)/(R1 + R2) = 10$ volts   4. $I_{E1} = V_{R3}/R3 = 1$ mA

For instance, with reference to the circuit of Figure 4-71, if $I_{E1} = 1$ mA, Vp = 6.7 volts, and C1 = 0.01 μF, then the frequency of oscillation is solved from:

$$f_o = \frac{1}{C1(Vp/I_{E1})} = \frac{I_{E1}}{VpC_1} = \frac{1 \text{ mA}}{6.7 \text{ V} \times 0.01 \text{ μF}} = \frac{10^{-3}}{6.7 \times 10^{-8}} = \frac{10^5}{6.7} \approx 14.9 \text{ kHz}$$

EXPERIMENTAL OBJECTIVES

To investigate the unijunction sawtooth oscillator.

LIST OF MATERIALS

1. Transistors: 2N2646 or 2N4871 and 2N3906 or equivalents
2. Resistors (one each):   100 Ω   330 Ω   4.7 kΩ   10 kΩ   27 kΩ   100 kΩ
3. Capacitor: 0.01 μF (one)

EXPERIMENTAL PROCEDURE

1. Connect the unijunction relaxation oscillator as shown in Figure 4-72.
2. Monitor the DC voltage drops around the circuit.
3. Insert the measured values into Table 4-18.

4. Calculate the frequency of oscillation from: $f_o = \dfrac{1}{R_3C_1 \ln(1/1 - \eta)}$ where: $\eta \approx 0.65$ for the 2N2646

255

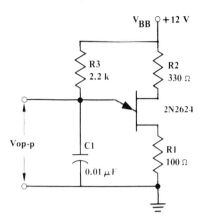

FIGURE 4-72

TABLE 4-18	V_E	VB #1	VB #2	f_o	Vp	V_v	V_{C1} Vop-p	Charge-Discharge
CALCULATED	///////	///////	///////			≈ 0.5 V		///////
MEASURED					///////	///////		

5. Calculate the magnitude of the waveshape from:

$$V_{C1} \approx Vp - V_v \quad \text{where:} \quad Vp \approx V_{D1} + \eta V_{BB} \quad \text{and} \quad V_v \approx 0.5 \text{ V}$$

6. Monitor the waveshape at the emitter of the device —across the C1 capacitor.
   a. Measure the magnitude.
   b. Measure the frequency of oscillation.
   c. Measure the charge-discharge ratio.
7. Monitor the waveshape at base #1 and at base #2.
8. Sketch the waveshapes, comparing the magnitudes and the timing of each of the pulses.
9. Insert the calculated and measured values into Table 4-18.
10. Connect the UJT relaxation oscillator with a constant current source as shown in Figure 4-73.
11. Monitor the DC voltage drops around the circuit.
12. Insert the measured values into Table 4-19.

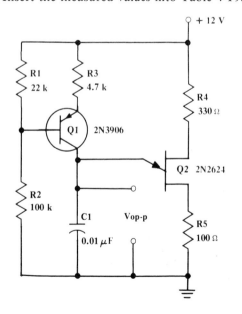

FIGURE 4-73

256

TABLE 4-19	V_{R1}	V_{R2}	V_{R3}	V_{C1}	VB #1	VB #2	f_o	V_{C1} Vop-p	Charge-Discharge
CALCULATED				///////	///////	///////		///////	///////
MEASURED									

13. Calculate the approximate frequency of oscillation from:

$$f_o \approx \frac{I_E}{VpC1} \quad \text{where:} \quad Vp = \eta V_{BB} \quad \text{and} \quad \eta \approx 0.65$$

NOTE: I_{E1} is solved from $V_R/R3$ and Vp can be anywhere from 0.5 V_{CC} to 0.7 V_{CC}, and with these kinds of variances to contend with, only "ballpark" calculations can be made.

14. Monitor the waveshape across the C1 capacitor:
   a. Measure the magnitude.
   b. Measure the frequency of oscillation.
   c. Note the linear ramp. Measure the charge-discharge ratio.
15. Monitor the waveshape at base #1, at base #2, and at the emitter of Q1.
16. Sketch the linear waveshape across capacitor C1 and the waveshape at base #1 of the UJT and compare the magnitude and timing.
17. Insert the calculated and measured values, as indicated, into Table 4-19.

PART 8 — 566 VOLTAGE CONTROLLED OSCILLATOR

GENERAL DISCUSSION

The 566 voltage controlled oscillator is another extremely versatile IC which can be used in a myriad of applications. As a function generator, the 566 provides both a triangle and a square wave at a frequency determined by the power supply, the biasing, and the RC components used. Additionally, the 566 can be used as a narrow band FM modulator, a clock generator, a signal generator, or a tone generator.

The 566 device contains a constant current source, a schmitt trigger, and two buffer amplifiers that serve to process the triangle and square waves to the output terminals. The amplitude of the square wave will be about one half of V_{CC} and the amplitude of the triangle wave will be about one fifth of V_{CC}. Construction technique requirements of the chip dictate the amplitudes. Essentially, the triangle wave charges and discharges linearly, while the Schmitt trigger is switched between two voltage extremes. The Schmitt trigger provides the square wave. An equivalent block diagram is shown in Figure 4-74.

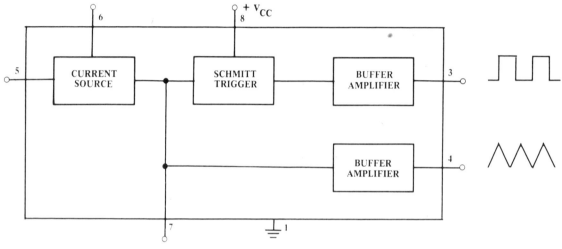

FIGURE 4-74

CIRCUIT OPERATION

Chip construction techniques of the 566 device require that Pin 5 be at a voltage level that is higher than three fourths V_{CC} and, therefore, biasing resistors R1 and R2 must be chosen to insure this device requirement. Additionally, load impedances much below 1 kΩ begin to load down the output waveshapes so output resistors of about 10 kΩ are used to make certain there is maximum output voltage amplitude. Load resistors that are too low (much below 500 Ω) will cause oscillations to terminate.

When the power is turned on to the circuits of Figure 4-75, timing capacitor C1 is charge and discharged by a constant current source which allows a sawtooth wave to be developed across the C1 capacitor. The amount of current provides the rate at which the capacitor charges and discharges and the current is determined by a combination of the V_{CC}, the voltage drop across timing resistor R1, and the biasing voltage of resistors R2 and R3. Because the charge-discharge time is essentially equal, the duty cycle remains at about 50% throughout the various R_1C_1 timing combinations. The representative circuits, shown in Figure 4-75, provide a constant frequency and a variable frequency, where either R1 or R2 (as an example), can be varied to change the oscillation frequency.

The approximate frequency of oscillation for the 566 device is solved from:

$$f_0 = \frac{2(V_{CC} - V[\text{Pin 5}])}{R_1 C_1 V_{CC}}$$

and if V_{CC} is 12 volts, R1 = 10 kΩ, and C1 = 0.02 μF, then:

$$V(\text{Pin 5}) = \frac{V_{CC} \times R3}{R2 + R3} = \frac{12\ V \times 10\ k\Omega}{2\ k\Omega + 10\ k\Omega} = 10\ \text{volts}$$

258

$$f_O = \frac{2(V_{CC} - V[\text{Pin 5}])}{R_1 C_1 V_{CC}} = \frac{2(12\text{ V} - 10\text{ V})}{10\text{ k}\Omega \times 0.02\ \mu\text{F} \times 12\text{ V}}$$

$$= \frac{2 \times 2\text{ V}}{10^4 \times 2 \times 10^{-8} \times 12\text{ V}} = \frac{4 \times 10^4}{24} \approx 1.67\text{ kHz}$$

Therefore, if a 100 kΩ variable resistor is substituted for resistor R1 and capacitor C1 is varied from 0.1 μF to 1000 pF, the frequency can be easily varied from below 100 Hz to above 100 kHz. The device has an upper limit of about 1 MHz. Additionally, if biasing resistors R1 and R2 are disconnected from V_{CC} and replaced with a (separate) variable supply that varies from greater than three quarters V_{CC} to slightly below V_{CC}, the frequency can be varied and, again, the duty cycle remains at about 50%. Also, the frequency will vary at a frequency rate as determined by the input signal rate if an input signal is provided to Pin 5 through a coupling capacitor. The effect is frequency modulation and the deviation can be as great as 10% of the operating frequency.

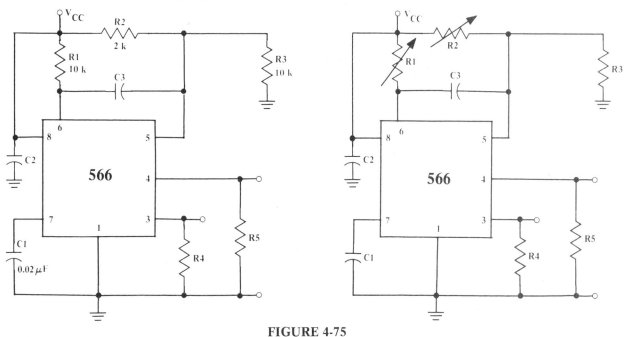

FIGURE 4-75

THE SCHMITT TRIGGER AND THE MONOSTABLE MULTIVIBRATOR

The main function of the Schmitt trigger and the monostable multivibrator is to provide an output square wave with a controlled duty cycle or "on" time. The Schmitt trigger circuit uses an input signal wave to turn on and off the device at a particular threshold voltage as determined by the circuit, while the monostable circuit uses an input signal to turn on the device and the circuit then control the length of time the device is kept on. Block diagrams of each circuit are shown in Figure 4-76.

FIGURE 4-76

Schmitt triggers are widely used in digital applications as pulse shapers or pulse restorers. They can be constructed from discrete devices or from monlythic devices such as the TTL series 7413, the 555 time, the μA710 comparator, or the LM301 op-amp.

The monostable multivibrator is also widely used in digital applications. And it is sometimes referred to as the one-shot, because the output pulse comes on for a certain length of time as determined by the multivibrators RC components. Also, since the output pulse is independent of the input pulse, it eliminates false triggering. The monostable multivibrator can be constructed from discrete transistors or from monolithic devices such as the TTL series 74123, the 555 timer, or the LM301 op-amp.

EXPERIMENTAL OBJECTIVES

To investigate the 566 voltage controlled oscillator in the astable mode.

LIST OF MATERIALS

1. Integrated Circuit: 566 (Mini-Dip Package)
2. Resistors: 1.8 kΩ (one)   4.7 kΩ (one)   10 kΩ (Three)   10 kΩ Potentiometer (one)
3. Capacitors (one each): 0.001 μF   0.01 μF   1 μF

EXPERIMENTAL PROCEDURE

1. Connect the 566 voltage controlled oscillator as shown in Figure 4-77.

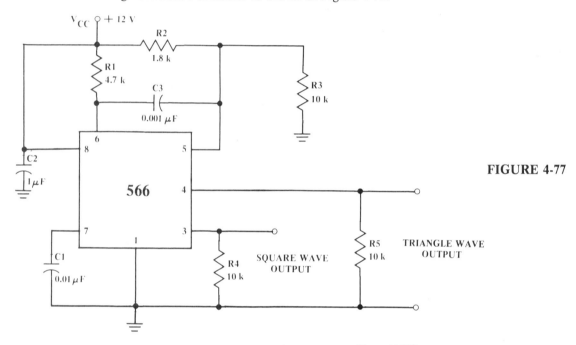

FIGURE 4-77

2. Calculate the approximate voltage at Pin 5 from: $V(\text{Pin } 5) = \dfrac{V_{CC} \times R3}{R2 + R3}$

3. Measure the voltage at Pin 5.

4. Calculate the frequency of oscillation from:

$$f_o = \frac{2(V_{CC} - V[\text{Pin } 5])}{R_1 C_1 V_{CC}} \qquad \text{where:} \quad V_{CC} = 12 \text{ V, } R1 = 4.7 \text{ k}\Omega \text{, and } C1 = 0.01 \text{ } \mu\text{F}$$

5. Monitor the **triangle wave across** timing capacitor C1 and then across the output resistor connected between Pin 4 and ground.
    a. Calculate the approximate amplitude of each from: Vp-p (triangle) ≈ $V_{CC}/5$.
    b. Measure the amplitude and frequency.
    c. Measure the duty cycle and the charge-discharge ratio.
6. Monitor the square wave across the output resistor, connecting between Pin 3 and ground.
    a. Calculate the approximate amplitude of each from: Vp-p (square) ≈ $V_{CC}/2$.
    b. Measure the amplitude and frequency.
    c. Measure the duty cycle.

260

7. Insert the calculated and measured values, as indicated, into Table 4-20.

TABLE 4-20	V (Pin 5)	f_0	Vop-p		Duty Cycle		VCO 10 kΩ Variable	
			Pin 4	Pin 3	Pin 4	Pin 3	MIN	MAX
CALCULATED					/////	/////	/////	/////
MEASURED								

8. Varying the frequency of oscillation:
   a. Vary either of the frequency determining RC components, R1 or C1.
   b. Vary the voltage at Pin 5 — to test the circuit as a voltage controlled oscillator.
      1. Do this by replacing R2 with a variable 10 kΩ resistor and varying the resistance, monitoring the frequency extremes (before oscillation ceases). A second method is to:
      2. Remove resistors R2 and R3 and connect another power supply, with the voltage set to about 10 volts. Vary the voltage ± 1 volts and monitor the frequency change.
9. Frequency modulating the device can be accomplished by appling and AC signal to Pin 5 through a 10 μF coupling capacitor, as shown in Figure 4-78. Use a low frequency to frequency-modulate the square wave output at a rate that will be slow enough to be followed, initally, on a scope. Higher frequencies can be applied later.

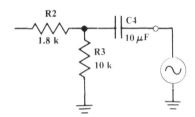

FIGURE 4-78

261

5 || INVERTERS, CONVERTERS, AND SWITCHING REGULATORS

GENERAL DISCUSSION

The conversion of AC power to DC power, in its simplest form, requires very few components. A step-up or a step-down transformer, a rectifier, a filter capacitor, and a load are all that are normally required. The DC to AC inverter, however, is a little more complex because the conversion of DC power to AC power must be accomplished with an oscillator and, then, the voltage is stepped up or stepped down by a transformer, before the output AC voltage at the secondary is rectified. Therefore, DC to AC involves power inverters and DC to DC involves power converters.

Switching regulators are also more complex than linear series pass voltage regulators because they use an oscillator to switch an in-line power transistor on and off. Since the device is either on or off (and not both at the same time), the power dissipated by the device is relatively low. Therefore, switching regulator techniques are considerably more efficient than linear series pass regulator techniques. However, because of the added compexity, the tradeoff of when to use a switching regulator over a linear series pass regulator is dictated by two factors: by the power that the regulator must deliver, and by the condition where a large voltage differenc exists between the input and output of the regulator.

For instance, if a large voltage difference exists between the load and the input voltage of the regulating circuitry, the switching regulator compensates for the difference voltage by controlling the duty cycle. Also, if power delivered to a load is above 100 watss, linear series pass techniques cause the devices to dissipate far too much power and switching regulator techniques should be used.

POWER INVERTERS

Inverters can be fixed frequency or free running. And inverters and converters almost always use the push-pull arrangement of power transistors and transformer proper. If the inverter is fixed, then an external frequency is used to turn on and turn off the power devices (alternately) and, if the inverter is free running, then the power devices, along with the transformer, are connected to insure oscillatory conditions.

The frequency at which power inverters operate varies with the application, but the ideal frequency is about 20 kHz, because it is just above the audio frequency (transformers will "sing"). Also, the transformers can be smaller and lighter than those used at 60 Hz and the efficiency of the inverter is still relatively high. However, above 20 kHz the device, because of the rise time required, become costly because transformer core loss becomes excessive and the efficiency of the circuit decreases. Hence, regardless of the ideal frequency of 20 kHz, most inverters operate lower than 20 kHz and the size of the transformer, the core, and the power transistors used, all play a role in the operating frequency. This is true, especially, with reference to free-running circuit connections.

PART 1 — THE FIXED FREQUENCY INVERTER

An example of the fixed frequency inverter is shown in Figure 5-1, where an astable multivibrator is used to drive push-pull transistors. Note that the signal off the collector of Q1 is 180° out of phase with the signal off the collector of Q2 and, therefore, the Q3 and Q4 push-pull transistors are turned on and off, alternately, causing the voltage across the Q3 and Q4 collectors to switch between $V_{CE(SAT)}$ and $2V_{CC}$, with respect to ground.

TRANSFORMER OPERATION

Capacitors store voltage and inductors store current and it is this storage of energy that creates larger voltage swing conditions, in some circuits, than the applied DC voltage. For instance, in the circuits of Figure 5-2, where a collector resistor is used, the peak-to-peak signal is approximately $V_{CC} = V_{CE(SAT)}$ and it is limited, before being distorted, by $V_{CE(SAT)}$ and V_{CC} of the power supply. When the transistor

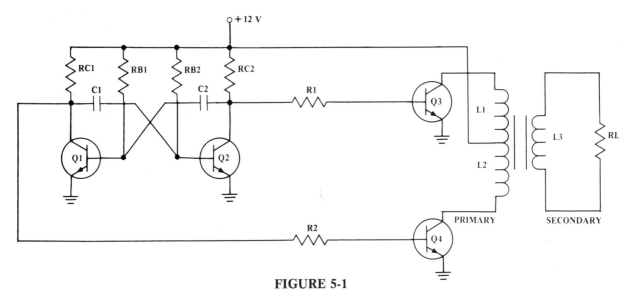

FIGURE 5-1

is switched by an input square wave, it switches between $V_{CE(SAT)}$ and V_{CC} or $V_{CC} - V_{CE(SAT)} = 12\ V - 0.3\ V = 11.7\ V$, where the average voltage at the collector for a 50% duty cycle is slightly less than 6 volts.

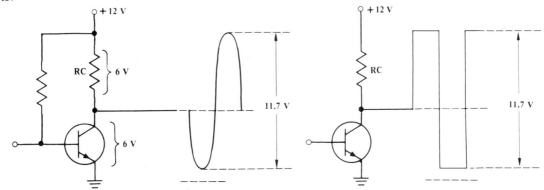

FIGURE 5-2

However, when the collector resistor is replaced by an inductor and, where the inductive reactance is at least equal to or greater than the output impedance of the device, then the voltage swing at the collector is approximately $2V_{CC}$ or, more accurately, $2V_{CC} - 2V_{CE(SAT)} = 2(12\ V - 0.3\ V) = 23.4\ Vp\text{-}p$. This phenomenon is also experienced in Class A power amplifiers and LC tuned circuits where, to the DC

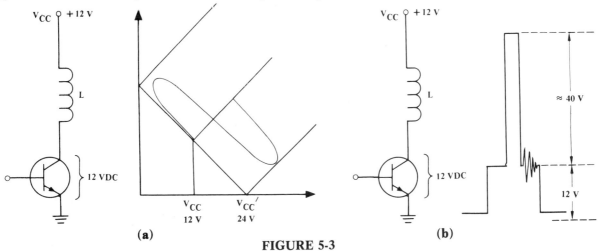

FIGURE 5-3

263

voltage, the inductor is a length of wire and the applied power supply voltage appears at the collector with respect to ground. Therefore, the V_{CC} voltage becomes the Q point of the transistor with the AC signal "riding" on it, and the transistor processes the negative-going peak and the inductor the positive-going peak. However, in the switching mode, the voltage at the collector can rise to five and six times the V_{CC} voltage amplitude, where the "on" pulse width will be proportionately narrower with regard to the total on time. Also, as shown in Figure 5-3(b), if the inductor is not critically damped, then ringing will occur during the off time of the transistor. These problems are cured, effectively, in Class B operations, because the transformer is being switch by both the input signals by way of the Q3 and Q4 transistors and the collapsing field of the alternate inductive winding and the device can be assumed to be either full on or full off.

In the Class B, push-pull operation, one transistor is in cutoff and the other is in saturation, or vice versa. Then, if we assume Q3 to be in saturation and Q4 to be in cutoff, as shown in Figure 5-4(a), then the full 12 volts of the V_{CC} is applied across the series $V_{CE(SAT)}Q3$ and the inductor L1. If $V_{CE(SAT)}Q3 \approx 0.3$ V, then $V_{L1} = 11.7$ V and, because of inductive coupling, $V_{L2} = 11.7$ V.

NOTE: A center-tapped transformer implies that the number of turns of L1 = L2 and, therefore, L1 = L2. Hence, any voltage developed across L1 is also developed across L2, assuming the coefficient of coupling is high.

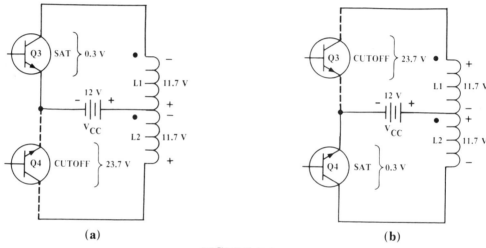

(a)   **(b)**

FIGURE 5-4

Therefore, in the condition of Q3 being saturated and Q4 being in cutoff, $V_{L1} = V_{L2} = 11.7$ V, the voltage at the collector of Q3 with respect to ground $V_{CE(SAT)}$ is 0.3 V, and the voltage at the collector of Q4 with respect to ground is $V_{CC} + V_{L2} = 12$ V $+ 11.7$ V $= 23.7$ V. This is shown in Figure 5-4(a). Likewise, when Q3 is in cutoff and Q4 saturated, the voltage at the collector of Q3 is approximately 23.7 volts and the voltage at the collector of Q4 is approximately 0.3 volts, all with respect to ground. This is illustrated in Figure 5-4(b).

Hence, where Q3 in saturation and Q4 in cutoff is one state and where Q3 in cutoff and Q4 in saturation is the next state, the two states develop approximately 23.7 volts at the collectors of Q1 and Q2, with respect to ground. Once the circuit is switching back and forth, a voltage developed across L1 is also developed across L2 and vice versa. So, the total voltage across the transformer primary is 47.4 Vp-p. However, both halves of the circuit are identical and the loading effects of the secondary on the primary are usually calculated on an individual Q1 or Q2 basis. Therefore, the voltage across the entire transformer is not necessary in making calculations.

In summation then, the square-wave, output signal at the collectors of both Q1 and Q2 switches between 0.3 volts and 24 volts, and the two transistors are 180° out of phase. If the voltage swing is monitored across the entire transformer (L1 + L2+ 2M), which is between the collectors of Q1 and Q2, the voltage should be approximately 2×23.7 V $= 47.4$ Vp-p.

CIRCUIT THEORY AND ANALYSIS

The fixed frequency inverter of Figure 5-5 uses the astable circuit to establish the driving frequency, the power transistors as the power switching devices, and the transformer to process the output square wave at the required voltage level. Additionally, the push-pull arrangement can handle about four times more power than the single-ended arrangement, given the same voltage ($V_{CC} = 12$ volts).

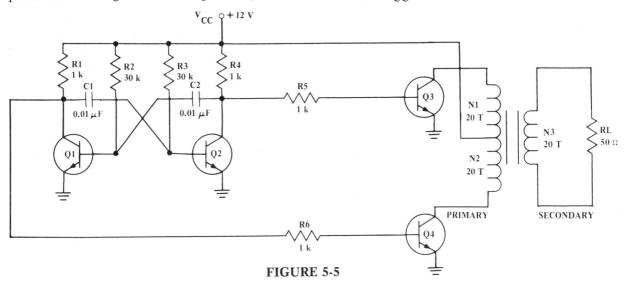

FIGURE 5-5

1. The frequency of oscillation for the circuit would be determined by the astable multivibrator, alone, if the loading effect were at least 10 times greater than the collector resistors. However, when the loading effect is effectively equal to the collector resistance, as shown in Figure 5-5, then the frequency is increased over that of the unloaded frequency.

a. Solving for the frequency of the unloaded multivibrator circuit:

$$f_o = \frac{1}{2(R_3 C_1 \times \ell n2)} = \frac{1}{2(R_2 C_2 \times \ell n2)} = \frac{1}{2(30 \text{ k}\Omega \times 0.01 \text{ }\mu F \times 0.693)} = 2404 \text{ Hz}$$

b. Solving for the frequency of the 1 kΩ multivibrator circuit, where R6 = R1 = R4 = R5 = 1 kΩ

$$f_o = \frac{1}{2R_3 C_1 \ell n(1 + [V_C/V_{CC}])} = \frac{1}{2R_2 C_2 \ell n(1 + [V_C/V_{CC}])}$$

$$= \frac{1}{2 \times 30 \text{ k}\Omega \times 0.01 \text{ }\mu F \times \ell n(1 + [6 \text{ V}/12 \text{ V}])} = \frac{1}{2 \times 3 \times 10^4 \times 10^{-8} \times \ell n1.5} \approx 4.11 \text{ kHz}$$

NOTE: The effect of loading the collector with a 1 kΩ resistor, when the device is in cutoff, is to halve the voltage at the collector and, hence, one half the voltage that the C1 and C2, 0.01 μF capacitors have to charge to. The net effect is that the charging time is decreased and the frequency is decreased. For instance, when Q1 and Q2 are unloaded, capacitors C1 and C2 (theoretically) follow a charge line from $-V_{CC}$ to V_{CC} and can be solved from $\ell n2$, when $V_C \approx V_{CC}$ from: $\ell n(1 + [V_C/V_{CC}] = \ell n(1 + 1)$. However, when V_C is loaded and one-half V_{CC} in voltage, the capacitors follow a charge line that (theoretically) goes from $-V_{CC}/2$ to V_{CC}, or -6 V to 12 V and the $\ell n1.5$ from $\ell n(1 + [6 \text{ V}/12 \text{ V})$. Additionally, if the load is decreased to 500 Ω, V_C drops to 4 V, and the $\ell n(1 + [V_C/V_{CC}]) = \ell n(1 + [4 \text{ V}/12 \text{ V}])$ $= \ell n1.33 = 0.2851$, and f_o increases to 5844 Hz.

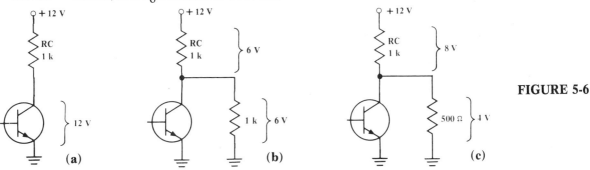

FIGURE 5-6

265

2. The amplitude of the voltage at the collector of Q1, and hence the collector of Q2, under loaded conditions of 1 kΩ is:

$$V_{C(Q1)}p\text{-}p = \frac{V_{C(Q1)}p\text{-}p\text{(unloaded)} \times R6}{R1 + R6} = \frac{\left(V_{CC} - V_{CE[SAT]}\right) \times 1 \text{ k}\Omega}{1 \text{ k}\Omega + 1 \text{ k}\Omega}$$

$$= \frac{(12 \text{ V} - 0.3 \text{ V}) \times 1 \text{ k}\Omega}{2 \text{ k}\Omega} = 5.65 \text{ Vp-p}$$

NOTE: Q2 is solved similarly and is not repeated.

3. The base current drive of transistors Q3 and Q4, during the on time of the device, is solved from:

$$I_{B(Q3)} \approx I_{B(Q4)} = \frac{V_{C(Q1)}p\text{-}p - V_{CE(SAT)}}{R6} = \frac{5.65 \text{ Vp-p} - 0.6 \text{ V}}{1 \text{ k}\Omega} \approx 5 \text{ mA}$$

a. Therefore, the average (DC) base current for a 50% duty cycle input square is solved from:

$$I_{B(av)} = \frac{I_{B(Q3)}p\text{-}p \times T_{on}}{T} = \frac{5 \text{ mA} \times (1.04 \times 10^{-4})}{2.08 \times 10^{-4}} = 2.5 \text{ mA}$$

where: $T = 1/f$ and $T = T_{on} + T_{off} = (1.04 \times 10^{-4}) + (1.04 \times 10^{-4})$

b. $I_{B(av)} = I_{B(Q3)}p\text{-}p \times 50\% = (5 \text{ mA} \times 50)/100 = 2.5 \text{ mA}$

NOTE: For the 50% duty cycle condition, the DC voltage is about one half the peak-to-peak amplitude. So, the DC or average current is one half the peak-to-peak "on" current condition. Also, the voltage drop across R5 (and R6) should be smaller than the approximate 5 volts, and it depends wholly on input resistance, and the rate of turn-on of the device used. Therefore, measuring Vp-p at the collector of Q1 and the base of Q4, and taking the difference, will provide $V_{R5p\text{-}p}$ exactly.

4. The amplitude of the voltage at the collector of Q3 is solved from:

$$V_{C(Q3)}p\text{-}p = V_{CC} + V_{L1} = 12 \text{ V} + 11.5 \text{ V} = 23.5 \text{ Vp-p}$$

where: $V_{L1} = V_{CC} = V_{BE(SAT \ Q3)} = 12 \text{ V} - 0.5 \text{ V} = 11.5 \text{ V}$

NOTE: Q4 is calculated similarly so it is unnecessary to repeat the calculations. However, further explanation can be obtained from the transformer operation section. Also, the $V_{CE(SAT)}$ of devices increases with increased collector current, hence $V_{CE(SAT)} = 0.5$ volts, instead of the 0.3 volts used at lower current (I_C) values.

5. The amplitude across the whole transformer primary is about twice the voltage at either of the Q3 or Q4 collectors, with respect to ground, if the transformer is center-tapped.

$$V(L1 + L2 + 2M)p\text{-}p = 2V_{C(Q3)}p\text{-}p = 2V_{C(Q4)}p\text{-}p = 2 \times 23.5 \text{ Vp-p} = 47 \text{ Vp-p}$$

6. Solving for the Vop-p to the load, assuming a transformer efficiency of 90% and (L1 + L2) : L3 turns ratio of 2 : 1, where: N1 = 20 T, N2 = 20 T, and N3 = 20 T.

a. $\text{Vop-p} = \dfrac{V(L1 + L2 + 2M)N3 \times \%eff}{N1 + N2} = \dfrac{47 \text{ Vp-p} \times 20 \text{ T} \times 0.9}{40 \text{ T}} = 21.15 \text{ Vp-p}$

b. $\text{Vop-p} = \dfrac{V_{L1} \times N3 \times \%eff}{N1} = \dfrac{23.5 \text{ Vp-p} \times 20 \text{ T} \times 0.9}{20 \text{ T}} = 21.15 \text{ Vp-p}$

NOTE: Either method is valid, because the reflected load impedance is handled by one transistor and then the other, alternately.

c. Therefore, the average voltage to the load for a 50% duty cycle is:

$$V(av) = \frac{Vop\text{-}p \times T_{on}}{T} = \frac{21.15\ V}{2} \approx 10.58\ VDC$$

If a full-wave bridge and a capacitor were used, this would be the approximate DC voltage value. Therefore, an estimation to use for a DC to DC situation where the V_{CC} of the inverter and the turns ratio are known, is to solve the output VDC from the turns ratio times V_{CC}. The 10.58 VDC divided by the efficiency factor is an example of this technique.

7. Solving for the power across a 50 Ω load resistor:

a. $Po = Vop\text{-}p^2 / 4RL = 21.15\ Vp\text{-}p^2 / (4 \times 50\ \Omega) \approx 2.24$ watts

b. $Po = V(av)^2 / RL = 10.58^2 / 50\ \Omega \approx 2.24$ watts

NOTE: Again, the $V(av) = VDC$ at one half the 50% duty cycle is solved like any DC power equation because doubling the voltage and squaring (2^2) provides the divide by 4.

8. Solving for the power delivered to the primary where the 90% efficiency factor is included.

$$Po(pri) = Po/\%eff = 2.24\ W/0.9 \approx 2.48 \text{ watts}$$

9. Solving for the approximate average (DC) current demand:
a. At the load from:

1. $I_{L(av)} = po/V(av) = 2.24\ W/10.58\ V \approx 211$ mA

2. $I_{L(av)} = 2Po/Vop\text{-}p = (2 \times 2.24\ W)/21.15\ Vp\text{-}p \approx 211$ mA

3. $I_{L(av)} = Vop\text{-}p/2RL = 21.15\ Vp\text{-}p/(2 \times 50\ \Omega) \approx 211$ mA

b. At the primary from

$$Ipri(av) = 2Po(pri)/Vop\text{-}p = (2 \times 2.48\ W)/23.5\ Vp\text{-}p \approx 211 \text{ mA}$$

c. And since the Q3 and Q4 devices each, alternately, supply the primary current, then:

1. $I_{C(Q3)} = I_{C(Q4)} = Ipri/2 = 211\ mA/2 = 105.5$ mA

2. $I_{C(Q3)} = I_{C(Q4)} = Po/VCp\text{-}p = 2.48\ W/23.5\ Vp\text{-}p \approx 105.5$ mA

NOTE: Ipri(av) at 211 mA is current that the power source must deliver along with the additional operating circuit currents. Therefore, IDC, the current out of the power source is slightly higher than Ipri(av).

10. DEVICE CHARACTERISTICS

a. The voltage breakdown for the Q3 and Q4 devices is calculated from $2V_{CC}$, approximately. However, because of spikes caused during rise and fall times, 3 V_{CC} is a considerabley safer design criteria. Therefore:

1. $V_{CED(min)} = 2V_{CC} = 2 \times 12\ V = 24\ V$, however

2. $V_{CED(safe)} = 3V_{CC} = 3 \times 12\ V = 36\ V$

b. Based on I_C and I_B calculations, the minimum beta the device should have is:

$$\beta(min) = I_C/I_B = 215\ mA/2.5\ mA = 86$$

c. The amount of power the devices dissipate is relatively small because the maximum current condition occurs when the device is in a saturated condition. The minimum current condition occurs when the maximum voltage is impressed across the device. In addition, rise time and fall time contribute to the power dissipation. If the device is slow in switching, and remains in the active region for too long, then the voltage times the maximum current for that time will provide power dissipation. Therefore, the faster the device, the less power consumed during turn-on and turn-off periods. Therefore:

1.   The power dissipated with Q3 in saturation is:

$$P_{Q3(SAT)} = T_{on}/T \times IDC \times V_{CE(SAT)} = (211 \text{ mA} \times 0.5 \text{ V})/2 \approx 52.8 \text{ mW, also}$$

$$P_{Q3(SAT)} = I_C \times V_{CE(SAT)} = 105.5 \text{ mA} \times 0.5 \text{ V}^2 \approx 52.8 \text{ mW}$$

2.   The power dissipated when Q3 is in cutoff, where the voltage across the device is approximately $2V_{CC}$ and I_{CEX} (the collector current with the base reverse biased) is given as 2 mA in manufacturer data sheets, is solved from:

$$P_{Q3(cutoff)} \approx T_{off}/T \times 2V_{CC} \times I_{CEX} = (2 \times 12 \text{ V} \times 2 \text{ mA})/2 = 48 \text{ mW}/2 = 24 \text{ mW}$$

3.   The power Q3 dissipates during the rise anf fall times is totally dependent on the devices. Therefore, assuming that $t_r = t_f = 2$ μsec and $T = 1/f = 1/5$ kHz $= 200$ μsec, the power dissipated for a worse case condition is calculated from:

$$P_{Q3}(t_r + t_f) = \frac{t_r + t_f}{T} \times \frac{V_{CC} \times IDC}{2} = \frac{4 \text{ μsec}}{200 \text{ μsec}} \times \frac{12 \times 211 \text{ mA}}{2} \approx 25.3 \text{ mW}$$

4.   Therefore, the total power dissipated by Q3 or Q4, for a 50% duty cycle condition, is:

$$P_{Q3} = P_{Q4} = P_{(SAT)} + P_{(cutoff)} + P_{(t_r + t_f)} = 52.8 \text{ mW} + 24 \text{ mW} + 25.3 \text{ mW} \approx 102 \text{ mW}$$

NOTE: I_{CEX} at 2 mA and $V_{CE(SAT)}$ at 0.5 volts are nominal values for devices processing higher collector currents, and where $V_{CE(SAT)}$ increases with increased current. Additionally, for the duty cycle of 50%, $T_{on} = T_{off} = 200$ μsec $= T/2$. Obviously, medium power devices (with heat sinks), such as a 2N5190, 2N6178, or 2N3053, can be used in this low power inverter.

11.   The peak current for the 50% duty cycle square wave inverter is approximately $2I_C$. Therefore, solving for the peak current of Q3 and Q4:

$$Ip(Q3) = Ip(Q4) \approx 2I_C \approx 2 \times 105.5 = 211 \text{ mA}$$

NOTE: Peak current does contribute to power loss and it is frequently used instead of average current in the design of inverters.

12.   The efficiency of a Class B, center-tapped, sine wave, amplifier circuit can be calculated to be about 78.5%, where the current flow out of the power source (neglecting the biasing currents) is calculated from: $IDC \approx I_C(Q3) + I_C(Q4)$. Also, this was seen in the center-tapped, full-wave, unfiltered, rectifier circuit, where each half of the circuit, alternately, contributed one half the current to the load. Therefore, in the 50% duty cycle, square-wave circuit, $I_C(Q3) = I_C(Q4) = IDC/2$. Hence, the current out of the power source (where power loss across the transformer, base circuit, and the astable multivibrator are neglected, initially, in the calculations) is calculated to be, theoretically:

$$IDC = 2I_{C(Q3)} = 2I_{C(Q4)} = 2 \times 105.5 \text{ mA} = 211 \text{ mA}$$

NOTE: IDC equals Ipri(av)

13.   Assuming there is about 39 mA of operating biasing currents because of the transformer, the base circuit, and the astable multivibrator, the $IDC \approx 250$ mA and the power provided by the power source is:

$$Pps = IDC \times V_{CC} = 250 \text{ mA} \times 12 \text{ V} = 3 \text{ watts}$$

14. The efficiency of the circuit is solved from:

$$\% \text{ efficiency} = Po/P_{PS} \times 100 = (2.24 \text{ V} \times 100)/3 \text{ W} \approx 74.6\%$$

NOTE: DC to DC conversion can be obtained if a full-wave bridge is connected in the secondary and adequate filtering is used.

TRANSFORMER CONSTRUCTION

If magnetic wire were wound on a cardboard form, the magnetic flux field would have to be created by the wire alone, because the cardboard form would be considered an air core with a permeability of unity (1). However, if core material composed of iron, steel, powdered iron, or ferrite were used, the permeability would be greater than unity, and the inductance would be greatly enhanced. Obviously, the higher the permeability of the core material, the greater the flux field, and large inductances with few turns can be achieved.

At power transformer frequencies of 60 Hz, core material in the form of I's and E's is used, because solid core material creates too much heat loss from eddy currents and hysteresis. And, if the frequency is increased to above 1 kHz, then the laminated core material exhibits excessive heat losses, so powdered iron and ferrite core materials are used. However, iron powder core material does not saturate easily, but ferrite material does, so at frequencies much above 1 kHz, ferrite core material is used in saturable inverter circuits. Still, ferrite material does not have the large permeability factors that silicon iron, silicon steel, or nickel iron does, so the trade-off of core loss versus the number of turns can sometimes be a design factor.

The main advantage of ferrite material is that it is highly resistive, because the structure of its atom is tightly bound and does not give up electrons easily — a necessity for conduction to occur. Therefore, eddy current losses are minimized and hysteresis losses, which are inevitable as the core is driven into saturation in one direction and then in the other, still occur. Additionally, the size of the core and the frequency play a role in core loss because increased frequency means that switching between positive and negative saturated conditions is increased, and the larger the core for the same material used means a larger flux field must be switched.

DESIGN TECHNIQUES

One formula used in the construction of 60 Hz power transformers is $V = 4.44 \times 10^{-8} A f_o B_m N S.F.$, and if the stacking factor is high, the formula can be further simplified to $V = 4.44 \times 10^{-8} A f_o B_m N$. Since the operating frequency is 60 Hz, the line voltage of power transformers (normally 120 V RMS), the area, the maximum flux density, and the number or primary turns are the unknowns. However, if an area of $A = 3$ square inches is assumed and a maximum magnetic flux density of 50,000 is chosen, which is nominal for silicon steel, then the number of turns can be calculated from:

$$N = \frac{V}{4.44 \times 10^{-8} A f_o B_m} = \frac{120 \text{ V RMS}}{4.44 \times 3 \times 60 \times 5 \times 10^4} = 300 \text{ turns}$$

and if the secondary were to be stepped down to 60 V RMS, the secondary would be wound with 150 turns.

TOROIDAL CONSTRUCTION

Ferrite core material comes in many forms, but the toroidal (doughnut shaped) core is popular because it has no air gaps and it can be easily saturated. The sizes of the cores vary but, if a core with an area of $3CM^2$ (which is relatively large) is used in a 12 volt power inverter, the number of turns on the primary is 40. Then, the modified formula:

$$N = \frac{V}{4 \times 10^{-8} A f_o B_m}$$

can be used where, besides changing the 4.44 to 4, the area is given as CM^2 rather than IN^2. Therefore, knowing the maximum flux density is 2000, the approximate operating frequency is:

$$f = \frac{V}{4 \times 10^{-8} A N B_m} = \frac{12 \text{ V}}{4 \times 10^{-8} \times 3 \times 40 \times 2 \times 10^3} = \frac{12 \text{ V}}{960 \times 10^{-5}} = 1250 \text{ Hz}$$

NOTE: If the number of turns, the area of the core, or the maximum flux density is increased, the operating frequency decreases.

The construction of the toroidal transformer begins by winding a 40 turn center tapped primary and then a 20 turn secondary. If the core is large, both can be wound as a single layer, but, if not, the primary is wrapped, then it is covered with a layer of insulation tape, and the 20 turn secondary is wrapped directly on top of the tape.

TRANSFORMER WINDING PROCEDURE (TOROIDAL CORES)

1. Wind the 40 turn center-tapped primary. Wind 20 turns, bring out the wire and twist two or three times to form the center tap, and continue winding the remaining 20 turns.

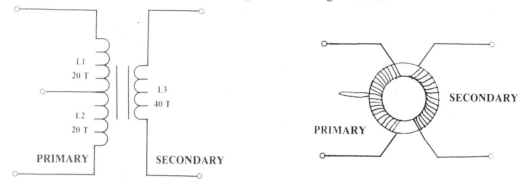

FIGURE 5-7

2. Wrap the 20 turn secondary as a single layer, as illustrated in Figure 5-7, and, if necessary because of a lack of room, proceed to wrap a layer of insulating tape and then wrap the 20 turn secondary directly on top of the tape. (This is only precautionary. Because of the low voltages involved, no arcing will occur.)

3. The size of the wire is flexible and can range from #24 to #20. Not larger than #20 is required, and the only reason for using the larger #22 or #20 is for convenience in wrapping or availability. Magnetic wire (formvar) is ideal.

4. Bipolar winding of the primary can be used in higher power circuits to cut down on stray inductances. For bipolar winding, two wires are wrapped together and then series-aiding connected, which requires one of the end wires to be connected with one of the beginning wires. Testing the full inductor on a bridge to insure series aiding is always a good policy.

5. The core materials are readily available. Arnold Engineering, Amidon Associates, and Ferroxcube Corporation are among the sources.

EXPERIMENTAL OBJECTIVES

To investigate the astable-driving inverter circuit.

LIST OF MATERIALS

1. Core: Ferrite Core Toroid
2. Wire: Magnetic Wire (formvar)
3. Transistors: 2N3904 (two)   2N3055 (two) or equivalents
3. Resistors: 47 Ω (one)   470 Ω (two)   1 kΩ (two)   1 kΩ potentiometer   27 kΩ (two)
5. Capacitors: 0.01 μF (two)

EXPERIMENTAL PROCEDURE

1. Wind a 40 turn, center-tapped primary and a 20 turn secondary on a toroidal (doughnut shaped) core. The core material is not critical providing the saturated flux density is nominally 2000 and the permeability is about 1000. Use the winding techniques discussed in the theory portion of this exercise.

2. Connect the circuit as shown in Figure 5-8. The bases of both Q3 and Q4 should be disconnected as shown in the illustration.

270

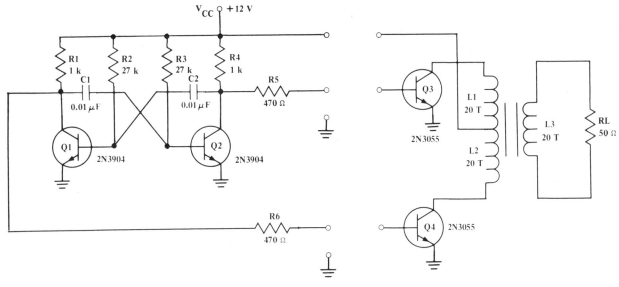

FIGURE 5-8

    a. Calculate the frequency of the unloaded astable multivibrator from:

$$f_o = 1/2(R_3C_1 \ln 2) \quad \text{where:} \quad R3 = 27 \text{ k}\Omega \quad \text{and} \quad C1 = 0.01 \text{ } \mu F$$

    b. Monitor the square wave output of either the Q1 or the Q2 collector and measure the approximate frequency. Also, measure the amplitudes at the bases of Q1 and Q2.

NOTE: If VBp-p is measured, the frequency can be (more) accurately calculated from $\ln(1 + [V_C/V_{CC}])$.

    c. Connect the R5 and R6, 470 Ω base resistors to ground, similar to connecting load resistors, and calculate the frequency of the loaded astable multivibrator. Calculate from:

$$f_o(\text{loaded}) = \frac{1}{2R_3C_1 \quad (1 + [V_C/V_{CC}])} \quad \text{where:} \quad V_C = R6/(R1 + R6)$$

    d. Monitor the square wave at the collector of Q1 or Q2 and measure the frequency. This will be the approximate operating frequency for the inverter circuit. Again, if VBp-p is measured f_o can be solved, comparatively accurately.

3. Connect the circuit as shown in Figure 5-9, where the R5, 470 Ω resistor is replaced by a 1 kΩ variable resistor and varied to about 470 Ω to insure circuit symmetry.

4. Monitor the waveshape across the 50 Ω load and adjust R5 for the cleanest, square-wave output. This will occur, approximately, at a 50% duty cycle. Also, note that the lowest operating current, overall, will occur when symmetry is achieved. Therefore, if the DC power source used has a voltage-current switch, monitor the current and the waveshape at the same time and adjust R5 for the cleanest square wave and lowest DC current. These conditions will occur almost simultaneously. Also, note that when symmetry is not achieved, one device will be turned on longer than the other and will run, therefore, hotter.

    a. Calculate the amplitudes of the waveshapes at the collectors of Q3 and Q4 to be approximately (slightly less than) 2 V_{CC}, and the amplitude between the collectors to be approximately 4V_{CC}.

    b. Calculate the amplitude of the waveshape across the load from:

$$\text{Vop-p(load)} = \frac{V_{C(Q3)}\text{p-p} \times N3}{N1} = \frac{V_{C(Q4)}\text{p-p} \times N3}{N2}$$

NOTE: N repesents the number of turns of the designated winding.

    c. Monitor the amplitudes of the square waves at the collectors of Q3 and Q4, with respect to ground.

    d. Monitor the amplitude of the square wave across the whole center-tapped inductor (collector of Q3

271

to the collector of Q4). $V_{C(Q3)} = V_{C(Q4)}$.

e. Monitor the amplitude of the square wave across the load.

NOTE: The output at the load will reflect the difference of the turns ratio of L1 to L3 or L2 to L3 and the coefficient of coupling, which should be higher than 90%.

f. Monitor the amplitude of the square wave at the collectors of Q1 and Q2, with respect to ground, and then at the bases of Q3 and Q4, with respect to ground.

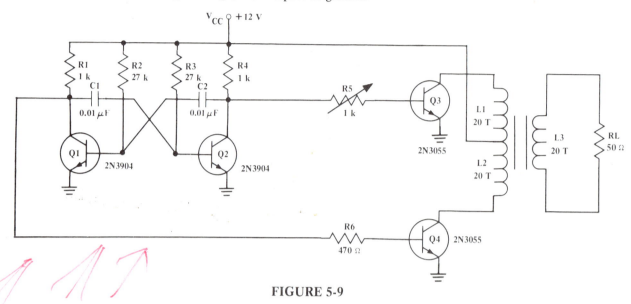

FIGURE 5-9

5. Insert the calculated and measured values, as indicated, into Table 5-1.

TABLE 5-1	f_o	V_Cp-p		V_Bp-p		Vop-p	V_Cp-p		Vp-p
		Q1	Q2	Q3	Q4	RL	Q3	Q4	$V_{C(Q3)} - V_{C(Q4)}$
NO LOAD CALCULATED		//////	//////	//////	//////	//////	//////	//////	//////
NO LOAD MEASURED						//////	//////	//////	//////
470 Ω LOAD CALCULATED		//////	//////	//////	//////				
470 Ω LOAD MEASURED									

6. Calculate the power developed across the load from: $Po = Vop\text{-}p^2/4RL$ where: $RL = 50\ \Omega,\ 5\ W$

NOTE: The power delivered to the primary will be slightly higher than that delivered to the secondary by a ratio of the coefficient of coupling. However, unless the exact turns ratio is known, the coefficient of coupling will be (unless large) difficult to obtain.

7. Calculate the average current through the load from: $I_L(av) = 2Po/Vop\text{-}p = Vop\text{-}p/2RL$

8. Calculate the average current delivered by the power source from:

$$IDC \approx 2Po/V_{C(Q3)} = 2Po/V_{C(Q4)}$$

9. Calculate the approximate I_C of the collectors of Q3 and Q4 from:

$$I_{C(Q3)} = I_{C(Q4)} = Po/V(Q3)p\text{-}p = Po/V(Q4)p\text{-}p$$

10. Measure $I_{C(Q3)}$, $I_{C(Q4)}$, and IDC

NOTE: Both the coefficient of coupling and $V_{CE(SAT)}$ were excluded in the calculations, so IDC should be slightly higher than calculated, if the turns ratio are precisely known and symmetry is maintained. Also, Po/Vop-p was derived from existing text information where, again, the efficiency of the transformer is neglected.

11. Use the measured Vp-p at the collectors of Q1 and Q2 and the bases of Q3 and Q4 and calculate the average base current from:

  a. $I_{B(Q3)} = \dfrac{V_{C(Q2)}\text{p-p} - V_{B(Q3)}\text{p-p}}{2 \times R6}$

  b. $I_{B(Q4)} = \dfrac{V_{C(Q1)}\text{p-p} - V_{B(Q4)}\text{p-p}}{2 \times R5}$

12. Measure the base currents of Q3 and Q4. Calculate $\beta(Q3) = \beta(Q4)$ from: I_C/I_B.

NOTE: If R5 is adjusted for symmetrical, clean square-wave conditions, the base currents should be close in value.

13. Increase the loading of Rl and monitor the cirucit currents. A 100 Ω resistor in parallel with the existing 50 Ω load will do. Monitor either the collector of Q3 or that of Q4, with respect to ground or measure across the load to make sure that the circuit continues to operate during this monitoring procedure. Also, since the transformer is audible, the "singing" will cease if the oscillator quits.
  a. While the 100 Ω resistor is in parallel with the load resistor, monitor the Q3 and Q4 collector currents and base currents.

NOTE: Place a DC ammeter in series with the base or collector leads.

  b. Calculate the beta requirements of Q3 and Q4 from $\beta(Q3) = \beta(Q4) = I_C/I_B$. Does beta change?
14. Monitor the collector currents of Q3 and Q4, the current flow (IDC) out of the power supply. Also, monitor the current flow out of the center-tap lead. Compare to IDC.
  a. For a 50 Ω load condition.
  b. For a load condition of 50 Ω and 100 Ω in parallel combination. Increased loading can also be used, but that is dependent on the power source available.
  c. Allow the load to "warm up" (to the touch) and touch the Q3 and Q4 devices. Are they cooler? If so, why?
15. For the loaded conditions of 50 Ω and the parallel combination of 50 Ω and 100 Ω, calculate the efficiency of the circuit from:

$$\% \text{ efficiency} = Po/P_{PS} \times 100$$

where: $P_{PS} \approx V_{CC}IDC$ and $Po = Vop\text{-}p^2/4RL$ for each of the loaded conditions. However, since Vop-p decreases slightly with increased loading, it should be monitored for no load and the two loaded conditions.
16. Insert the calculated and measured values, as indicated, into Table 5-2.

TABLE 5-2	R_L		I_C		IDC	I_B		Vop-p	P_{PS}	% eff
	Po	I(av)	Q3	Q4		Q3	Q4			
50 Ω LOAD CALCULATED								///////		
50 Ω LOAD MEASURED	///////	///////							///////	///////
50 Ω // 100 Ω CALCULATED			///////	///////	///////	///////	///////	///////		
50 Ω // 100 Ω MEASURED	///////	///////							///////	///////

273

PART 2 — THE FREE-RUNNING INVERTER

GENERAL DISCUSSION

The main drawback to the driven "fixed frequency" DC to AC inverter, in which discrete component astable multivibrators are used, is that the astable circuit is limited by the amount of current it can deliver to the bases of push-pull inverters. This, in turn, limits the collector current. Therefore, large power requirement are not attained, normally, by using this circuit, and only increased circuit compexity, in achieving lower driving impedances, will increase the available power that can be delivered to the load.

A better method for increasing available power to the load is to use the free-running inverter circuit shown in Figure 5-10. In this circuit, the transformer is alternately driven into saturation, first in one direction as Q1 saturates and then in the other direction as Q2 saturates, and the maximum value of base current (and hence collector current) is limited only by the induced voltage of the base windings and the value of the base resistors. Therefore, at high power and high frequency conditions, the choice of devices becomes critical.

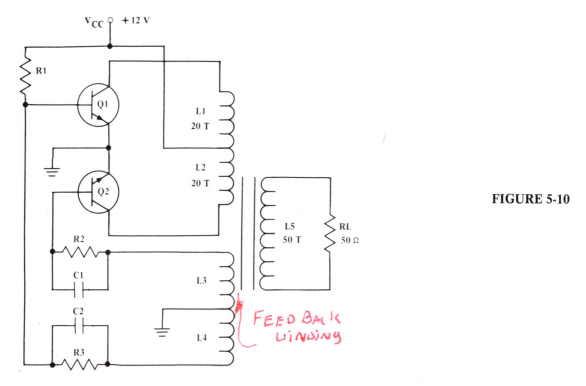

FIGURE 5-10

NOTE: The commutating capacitors C1 and C2 are connected in parallel with base resistors R2 and R3 to reduce transition time and to aid turn-on and turn-off time of the transistors.

THEORY OF OPERATION

When power is applied to the circuit shown in Figure 5-10, Q1 begins to conduct, because of biasing resistor R1, and drives itself into a saturated condition, and approximately $V_{CC} - V_{CE(SAT)}$ is applied across the upper L1 winding.

NOTE: Resistor R1 serves only as a starting resistor. Once the circuit is in operation, it can be removed and oscillations will continue.

Since the voltage applied to the L1 winding is constant, a constant DC current will flow. In addition, a linearly increasing (di/dt) current will flow through the inductor and, hence, through R1, which increases the voltage across L1 to L1di/dt. Also, this additional voltage is induced across the other L2, L3, and L4 windings.

Since the polarity of the increasing voltage across feedback winding L4 is such that it increases the Q1

274

base current flow, then Q1 is forced into further saturation. Too, the polarity across both L2 and L3 insures that transistor Q2 will remain, momentarily, in cutoff. However, with Q1 in saturation, the primary voltage build-up can no longer increase, but Q1 can continue to supply collector current to the "load" as long as L1 is not saturated and the voltage developed across L1, during the build-up time, is maintained. But, during this constant voltage across L1 condition, the magnetic flux coninues to increase ($V = N \, d\Phi/dt$) and, when it ceases to increase, L1 is in saturation.

Essentially, as long as L1 is not in saturation, the rate of change of the collector current (di/dt) remains constant and the voltage across $L_1 V_{L_1} = L_1 di/dt$ remains constant. But as soon as L1 goes into saturation, the entire transformer goes into saturation, and the rate of change of collector current, because it is not restricted by the inductance of L1, increases rapidly.

NOTE: The increased collector current is a result of decreasing inductance, but the decreasing inductance of the collapsing field results in less and less base current drive and Q1 comes out of saturation.

As the flux field collapses, Q2 begins to conduct, the polarities of the voltages across the windings are reversed, the core is driven into negative saturation, the flux field collapses toward zero, and the cycle begins again. The polarity for both the positive saturated condition, where Q1 is in saturation and Q2 is in cutoff and the negative saturated condition, where Q2 is in saturation and Q1 is in cutoff, is shown in Figure 5-11.

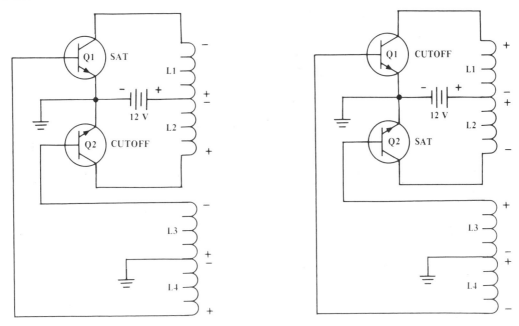

FIGURE 5-11

TRANSFORMER CONSTRUCTION AND DESIGN

The analysis of the transformer used in the push-pull, free-running circuit of Figure 5-10 is straight-forward. If the turns of the primary, secondary, and feedback windings, alone, are known, and knowing the V_{CC}, all the other circuit parameters can be estimated. However, the number of turns, the size of the core, the operating frequency, and the type of core used are all part of the design input and should be investigated to support the circuit analysis.

NOTE: The design approach is only, on rare occasions, more accurate than analysis techniques, because analysis techniques begin with the finished product, while design techniques begin at "square one".

TRANSFORMER CONSTRUCTION

Transformer construction techniques merely involve the duplication in hardware of the schematic diagram of Figure 5-12(a). Use magnetic wire and (usually) a ferrite toroidal core.

The primary is wound first, the feedback winding second and, if there is room, the secondary winding is added. A representative winding technique is illustrated in Figure 5-12(b). Therefore, assuming the primary and feedback windings as a first layer and the secondary winding as a second layer, the winding procedure is as follows:

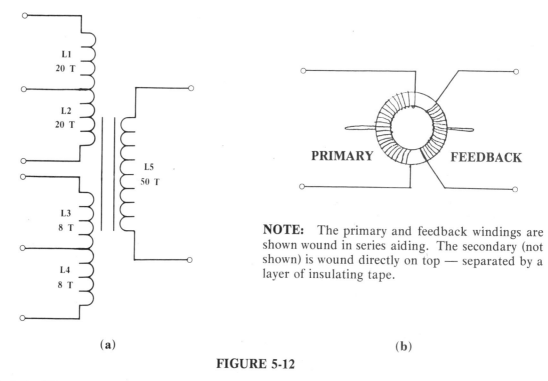

NOTE: The primary and feedback windings are shown wound in series aiding. The secondary (not shown) is wound directly on top — separated by a layer of insulating tape.

(a)   (b)

FIGURE 5-12

1.   Wind the 40 turns, center-tapped primary. Wind 20 turns, bring out the wire, twist two or three times, and continue winding the remaining 20 turns.
2.   Wind the 16 turn, center-tapped, feedback winding and maintain the same direction of winding so there is an effective series aiding condition. Wind 8 turns, bring out the wire, twist two or three times, and continue winding the remaining 8 turns.
3.   Wrap a layer of insulating tape over the windings. Wrap 50 turns of secondary wire directly over the tape. The direction of winding is not critical. Identify the leads and use tape to secure the secondary in place. Do not overtape because, when the power delivered to the load is high, the core can get quite warm. And too much tape will trap the heat and create further heat losses.
4.   The wire size used is also a consideration. Too large a wire size will limit the number of turns for the core used, and too small a wire size can cause excessive copper losses contributing to a possible overheating problem of the core material.
5.   Ferrite core material is generally used in free-running, saturating, power transformers because ferrite material saturates easily. Also, it will operate at frequencies above 1 kHz up to 20 kHz with moderate core loss. Above 20 kHz, core material and the size of the core, for the power handled, becomes critical. They must be carefully selected by the designer.

CORE SIZE

To calculate transformer equations, it is necessary to know the transformer cross sectional area and the mean length. These parameters are given in CM² and CM respectively. However, some manufacturers give the diameters of toroids (inner and outer) and the height of the core in inches, which have to be converted to centimeters. To find the cross sectional area and the mean length, in centimeters, of the toroid shown on the next page, where the measurements are in inches, is as follows:
1.   The cross sectional area of the toroid is found by multiplying the height in centimeters by the width in centimeters. Therefore, the height of 1/2 inches converted to centimeters is $0.5 \times 2.54 = 1.27$ CM, and the width of 1/4 inches, converted to centimeters, is $0.25 \times 2.54 = 0.635$ CM. Therefore, the cross sectional area is: 1.27 CM $\times$ 0.635 CM $\approx$ 0.81 CM².
2.   The mean lenght (ℓ) of the toroid is found by taking the centimeter average of the outer and inner

circumferences.

   a. Outer Circumference

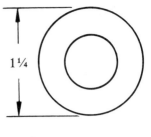

$$\text{In Inches:} \quad C = 2\pi R = 2\pi D/2 = 6.28 \times 1.25/2 = 3.93 \text{ inches}$$

$$\text{In Centimeters:} \quad C = 3.93 \text{ inches} \times 2.54 = 9.97 \text{ CM}$$

   b. Inner Circumference

$$\text{In Inches:} \quad C = 2\pi R = 2\pi D/2 = 6.28 \times 0.75/2 = 2.36 \text{ inches}$$

$$\text{In Centimeters:} \quad 2.36 \text{ inches} \times 2.54 = 5.98 \text{ CM}$$

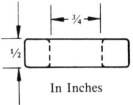

In Inches

   c. Mean Length $= (OC + IC)/2 = (9.97 \text{ CM} + 5.98 \text{ CM})/2 = 7.98 \text{ CM}$

CIRCUIT ANALYSIS

The circuit of Figure 5-14 is analyzed much the same as the astable multivibrator driven inverter circuit. The main difference is that the feedback windings have replaced the astable multivibrator. However, the operating frequency is not as easily found because maximum magnetic flux density transformer core data and the area (size) of the transformer core are needed. Also, the V_{CC} and the number of turns on the primary must be known. So, if $V_{CC} = 12$ V, N(pri) $= 40$ T, $B_m = 2000$, and the core area is 0.81 CM²:

$$f_o = \frac{V_{CC}}{4 \times 10^{-8} \, A \, N(pri) B_m} = \frac{12 \text{ V}}{4 \times 10^{-8} \times 8.1 \times 10^{-1} \times 40 \times 2 \times 10^3} = \frac{12 \text{ V}}{2.592 \times 10^{-3}} \approx 4.63 \text{ kHz}$$

NOTE: Obviously, if the V_{CC} is increased, or the area of the core of the number of turns is decreased, the operating frequency will be increased proportionately.

Also, if the mean length of the core is 7.98 CM and the permeability is 1000, then the inductance of the 40 T primary of the transformer can be calculated from:

$$L(pri) = \frac{4\pi N(pri)^2 \times 10^{-9} \times \mu A}{\ell} = \frac{4\pi \times 40^2 \times 10^{-9} \times 10^3 \times 8.1 \times 10^{-1}}{7.98}$$

$$\approx \frac{12.57 \times 1.6 \times 10^3 \times 10^{-9} \times 10^3 \times 8.1 \times 10^{-1}}{7.98} = \frac{162.86 \times 10^{-4}}{7.98} \approx 2.04 \text{ mH}$$

NOTE: To increase the inductance, the number of turns or the permeability should be increased. Increasing the area, also increases the mean length (usually), but it is the least desirable method to use. Also, increasing the number of turns is limited by the size of the core. Therefore, increasing the permeability (coil selection) is the best solution to increase the inductance.

1. The peak-to-peak square waves at the collectors of Q1 and Q2 are approximately $2V_{CC} = 2 \times 12$ V $= 24$ Vp-p or, more accurately:

$$V_{C(Q1)}\text{p-p} = V_{C(Q2)}\text{p-p} = V_{CC} - V(\text{SAT Q1}) + V_{L1} = (12 \text{ V} - 0.5 \text{ V}) + 12 \text{ V} = 23.5 \text{ Vp-p}$$

2. The voltage developed across the secondary winding L5 can be solved in two ways: by the ratio of N1 to N5 or N2 to N5, where the peak-to-peak voltage at either collector and associated winding is coupled, alternately, to the L5 secondary winding.

$$V(\text{sec})\text{p-p} = \frac{V_{C(Q1)}\text{p-p} \times N5}{N1} = \frac{V_{C(Q2)}\text{p-p} \times N5}{N2} = \frac{23.5 \text{ Vp-p} \times 50 \text{ T}}{20 \text{ T}} = 58.75 \text{ Vp-p}$$

or by the ratio (L1 + L2) to L5 where the total voltage across (L1 + L2) is coupled to L5:

$$V(\text{sec})\text{p-p} = \frac{V(L1 + L2)\text{p-p} \times L5}{N1 + N2} = \frac{47 \text{ Vp-p} \times 50 \text{ T}}{40 \text{ T}} = 58.75 \text{ Vp-p}$$

3. The voltage developed across the center-tapped, feedback winding also can be solved in two ways: by the ratio of N1 to N4, or by the ratio of N2 to N3, since they switch alternately,

$$V_{L4} = \frac{V_{C(Q1)}\text{p-p} \times N4}{N1} = \frac{23.5 \text{ Vp-p} \times 5 \text{ T}}{20 \text{ T}} \approx 5.88 \text{ Vp-p}$$

$$V_{L3} = \frac{V_{C(Q1)}\text{p-p} \times N3}{N2} = \frac{23.5 \text{ Vp-p} \times 5 \text{ T}}{20 \text{ T}} \approx 5.88 \text{ Vp-p}$$

or, by the ratio of (N1 + N2) to (N3 + N4) and then halving (N3 + N4), since N3 = N4 = (N3 + N4)/2,

$$(V_{L4} + V_{L3}) = \frac{V(L1 + L2) \times (N3 + N4)}{N1 + N2} = \frac{47 \text{ Vp-p} \times 10 \text{ T}}{40 \text{ T}} = 11.75 \text{ Vp-p}$$

$$V_{L3} = V_{L4} = (V_{L4} + V_{L3})/2 = 11.75 \text{ Vp-p}/2 \approx 5.88 \text{ Vp-p}$$

4. Calculating the power developed across a 50 Ω load, where a coefficient coupling of 0.96 is assumed:
   a. Solving for the V_{RL}p-p, where K = 0.96 is assumed:

$$V\text{op-p} = V_{RL}\text{p-p} = V(\text{sec})\text{p-p} \times 0.96 = 58.75 \text{ Vp-p} \times 0.96 \approx 56.4 \text{ Vp-p}$$

   b. $Po = V\text{op-p}^2/4RL = 56.4^2/(4 \times 50 \text{ Ω}) = 15.9$ watts

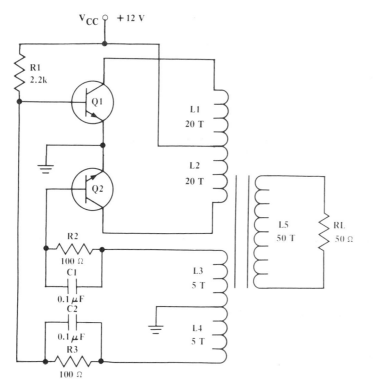

FIGURE 5-13

5. Calculating the power delivered to the primary windings to provide 15.9 watts at the load, assuming a 0.96 coefficient of coupling:

$$P(\text{pri}) = Po/K = 15.9 \text{ W}/0.96 \approx 16.56 \text{ watts}$$

6. Calculating the average (DC) current that the supply voltage must provide to insure approximately 16.56 watts of power to the primary windings:

$$IDC = P\text{pri}/V_{CC} = 16.56 \text{ W}/12 \text{ V} = 1.38 \text{ A}$$

278

NOTE: Since Q1 and Q2 supply current to the load, alternately, the current (1.38 A) will be shared by both devices.

7. Calculating the approximate average collector current of the Q1 and Q2 devices:

$$I_{C(Q1)} = I_{C(Q2)} = IDC/2 = 1.38 \text{ A}/2 = 690 \text{ mA}$$

8. Calculating the approximate average base current the devices must have (assuming a beta of 50 for power transistors):

$$I_{B(Q1)} = I_{B(Q2)} = I_C/B = 690 \text{ mA}/50 = 13.8 \text{ mA}$$

NOTE: Designers will normally derate the beta by 50% to insure a saturated condition, making the base current as high as 27.6 mA.

9. Solving for the approximate voltage drop across base resistors R2 and R3, where $V_{L3}\text{p-p} = V_{L4}\text{p-p} \approx$ 5.88 Vp-p and the voltage at the bases of Q1 and Q2 is assumed to be 1.88 Vp-p:

$$V_{R2}\text{p-p} = V_{L3}\text{p-p} = V_{B(Q1)}\text{p-p} = 5.88 \text{ Vp-p} - 1.88 \text{ Vp-p} = 4 \text{ Vp-p}$$

$$V_{R3}\text{p-p} = V_{L4}\text{p-p} = V_{B(Q2)}\text{p-p} = 5.88 \text{ Vp-p} - 1.88 \text{ Vp-p} = 4 \text{ Vp-p}$$

NOTE: The voltage developed at the bases of Q1 and Q2 is dependent on the collector current, the beta of the transistors, and the series bases resistors.

10. Solving for the average (DC) base resistor voltage across the R2 and R3 resistors:

$$V_{R2} = V_{R2}\text{p-p}/2 = 4 \text{ Vp-p}/2 = 2 \text{ VDC}$$

$$V_{R3} = V_{R3}\text{p-p}/2 = 4 \text{ Vp-p}/2 = 2 \text{ VDC}$$

11. Solving for the approximate R2 and R3 base resistor values based on a nominal base current value of 20 mA, which is between the calculated 13.8 mA and the (design) upper limit of 27.6 mA:

$$V_{R2} = V_{R3} = V_{R2}/I_{B(Q2)} = V_{R3}/I_{B(Q3)} = 2 \text{ VDC}/20 \text{ mA} = 100 \text{ }\Omega$$

12. **DEVICE CHARACTERISTICS**

a. The voltage breakdown for the power devices is calculated from $2V_{CC}$ but, because of spikes, a safe design value is $3 V_{CC}$. Therefore:

$$V_{CEo} = 3V_{CC} = 3 \times 12 = 36 \text{ volts}$$

b. Based on the nominal values of 20 mA of base current and 690 mA of collector current, the minimum beta for the Q3 and Q4 transistors is:

$$\beta(\text{min}) = I_C/I_B = 690 \text{ mA}/20 \text{ mA} \approx 35$$

c. The power dissipated by the devices is the sum of the power dissipated during saturation when the voltage dropped across the device is at its lowest and the current through the device is at its highest, during cutoff when the voltage across the device is at the highest but the current through the device is at its lowest, and during the turn-on and turn-off times, when both maximum voltage and current conditions are assumed for that (hopefully) small time with regard to the total time. Therefore:

1. The power dissipated by Q1 when it is in saturation:

$$P_{Q1(SAT)} = T_{on}/T (IDC \times V_{CE[SAT]}) = (1.38 \text{ A} \times 0.5 \text{ V})/2 = 345 \text{ mW}$$

2. Power dissipated by Q1 when it is in cutoff:

$$P_{Q1(\text{cutoff})} = T_{off}/T (2V_{CC} \times I_{CEX}) = (2 \times 12 \text{ V} \times 5 \text{ mA})/2 = 60 \text{ mW}$$

3. Power dissipated by Q1 durng rise and fall times where: $t_r = t_f = 2$ μsec. Therefore, since $T = 1/F = 1/1$ kHz $= 1$ msec, the power dissipated is:

$$P_{Q1}(t_r + t_f) = (t_r + t_f)/T \times (V_{CC} \times I_C)/2 = 4 \text{ μsec}$$

$$4 \text{ μsec}/1000 \text{ μsec} \times (12 \text{ V} \times 1.38 \text{ A})/2 \approx 33 \text{ mW}$$

4. Therefore, the total power dissipated by Q1, and hence Q2, is:

$$P_{Q1} = P_{Q2} = P(SAT) + P(\text{cutoff}) + P(t_r + t_f) = 345 \text{ mW} + 60 \text{ mW} + 33 \text{ mW} = 438 \text{ mW}$$

NOTE: The power dissipated by the devices is on the borderline between using medium power or higher power devices. For instance, the switching frequency at 1 kHz is not critical so either a medium power 2N3053 or a higher power 2N3055 can be used. The medium power device, because it has a lower $V_{CE(SAT)}$, will dissipate less power during saturation and, because it has a lower I_{CEX}, it will dissipate less power during cutoff. However, the 2N3055 is capable of delivering considerably more output power and can be used. The trade-off is a matter of design.

EXPERIMENTAL OBJECTIVES

To investigate the free-running, transformer-feedback inverter.

LIST OF MATERIALS

1. Core Material: Ferrite Toroidal Material
2. Magnetic Wire: 20 to 24 gauge
3. Transistors: 2N3053 or 2N3055 (two) or equivalents
4. Resistors: 50 Ω (one)   100 Ω (two)   1 kΩ (one)
5. Capacitors: 0.1 μF (two)

EXPERIMENTAL PROCEDURE

1. Wind a 40 turn center-tapped primary, a 12 turn center-tapped feedback winding, and a 20 turn secondary winding. Make sure the feedback winding is wound in the same direction as the primary winding, as previously mentioned. The direction of the secondary winding is not critical.

2. Connect the circuit, as shown in Figure 5-14, where two 2N3055's are used to minimize the possibility of the destruction of the device. However, medium power transistors can be substituted for the 2N3055 if they are mounted on heat sinks and adequate construction care is taken.

3. Calculate the square-wave amplitudes at the collectors of Q1 and Q2 to be approximately $2V_{CC}$.

4. Calculate the square wave amplitude across the load-secondary, L5 winding from:

$$V_{op\text{-}p}(\text{load}) = (V_{C[Q1]}\text{p-p} \times N5)/N1 = (V_{C[Q2]}\text{p-p} \times N5)/N2$$

NOTE: · N represents the number of turns of the designated winding.

5. Calculate the square wave amplitude across the L3 and L4 windings from:

    a. $V(L3)\text{p-p} = (V_{C[Q1]}\text{p-p} \times N3)/N2$

    b. $V(L4)\text{p-p} = (V_{C[Q2]}\text{p-p} \times N4)/N1$

6. Monitor the amplitude of the square wave:
    a.   At the collector of Q1 with respect to ground.
    b.   At the collector of Q2 with respect to ground.
    c.   Across the load resistor-secondary winding L5.
    d.   Across the feedback winding L3.
    e.   Across the feedback winding L4.
    f.   At the base of Q1 with respect to ground.
    g.   At the base of Q2 with respect to ground.

7. Monitor the frequency.
   a. Monitor the on-time (T_{on}).
   b. Monitor the off-time (T_{off}).
8. Calculate the power across the load from:

$$Po = Vop\text{-}p^2 / 4RL$$

9. Calculate the average current:
   a. Through the load from:

$$I_L(av) = 2Po/Vop\text{-}p = Vop\text{-}p/2R_o$$

   b. Delivered by the power supply:

$$IDC \approx 2Po/V_{C(Q1)}p\text{-}p = 2Po/V_{C(Q2)}p\text{-}p$$

   c. Through the Q1 and Q2 transistors:

$$I_{C(Q1)} = I_{C(Q2)} \approx IDC/2$$

$$I_{C(Q1)} = I_{C(Q2)} \approx Po/V_{C(Q1)}p\text{-}p = Po/V_{C(Q2)}p\text{-}p$$

NOTE: Base current flow and I^2R power losses are not included in the calculations and, therefore, IDC will always be slightly higher than $(I_{C[Q1]} + I_{C[Q2]})$.

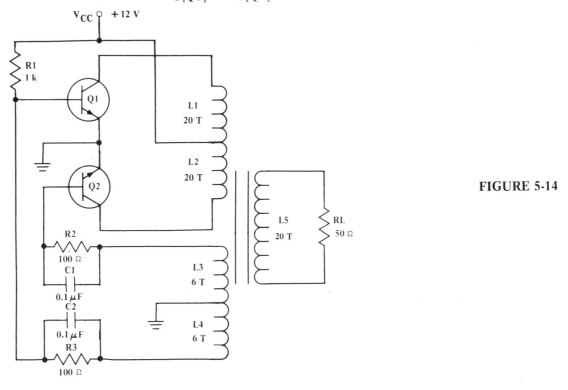

FIGURE 5-14

10. Calculate the (average) base currents of Q1 and Q2 from:

$$I_{B(Q1)} = \frac{V(L4)p\text{-}p - V_{B(Q1)}p\text{-}p}{2R_3} \quad or \quad I_{B(Q2)} = \frac{V(L3)p\text{-}p - V_{B(Q2)}p\text{-}p}{2R_2}$$

NOTE: The difference of the two voltages provides the peak-to-peak amplitude and this value, divided by two, provides the (DC) average current.

11. Measure the IDC, $I_{C(Q1)}$, $I_{C(Q2)}$, $I_{B(Q1)}$, and $I_{B(Q2)}$.

12. Calculate the power delivered from the source from: $P_{PS} = V_{CC}IDC$

13. And knowing Po, calculate the efficiency from: % efficiency $= Po/P_{PS} \times 100$

NOTE: The load resistor of 50 Ω can be increased or decreased, depending on the availability of the power supply source, to supply the necessary power.

14. Touch the load resistor and the power transistors to test the 50 Ω loaded condition for heat. A thermometer can also be used, obviously, but this test is relative only. It demonstrates a condition of high efficiency circuits, where the load "runs" warm and the transistors cool.

15. Insert the calculated and measured values, as indicated, into Table 5-3 and Table 5-4.

TABLE 5-3	f_o	V_Cp-p		Vp-p		V_Bp-p		Vop-p	T_{on}	T_{off}	Po
		Q1	Q2	L3	L4	Q1	Q2	L5			
CALCULATED	//////					//////	//////		//////	//////	
MEASURED											//////

TABLE 5-4	I_L(av)	IDC	IDC		I_B(DC)		P_{PS}	%EFF
			Q1	Q2	Q1	Q2		
CALCULATED								
MEASURED	//////						//////	//////

PART 3 — SWITCHING REGULATORS

GENERAL DISCUSSION

Switching regulators, or switchers as they are sometimes referred to, are primarily used in higher power applications to take advantage of high circuit efficiencies. They can be constructed with all discrete devices or they can be constructed with monolithic devices, where discrete devices are then used to handle the power. The discrete-device-only circuitry is primarily used in high-power, high voltage circuits, where the voltage levels of the monolithics are exceeded. However, in lower-power, low-voltage circuitry, monolithic devices minimize the number of components used and greatly simplify the circuitry.

The building blocks of most switching regulators are the reference voltage, the differential amplifier (comparator), the series pass circuitry, and the LC filter. A basic, but not very practical circuit configuration, is shown in Figure 5-16. In this circuit, it is assumed that the output of the op-amp will turn on and turn off the series pass device. Note, however, that it is necessary to provide a power dissipating collector resistor to make sure that the Q1 series pass transistor will saturate. Otherwise, the collector is connected directly to the low resistance of the power source and Q1 cannot fully turn on. Therefore, if the power dissipation of the R1 resistor is ignored and Vin (only) is considered, an introduction to the basic switching regulator can be made.

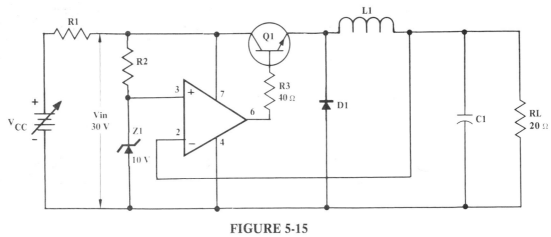

FIGURE 5-15

CIRCUIT OPERATION

When power is applied, initially, to the circuit of Figure 5-16, the voltage at Pin 3 of the op-amp will be established at 10 volts. Then, since the output filter capacitor is not charged (theoretically), Pin 2 is lower in voltage value than Pin 3, and the comparator output voltage switches to approximately Vin, causing Q1 to conduct. Therefore, with Q1 in saturation, the filter capacitor attempts to charge toward Vin = V_{CC} = 30 V, but as soon as the voltage across capacitor C1 is slightly larger than 10 volts, the voltage at Pin 2 is slightly higher in value than that at Pin 3, and the output of the comparator switches to approximately ground condition, forcing Q1 into cutoff. Obviously, with the base at a near zero volt condition and the emitter at about 10 volts, Q1 is, indeed, cut off.

Therefore, Pin 2 of the op-amp samples the output voltage, compares it to the zener voltage at Pin 3, and the output of the comparator (op-amp) switches "on" and "off" to provide a regulated 10 volts to the output load. Hence, the waveshape at the output of the comparator (a square wave) switches between approximately zero volts and Vin at a frequency that is determined by Vin, Vo, L, C, and the charging current through the inductor.

The duty cycle of the circuit is determined by the voltage out as compared to the voltage in. If no DC voltage drop across the transistor or no DC resistance of the inductor is assumed (to simplify the mathematics), then a 50% duty cycle occurs when the input voltage is 20 volts, a 40% duty cycle occurs when the input voltage is 25 volts, and a 60% duty cycle occurs when Vin is 16.67 volts. Therefore, the average (or DC) output volts, for each of the conditions of input voltage and duty cycle, is 10 volts. The three conditions are shown in Figure 5-17.

Essentially the series pass device is switched on and off to meet the needs of the voltage-out with reference to the voltage-in condition. If the input voltage decreases, the duty cycle increases and, if the input

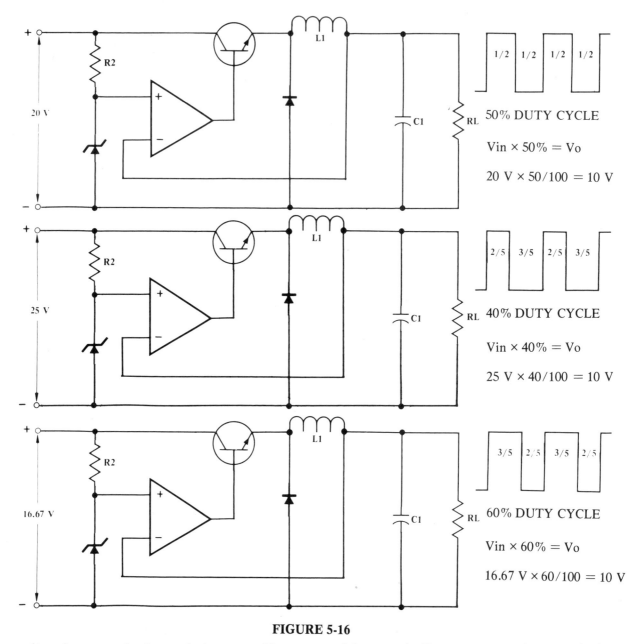

FIGURE 5-16

voltage increases, the duty cycle decreases. The average of the on-and-off, square-wave voltage equals the output voltage. Switches can be designed to provide a 10% to 90% duty cycle.

An obvious advantage of the switches is that large voltage differences can exist between the input and output with considerably less power dissipated by the devices than with a comparable linear, series-pass circuit. The reason is that maximum current conditions exist when the voltage across the device is at its lowest (in saturation), and the minimum current conditions exist when the voltage across the device is at its highest (in cutoff). For the linear, series-pass configuration, the device can operate at maximum voltage and maximum current conditions.

Certainly, switching regulators are more efficient than comparable linear circuits, they can operate from a DC source or a rectified bulk supply and, because the operating frequency can be made considerably higher than the 60 Hz line frequency, the LC components can be, generally, smaller in size and lighter in weight.

However, there are some drawbacks to switching regulators. At operating frequencies much beyond 20 kHz, the devices turn-on and turn-off time becomes critical. The reason for this is that during rise and fall times, the full load current flows through the device. This causes the device to dissipate needless power if the time to turn on and the time to turn off are too long. Also, switching large amounts of current

284

produces spikes and electromagnetic interference, so the EMI must be minimized by using good RF practices in circuit layout and minimum wire length. Too, voltage (current) spikes induce further circuit noise which must be limited, because spikes can destroy the circuit devices through high and second voltage breakdown. A commutating, free-wheeling diode at the junction of L1 and the emitter of Q1 will minimize large spike possibilities when Q1 is in cutoff, but at frequencies approaching 20 kHz, a fast diode, such as the Schotsky barrier diode, must be used.

DIRECT CURRENT ANALYSIS

The DC analysis of the switching regulator, shown in Figure 5-16, begins with the zener voltage and, hence, the output voltage. Then, the voltage drops of the circuit are calculated, providing I_C and Beta are known. The voltage at the output of the comparator will set the stage for the initial duty cycle, where the op-amp switches between zero and 30 volts. The 12 volt output must be the average, or two-fifths, of 30 volts. Therefore, the DC sets the conditions of switching based on input and output voltages.

1. Solving for the DC voltage drops around the circuit:

    a. $V_o = V(zener) = V(Pin\ 3) = V(Pin\ 2) = 10$ volts

    b. $I_{RL} = V_o/RL = 10\ V/20\ \Omega = 500$ mA

    c. $V_{L1}(DC) = I_{RL} \times R_{L1} = 500\ mA \times 2\ \Omega = 1$ volt

    d. $V_{BE(Q1)} = 0.6$ volts

    e. $V_{R3} = (I_{RL}/\beta[Q1]) \times R3 = (500\ mA/50) \times 40\ \Omega = (5 \times 10^{-1} \times 40)/(5 \times 10^{1}) = 0.4$ volts

2. Solving for the voltages taken with respect to ground:

    a. $V_o = 10$ volts

    b. $V_{E(Q1)} = V_o + V_{L1}(DC) = 10\ V + 1\ V = 11$ volts

    c. $V_{B(Q1)} = V_{E(Q1)} + V_{BE(Q1)} = 11\ V + 0.6\ V = 11.6\ V$

    d. $V_{Pin\ 6(op\text{-}amp)} = V_{B(Q1)} + V_{R3} = 11.6\ V + 0.4\ V = 12$ volts

    e. $V_{in} = 30$ volts

ALTERNATING CURRENT ANALYSIS

Designers normally make sure that the charging and discharging inductor current is not more and 10% of the maximum load condition so the current flow through the inductor is continuous. Therefore, if the load current is 0.5 amperes, then the charging and discharging current should be 50 mA. However, capacitor C1 also helps in providing current to the load and the larger the value of C1 (within functional limits), the less the current demand on the inductor.

NOTE: If the inductor used is not gapped, then the core material can saturate if the current "demand" is too high. Capacitor C1 minimizes this possibility.

Essentially, the frequency of oscillation is determined by the DC parameters of the circuit. For instance, if the frequency of oscillation is selected at 10 kHz and, since the selected DC voltage level controls the duty cycle, then 12 volts will be needed at the output of the comparator if 10 volts is required at the load. Then, the ratio of the input voltage to the comparator output voltage, with regard to the operating frequency, and hence the time of one complete cycle, provides the duty cycle or turn-on time. Therefore:

1. $T = 1/f = 1/10\ kHz = 10^{-4} = 100\ \mu sec$

2. $T_{on} = V_o/V_{in} \times T = 12\ V/30\ V \times 10^{-4} = 40\ \mu sec$

3. $T_{off} = T - T_{on} = 100 \ \mu sec - 40 \ \mu sec = 60 \ \mu sec$

$\therefore T = T_{on} + T_{off} = 40 \ \mu sec + 60 \ \mu sec = 100 \ \mu sec$

4. Duty Cycle $= T_{on}/T \times 100 = 40 \ \mu sec/100 \ \mu sec \times 100 = 40\%$

Furthermore, the voltage out of the comparator, in this case, is a function of the on-time and the input voltage versus the total time of one complete cycle. Hence:

$$V(\text{Pin } 6) = \frac{V_{in}T_{on}}{T_{on} + T_{off}} = \frac{V_{in}T_{on}}{T} = \frac{30 \ V \times 40 \ \mu sec}{100 \ \mu sec} = \frac{30 \times 4 \times 10^{-5}}{10^{-4}} = 12 \text{ volts}$$

which is required to provide 10 V DC to the load, where:

$$V_o = V_{\text{Pin } 6} - [V_{R1} + V_{BE(Q1)} + V_{L1}(DC)] = 12 \ V - [0.4 \ V + 0.6 \ V + 1 \ V] = 10 \text{ volts}$$

Since the amplitude of the 40% duty cycle at the output of the comparator (Pin 6) switches between approximately zero volts and Vin = 30 volts, the peak-to-peak, square-wave amplitude is:

Vop-p (Pin 6) $\approx$ 30 Vp-p

Then, because the average of the 30 Vp-p, 40% duty cycle, square wave (as illustrated in Figure 5-18) is 12 volts, the transistor is turned on for 2/5 of the total time and turned off for 3/5 of the total time, or it is on for 40 μsec and off for 60 μsec for a total time of 100 μsec.

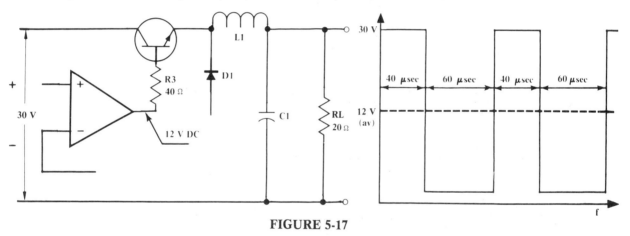

FIGURE 5-17

Therefore, during the "ON TIME" the load gets 10 volts and, if we assume (for simplicity) that the saturated transistor and the DC resistance of the coil drop the remaining 2 volts, then the voltage drop across the inductance is 18 volts. So in order to maintain a charging and discharging continuous current flow through the inductor or 50 mA, the inductor value must be:

$$L = V_L/\Delta I_L \times T_{on} = (18 \ V \times 40 \ \mu sec)/50 \ mA = (18 \times 4 \times 10^{-5})/(5 \times 10^{-3}) = 14.4 \text{ mH}$$

Also, the inductance can be solved if the device is considered to be in cutoff, where the output voltage is 10 volts. And, if the additional 2 volts of circuit drop is included in the calculations (to keep the mathematics simple), the inductance is:

$$L = \frac{(V_o + 2 \ V)T_{off}}{\Delta I_L} = \frac{(10 \ V + 2 \ V)60 \ \mu sec}{50 \ mA} = \frac{12 \ V \times 6 \times 10^{-5}}{5 \times 10^{-3}} = 14.4 \text{ mH}$$

NOTE: Initial design practices usually involve the use of ideal, no-loss components to minimize the complexity of the mathematics. Therefore, in this situation, the 2 volts of circuit voltage would not be in question and $L = ([V_{in} - V_o]T_{on})/\Delta I_L$ would equal $V_oT_{off}/\Delta I_L$ exactly.

286

If ideal, no-loss components were used, 12 volts would appear at the load and ΔI_L can be solved for a Q1 that is in cutoff and in saturation. Therefore, the voltage developed across the inductor for:

1. Q1 in saturation:

   a. $\Delta I_L = \dfrac{(V_{in} - V_o)T_{on}}{L1} = \dfrac{V_{L1}T_{on}}{L1} = \dfrac{(30\ V - 12\ V)40\ \mu sec}{14.4\ mH} = \dfrac{18 \times 4 \times 10^{-5}}{14.4 \times 10^{-3}} = 50\ mA$

   b. $V_{L1} = \dfrac{L1 \times \Delta L1}{T_{on}} = \dfrac{14.4\ mH \times 50\ mA}{40\ \mu sec} = \dfrac{14.4 \times 10^{-3} \times 5 \times 10^{-2}}{4 \times 10^{-5}} = 18\ volts$

2. Q1 in cutoff:

   a. $\Delta I_L = \dfrac{V_o T_{off}}{L1} = \dfrac{12\ V \times 60\ \mu sec}{14.4\ mH} = \dfrac{12 \times 6 \times 10^{-5}}{14.4 \times 10^{-3}} = 50\ mA$

   b. $V_{L1} = V_{RL} = \dfrac{L1 \times \Delta I_L}{T_{off}} = \dfrac{14.4\ mH \times 50\ mA}{60\ \mu sec} = \dfrac{14.4 \times 10^{-3} \times 5 \times 10^{-2}}{6 \times 10^{-5}} = 12\ volts$

To solve for the approximate ripple voltage to the output, assuming no loss across the ideal components and using a 100 μF capacitor:

1. $\Delta V_o = \dfrac{(V_{in} - V_o)(T_{on})^2}{LC} = \dfrac{(30\ V - 12\ V)(40\ \mu sec)^2}{14.4\ mH \times 100\ \mu F} = \dfrac{18\ V (4 \times 10^{-5})^2}{14.4 \times 10^{-3} \times 10^{-4}}$

   $= \dfrac{18 \times 16 \times 10^{-10}}{14.4 \times 10^{-7}} = \dfrac{288 \times 10^{-3}}{14.4} = 20\ mV$

2. $\Delta V_o = \dfrac{L \times \Delta I_L^2}{(V_{in} - V_o)C} = \dfrac{14.4\ mH \times 50\ mA^2}{(30\ V - 12\ V)100\ \mu F} = \dfrac{14.4 \times 10^{-3}(5 \times 10^{-2})^2}{18 \times 10^{-4}}$

   $= \dfrac{14.4 \times 10^{-3} \times 25 \times 10^{-4}}{18 \times 10^{-4}} = \dfrac{360 \times 10^{-3}}{18} = 20\ mV$

3. $\Delta V_o = \dfrac{V_{in} - V_o}{LC}\left(\dfrac{V_o}{V_{in}f}\right)^2 = \dfrac{30\ V - 12\ V}{14.4\ mH \times 100\ \mu F}\left(\dfrac{12\ V}{30\ V \times 10\ kHz}\right)^2$

   $= \dfrac{18\ V}{14.4 \times 10^{-3} \times 10^{-4}}\left(\dfrac{0.4}{10^4}\right)^2 = \dfrac{18}{14.4 \times 10^{-7}} \times (0.16 \times 10^{-8}) = 20\ mV$

NOTE: There are several more formula combinations that could be presented to solve for the approximate ripple voltage at the output, but merely three are given to show that algebra and substitution can provide many solutions to the problem.

Then, solving for the frequency using several methods of solution:

1. $f_o = \dfrac{V_o}{V_{in}} \times \dfrac{1}{T_{on}} = \dfrac{12\ V}{30\ V} \times \dfrac{1}{40\ \mu sec} = \dfrac{0.4}{4 \times 10^{-5}} = 10\ kHz$

2. $f_o = 1 - \dfrac{V_o}{V_{in}} \times \dfrac{1}{T_{off}} = 1 - \dfrac{12\ V}{30\ V} \times \dfrac{1}{60\ \mu sec} = \dfrac{1 - 0.4}{6 \times 10^{-5}} = \dfrac{0.6 \times 10^5}{6} = 10\ kHz$

287

3. $f^2 = \dfrac{V_{in} - V_o}{\Delta V_o \times LC}\left(\dfrac{V_o}{V_{in}}\right)^2 = \dfrac{30\,V - 12\,V}{20\,mV \times 14.4 \times 10^{-3} \times 10^{-4}}\left(\dfrac{12\,V}{30\,V}\right)$

$$= \dfrac{18\,V}{20 \times 10^{-3} \times 14.4 \times 10^{-7}} \times 0.4^2 = \dfrac{18 \times 0.16}{288 \times 10^{-10}} = \dfrac{2.88 \times 10^{10}}{288} = 10^8$$

and $\quad f = \sqrt{10^8} = 10^4 = 10\ kHz$

Ideal, no-loss, component conditions were used so that the exact circuit parameter calculations would occur for any number of algebraically manipulated formulas. However, if the actual output voltage of 10 volts is used, the results will be only slightly different, because the 12 volts at the output of the comparator dictates the duty cycle, and that will remain basically the same. Only the voltage drop of $V_{in} - V_o$ will change. But even that will not be dramatic since the difference between the ideal and the actual is only 0.5 volts, if $V_{CE}(SAT)$ for the ideal is 0.5 volts and the actual is $V_{CE}(SAT) = 1$ volts. The $V_{CE}(SAT)$ of power devices does increase with increased collector current and, for the 2N3055, the $V_{CE}(SAT)$ is given in data sheets as 1.1 volts, when $I_C = 4$ amperes.

Therefore, the circuit acts like a switching regulator except for the power loss across the R1 resistor, which guarantees Q1 saturation but, then, it dissipates as much, or more, power in the process of providing a solution to the original problem. And, if R1 were removed and V_{CC} adjusted to $V_{in} = 30$ volts, then Q1 would not turn on fully, and it would have to dissipate the power. Therefore, the efficiency of the circuit is poor. It is used only to introduce switching regulators but it is not a practical switching regulator. Therefore, no laboratory exercise is included with this discussion.

INDUCTORS FOR SWITCHING REGULATORS

High-permeability toroidal cores provide large inductances with few turns, but they are susceptible to being saturated. And, while this is exactly what was needed for the inverter transformers, it is exactly the opposite requirement for inductors used in switching regulators, because switching regulators must be kept out of saturation in order to maintain a continuous current flow to the load.

The ideal inductor for use in switching regulators is pot-core constructed, as shown in Figure 5-18, because ferrite material is used to enhance the inductance, and the construction techniques provide natural "air gaps that minimize magnetic saturation. Cylindrical constructed inductors rank next for use in switching regulator circuits because, while usually more bulky for the amount of inductance obtained, their construction is simple, with many turns wound around a ferrite rod. However, toroid construction can be used if the core material does not saturate easily. For instance, powdered iron core can be used and, in some instance, a ferrite toroid with an air gap. However, the permeability of a "gapped" ferrite toroid is too dependent on the size of the gap, making it a less likely choice.

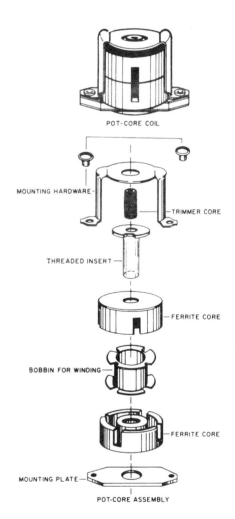

POT-CORE COIL

MOUNTING HARDWARE

TRIMMER CORE

THREADED INSERT

FERRITE CORE

BOBBIN FOR WINDING

FERRITE CORE

MOUNTING PLATE

POT-CORE ASSEMBLY

FIGURE 5-18: Construction of pot-core coil.

Courtesy of J.W. Miller Company

PART 4 — MONOLITHICALLY CONSTRUCTED SWITCHING REGULATORS

GENERAL DISCUSSION

The operation of a monolithic device in a switching regulator is not much different from the operation of the previously analyzed op-amp driven circuit. For instance, the basic building block of all monolithic regulators are still the reference voltage, the differential amplifier (comparator), and the output circuit. In the op-amp driven circuit, the zener diode and the op-amp served these functions and, in the monolithic switching regulator, the monolithic device does the same thing. The only difference is that a PNP series pass transistor is used to switch the load on and off. Since the load is connected to the collector of the transistor, the transistor does not require a power dissipating series resistor to insure that it goes into saturation. Therefore, the efficiency of the monolithic regulator is high — greater than 70%.

Both the LM100 and the μA723 are widely used in monolithic switching regulators. However, the former device has fewer pin connections than the μA723 and the reference voltage is lower, so it gives more flexibility in varying the output voltage. The LM105 is an improved version of the LM100 and either device can be used as the basis of analysis and for application in this exercise.

DIRECT CURRENT CIRCUIT ANALYSIS

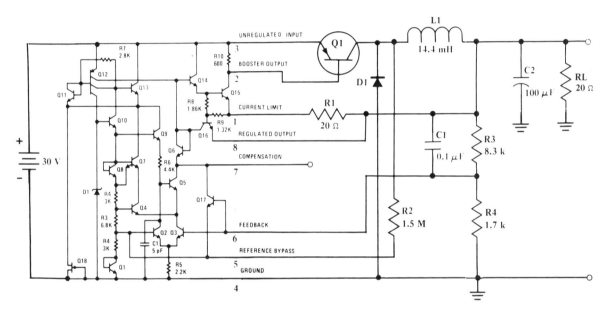

Courtesy of National Semiconductor Corporation

FIGURE 5-19

The direct current analysis begins by knowing the reference voltage at the base of Q2 (approximately 1.7 volts) and, therefore, at the base of Q3 — output of Pin 6, where the voltage developed across resistor R4 is multiplied by the ratio of (R3 + R4)/R4 in providing the output voltage. Then, the voltage at the emitter of Q1 is the output voltage plus the DC voltage drop across L1. Therefore:

a. $V_{B(Q2)} = V_{B(Q3)} = V(\text{Pin } 6) = V_{R4} \approx 1.7$ volts

b. $V_o = \dfrac{V_{R4} \times (R3 + R4)}{R4} = \dfrac{1.7 \text{ V} \times (8.3 \text{ k}\Omega + 1.7 \text{ k}\Omega)}{1.7 \text{ k}\Omega} = 10$ volts

c. $I_{RL} = \dfrac{V_o}{RL} = \dfrac{10 \text{ V}}{20 \text{ }\Omega} = 500$ mA

d. $V_{L1}(\text{DC}) = I_{RL} \times L1 = 500 \text{ mA} \times 4 \text{ }\Omega = 2$ volts

289

NOTE: The DC resistance of L1 is purposely changed to 4 ohms for the convenience of mathematical calculations in the AC section.

2. The voltage at the emitter of Q1 is V_{CC} and the voltage at the base of Q1 should be approximately 0.6 volts lower. External resistor R1 provides current limiting, and it also has some control over the base current drive series pass transistor Q1. For example, a 20 ohm R1 resistor will develop from 4-5 mA of current, and lowering the value of R1 will provide further current increases that limit out at about 10 mA. Therefore, the base drive (and hence the output current) is effectively limited to beta times I_B or to about 500 mA. Consequently, the voltage at Pin 1 should be about 0.1 volts less than the output voltage and the voltage at Pin 8 should equal the output voltage. Pin 5, the reference bypass terminal, should be approximately equal to the reference voltage of 1.7 volts, Pin 3 is tied directly to V_{CC}, and Pin 7, the compensation terminal, is left uncompensated for switching applications and it is at a voltage equal to, or a few tenths of a volt higher than, Pin 8.

    a. $V_{E(Q1)} = V_{CC} = 30$ volts

    b. $V_{B(Q1)} = V_{CC} - V_{BE(Q1)} = 30\ V - 0.6\ V = 29.4$ volts

    c. $V_{C(Q1)} = V_o + V_{L1}(DC) = 10\ V + 2\ V = 12$ volts

    d. $V(Pin\ 1) = V_o + V_{R1} = 10\ V + 0.1\ V = 10.1$ volts

    e. $V(Pin\ 8) = V_o = 10$ volts

    f. $V(Pin\ 5) \approx V(ref) = 1.7$ volts

    g. $V(Pin\ 3) = V_{CC} = 30$ volts

ALTERNATING CURRENT ANALYSIS

The advantage of the PNP connection of the monolithic switching regulator is that the series pass device will be turned on and off without the need for an in-series resistor, making it considerably more efficient than the op-amp driven circuit. However, in all other respects the two circuits work in a similar manner. The voltage at Pin 6 samples the output voltage and compares it to the reference voltage: if the output voltage is low, the Q1 transistor is turned on and, if the output voltage is high, Q1 is cut off.

1. The peak-to-peak voltage at the collector of Q1 is about 30 Vp-p and the duty cycle is determined by the ratio of the input voltage to the voltage at the emitter of Q1. Therefore, if the output voltage is 10 volts, the voltage at the emitter is 12 volts, and the operating frequency is 10 kHz, then:

    a. $T = 1/f = 1/10\ kHz = 10^{-8} = 100\ \mu sec$

    b. $T_{on} = V_o/Vin \times T = 12\ V/30\ V \times 10^{-4} = 40\ \mu sec$

    c. $T_{off} = T - T_{on} = 100\ \mu sec - 40\ \mu sec = 60\ \mu sec$

    d. Duty Cycle $= T_{on}/T \times 100 = 40\ \mu sec/100\ \mu sec \times 100 = 40\%$

2. Therefore, the voltage at the collector of the series pass transistor Q1 and at the output can be mathematically verified from the turn-on and turn-off periods or from the duty cycle.

    a. $V_{C(Q1)} = \dfrac{Vin \times T_{on}}{T_{on} + T_{off}} = \dfrac{Vin \times T_{on}}{T} = \dfrac{30\ V \times 40\ \mu sec}{100\ \mu sec} = 12$ volts

        $V_{C(Q1)} =$ Duty Cycle $\times$ Vin $= 0.4 \times 30\ V = 12$ volts

    b. $V_o = V_{C(Q1)} - V_{L1}(DC) = 12\ V - 2\ V = 10\ V$

3. The inductance is calculated, like that of the op-amp driven circuit, where 18 volts and then 12 volts are alternately impressed across the inductor during saturation and cutoff, respectively. And, for the

convenience of the mathematics, the inductor drops 2 volts and VCE(SAT) is considered to be zero volts.

a. $L = \dfrac{V_{L1} \times T_{on}}{I_L} = \dfrac{18\ V \times 40\ \mu sec}{50\ mA} = \dfrac{18 \times 4 \times 10^{-5}}{5 \times 10^{-3}} = 14.4\ mH$

b. $L = \dfrac{V_{C(Q1)} \times T_{off}}{I_L} = \dfrac{12\ V \times 60\ \mu sec}{50\ mA} = \dfrac{12 \times 6 \times 10^{-5}}{5 \times 10^{-3}} = 14.4\ mH$

NOTE: The charging current at 50 mA and the voltage developed across L1 for cut-off and saturated conditions were all verified for the op-amp driven circuit. So, they are not repeated. However, note that $V_{C(Q1)} = 12$ volts is used instead of the idealistic value of $V_o = 12$ volts.

4. The ripple voltage to the output can be solved in several ways. Two ways are:

a. $\Delta V_o = \dfrac{(V_{in} - V_{C[Q1]})(T_{on})^2}{L_1 C_1} = \dfrac{(30\ V - 12\ V)(40\ \mu sec)^2}{14.4\ mH \times 100\ \mu F} = 20\ mV$

b. $\Delta V_o = \dfrac{L \times I_L{}^2}{(V_{in} - V_{C[Q1]})C} = \dfrac{14.4\ mH \times 50\ mA^2}{(30\ V - 12\ V)100\ \mu F} = 20\ mV$

5. Resistor R2 is part of the controlled feedback path in this monolithic regulator circuit and it can control, to some degree, the output ripple voltage and the operating frequency. Its value can be approximated if the input impedance to the base of Q2 is known (at approximately 1 kΩ), the peak-to-peak voltage at the collector of the external, series-pass transistor is known (at approximately 30 Vp-p), and the output ripple voltage is known (at approximately 20 mVp-p). Therefore, by rearranging and simplifying the voltage divider equation of:

$$\Delta V_o = \dfrac{V_{in} \times Z_{in}(Base\ Q2)}{R2 + Z_{in}(Base\ Q2)} \approx \dfrac{V_{in} \times Z_{in}(Base\ Q2)}{R2} \quad \text{where } R2 \gg Z \text{ in (base Q2),} \quad \text{then:}$$

$$R2 = \dfrac{V_{in} \times Z_{in}(Base\ Q2)}{\Delta V_o} = \dfrac{30\ Vp\text{-}p \times 1000\ \Omega}{20\ mV} = 1.5\ M\Omega$$

NOTE: Increasing the value of R2 will decrease the ripple voltage, but it will also decrease the operating frequency.

6. The frequency of the circuit can be verified from several formulas. Two are:

a. $f_o = V_{C(Q1)}/V_{in} \times 1/T_{on} = 12\ V/30\ V \times 1/40\ \mu sec = 10\ kHz$

b. $f_o = \left(1 - \dfrac{V_{C(Q1)}}{V_{in}}\right) \times \dfrac{1}{T_{off}} = \left(1 - \dfrac{12\ V}{30\ V}\right) \times \dfrac{1}{60\ \mu sec} = 10\ kHz$

7. POWER DISSIPATION

a. The power dissipated by the load:

$$P_{RL} = V_o{}^2 / RL = 10\ V^2 /20\ \Omega = 100\ V/20\ \Omega = 5\ watts$$

b. The power dissipated by the DC resistance of L1:

$$P_{L1} = V_{L1}(DC) \times I_{RL} = 2\ V \times 0.5\ A = 1\ watt$$

c. The power dissipated by the Q1 series-pass device:
   1. With Q1 in saturation:

$$P(Q1\ SAT) \approx T_{on}/T \times I_{RL} \times V_{CE(SAT)}$$

$$= 40\ \mu sec/100\ \mu sec \times 500\ mA \times 0.5\ V = 0.4 \times 0.25\ W = 100\ mW$$

2. With Q1 in cutoff:

$$P(Q1\ Cutoff) = T_{off}/T \times (V_{CC} - V_{C[Q1]}) \times I_{CEX}$$

$$= 60\ \mu sec/100\ \mu sec \times (30\ V = 12\ V) \times 5\ mA = 0.6 \times 18\ V \times 5\ mA = 54\ mW$$

3. During rise and fall times of 2 μsec each:

$$PQ1(t_r + t_f) = (t_r + t_f)/T \times V_{CE} \times I_C$$

$$= (2\ \mu sec + 2\ \mu sec)/100\ \mu sec \times 18\ V \times 500\ mA = 360\ mW$$

4. Total power:

$$P_{Q1} = P(Q1\ SAT) + P(Q1\ Cutoff) + P(t_r + t_f)$$

$$= 100\ mW + 54\ mW + 360\ mW = 514\ mW$$

NOTE: $V_{CE(SAT)}$ at 0.5 volts, I_{CEX} at 5 mA, and rise and fall times of 2 μsec, respectively (or 4 μsec total) are good specifications for a non-switching device. Therefore, if the rise time and fall time were double, the device would "cook" at slightly less than 1 watt of power. Also, note that the rise and fall time relates to the "on" pulse only and, therefore, the smaller the duty cycle at increased frequencies, the narrower the "on time" and the faster the switching device must be. Obviously, the switching time of the device is one of the most critical parameters of switching regulators.

d. The power delivered by the power source is given, theoretically, as:

$$P_{PS} = V_{CC} \times I_{RL} \times T_{on}/T = 30\ V \times 500\ mA \times 40\ \mu sec/100\ \mu sec = 6\ watts$$

NOTE: The P_{PS} at 6 watts assumes ideal components with no loss. However, the power dissipated across L1 is excessive and the resistance of 4 ohms is not practical. It was used to simplify the mathematics. In a practical circuit, the DC resistance of L1 would be less than 1 ohm.

8. The percent efficiency of the circuit, based on an output power of 5 watts and a power supply drain of 6 watts, is:

$$\% \text{ efficiency} = P_o/P_{PS} \times 100 = 5\ W/6\ W \times 100 = 83.3\%$$

NOTE: Power losses across L1 and the associated circuit components will drop efficiency to less than 80%, where 70% for switching regulators is nominal efficiency.

EXPERIMENTAL OBJECTIVES

To investigate the LM100, monolithic-driven, switching regulator.

LIST OF MATERIALS

1. Monolithic Regulator: LM100 or LM105 or equivalent
2. Transistor: MJ2955 or equivalent
3. Inductor: > 1 mH (non-saturable, air-gap, pot core) or equivalent
4. Resistors: 22 Ω (one)   50 Ω, 5 W (one)   1 kΩ (one)   4.7 kΩ (one)   1.5 MΩ
5. Capacitors: 0.1 μF (one)   100 μF (one)

EXPERIMENTAL PROCEDURE

1.   Connect the circuit as shown in Figure 5-27. Initially, set Vin to 24 V DC. Use a non-saturable core inductor of about 1 mH or greater. Also, use the MJ2955, even though it is higher powered than is necessary, and it has relatively poor rise and fall times. Using the higher power capabilities will provide protection against a non-working circuit, and the device can easily dissipate the additional heat that results from the poor rise and fall times. (However, if a faster PNP device is readily available, use it.)

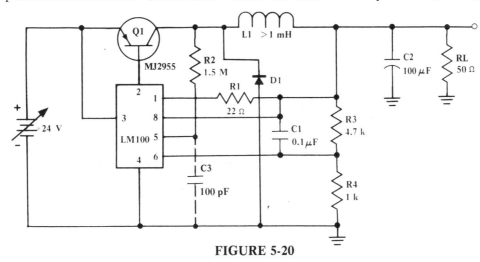

FIGURE 5-20

NOTE: A non-saturated core is required for the inductor to insure coninuous load current flow. A pot core construction with an air gap is ideal. However, any core material that does not saturate easily, such as iron powder core material or standard cylindrical construction can be used. (C3 minimizes oscillations.)

2.   Measure the DC reference voltage at Pin 6. It should be about 1.7 volts. Then, using the measured reference voltage:

a.   Calculate the DC output voltage from: $V_O = (V[Pin\ 6] \times [R3 + R4])/R4$

b.   Calculate the DC voltage at Pin 8 from: $V(Pin\ 8) = V_O$

c.   Calulate the DC voltage at Pin 1 from: $V(Pin\ 1) = V(Pin\ 8) - V_{R1}$

NOTE: The current flow through resistor R1 is nominally maintained at about 5 mA, and it reaches a maximum **current** condition of about 10 mA when R1 is reduced to zero ohms. Therefore, for an R1 of 22 ohms, V_{R1} can be estimated at 0.1 volts.

d.   Calculate the DC voltage at Pin 5 from: $V(Pin\ 5) \approx V(ref)$

3.   Measure the DC voltages around the circuit.
a.   At the junction of resistors R3 and R4.

293

b. At the output.

c. At Pin 1.

d. At Pin 5.

e. At the collector of Q1.

f. At the base of Q1.

g. At the emitter of Q1 — Pin 3 of the monolithic device.

4. Calculate:

a. The load current from: $I_{RL} = V_o/RL$

b. The DC resistance of inductor L1 from: $R_{L1(DC)} = (V_{C[Q1]} - V_o)/I_{RL}$

c. The voltage at the base of Q1 from: $V_{B(Q1)} = V_{E(Q1)} - V_{BE(Q1)}$, where: $V_{E(Q1)} = Vin$

d. The DC voltage drop across the collector-emitter of Q1 from: $V_{CE(Q1)} = V_{E(Q1)} - V_{C(Q1)}$

5. Insert the calculated and measured values into Table 5-5. Insert I_{RL} into Table 5-6.

TABLE 5-5	V(Pin 6)	V_o	V (Pin 8)	V (Pin 5)	R_{L1} DC	$V_{C(Q1)}$	$V_{E(Q1)}$ − V(Pin 3)	$V_{B(Q1)}$	$V_{CE(Q1)}$
CALCULATED	/////					/////			
MEASURED									

6. PEAK-TO-PEAK SQUARE WAVE

a. Calculate the amplitude of the peak-to-peak square wave at the collector of Q1 from:

$$V_{C(Q1)}\text{p-p} \approx Vin$$

b. Calculate the duty cycle of the square wave at the collector of Q1 from:

% Duty Cycle $= (V_{C[Q1]}/Vin) \times 100$

c. Monitor the peak-to-peak square wave at the collector of Q1 and at the base of Q1.

7. RIPPLE VOLTAGE

a. Calculate the ripple voltage at the output from: $\Delta V_o \approx (Vin \times 1\ k\Omega)/R2$

b. Measure the ripple voltage at the load.

8. FREQUENCY — TIME

a. Monitor the square wave at the collector of Q1 and measure T_{on}, T_{off}, and T.

b. Calculate the frequency from: $f = 1/T = 1/(T_{on} + T_{off})$

c. Verify the (percent) duty cycle from: % Duty Cycle $= T_{on}/T \times 100$

9. POWER DISSIPATION

a. Calculate the power dissipated by the load from:

$$P_o = P_{RL} = V_o^2/RL, \quad \text{where:} \quad V_o \text{ is measured in DC volts}$$

b. Calculate the power delivered by the power supply from:

$$P_{PS} = V_{CC} \times I_{DC} \approx V_{CC} \times I_{RL} \times T_{on}/T$$

NOTE: Most power supplies have a voltage-current switch so the voltage or current can re readily monitored. Therefore, when the current out of the power supply is being monitored, it will be less than the current into the load, and about directly proportional to the duty cycle percentage.

c. Calculate the power dissipated by the Q1 transistor from:

$$P(Q1) = P(Q1\ SAT) + P(Q1\ Cutoff) + P(Q/[t_r + t_f])$$

where: $P(Q1\ SAT) = V_{CE(Q1\ SAT)} \times I_{RL}$

$$P(Q1\ Cutoff) = V_{CE(Q1)} \times I_{CEX}$$

$$P(Q1/[t_r + t_f]) = V_{CE(Q1)} \times I_{RL} \times \frac{t_r + t_f}{T}$$

10. Calculate the efficiency of the switching regulator from:

$$\%\ efficiency = P_o/P_{PS} \times 100 \quad where: \quad Vin = 24\ volts$$

NOTE: One major advantage of switching regulators over linear regulators is that the difference voltage between the input and output can be high (across the device) with minimum power dissipation by the device. Therefore, when the input voltage is increased, only slight power dissipation by the device will be noted. In the linear regulators, the device would have to absorb all of the increased voltage and hence dissipate power.

11. Monitor the input voltage and current. Increase and decrease the input voltage.
   a. Does the input current increase with decreased input voltage?
   b. Does the input current decrease with increased input voltage?
   c. Should this happen? **Does IDC $= I_{RL} \times T_{on}/T$?**
12. The efficiency formula can be used to prove your monitored observations.
   a. Decrease the input voltage to 18 volts and monitor the input current. Calculate the "new" power supplied by the power supply from:

$$P_{PS} = V_{CC} \times I_{DC} \approx V_{CC} \times I_{RL} \times T_{on}/T \quad where: \quad V_{CC} = 18\ volts$$

   b. Monitor the voltage at the load and the current to the load. Confirm that the output power remains constant with the decrease in input voltage.
   c. Calculate the "new" circuit efficiency for the Vin $= V_{CC}$ of 18 volts from:

$$\%\ efficiency = P_o/P_{PS} \times 100 \quad where: \quad Vin = 18\ volts$$

13. Insert the calculated and measured values, as indicated, into Table 5-6.

TABLE 5-6	$V_{C(Q1)}$	Vp-p $V_{B(Q1)}$	Duty Cycle	ΔV_o	T_{on}	Time-Frequency T_{off}	T	f	P_o	P_{PS}	I_{DC}
CALCULATED		/////			/////	/////	/////				
MEASURED								/////			

	I_{RL}	P(Q1)	Q1 DISSIPATION P(Q1) Cutoff	P(Q1) SAT	P(Q1) $t_r + t_f$	% EFF.	Vin = 18 V P_o	P_{PS}	% EFF
CALCULATED									
MEASURED		/////	/////	/////	/////	/////	/////	/////	/////

295

APPENDIX A: PRACTICAL FET MEASUREMENTS

Field effect transistors are square law devices and several techniques can be used to measure the I_{DSS}, I_D, and V_{GS} of the device. However, the pinch-off voltage Vp must be calculated. That is, it must be calculated if some degree of accuracy is to be achieved.

A simple, but relatively accurate technique for measuring I_{DSS}, I_D, and V_{GS}, for a given RS resistor, is shown in Figure A-1, and when these three parameters are known, Vp can be calculated. The technique uses a milliammeter, a source resistor, and a 12 volt supply to measure I_D on the milliammeter, V_{GS} across the source resistor RS, and then I_{DSS}, when the RS resistor is shorted. The pinch-off voltage Vp is calculated from:

$$Vp = \frac{V_{GS}}{1 - \sqrt{I_D/I_{DSS}}}$$

This technique can be used for JFET's and depletion type MOSFET's, including the dual-gate MOSFET, where one gate is connected to 4 volts and the other to ground, as shown in Figure A-1. The formula for solving for Vp is derived from the standard $I_D = I_{DSS}(1 - V_{GS}/Vp)^2$ formula as follows:

$$I_D = I_{DSS}(1 - VGS/Vp)^2 \qquad \text{Transposing } I_{DSS}$$

$$\frac{I_D}{I_{DSS}} = (1 - V_{GS}/Vp)^2 \qquad \text{Taking the square root of both sides}$$

$$\sqrt{I_D/I_{DSS}} = (1 - V_{GS}/Vp) \qquad \text{Transposing}$$

$$\frac{V_{GS}}{Vp} = 1 - \sqrt{I_D/I_{DSS}} \qquad \text{Solving in terms of Vp}$$

$$Vp = \frac{V_{GS}}{1 - \sqrt{I_D/I_{DSS}}}$$

Therefore, to make practical JFET or dual-gate MOSFET measurements:
1. Connect the JFET or dual-gate MOSFET as shown in Figure A-1.
2. Measure the I_D on the milliammeter and V_{GS} across the RS resistor or gate-to-source junction.

NOTE: A nominal value of RS for the JFET is selected at 1 kΩ, but it is not critical. Likewise, RS at 330 ohms is a nominal value for the dual-gate MOSFET. It, too, is not that critical.

3. Short the source resistor RS and monitor the I_{DSS} on the milliammeter. Calculate the pinch-off voltage Vp from: $$Vp = \frac{V_{GS}}{1 - \sqrt{I_D/I_{DSS}}}$$

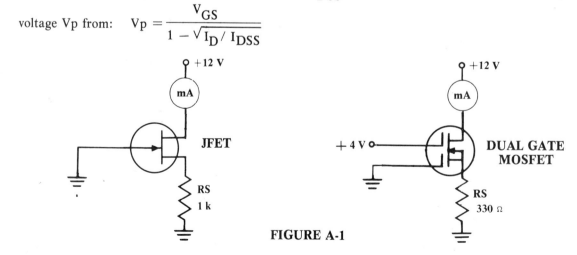

FIGURE A-1

APPENDIX B: COMPONENTS REQUIRED TO WORK ALL THE EXPERIMENTS

ACTIVE DEVICES (One each except where indicated):

2N3904 or equivalent (two)
2N3906 or equivalent
2N3819, or 2N5951, or equivalent
CA40673 or equivalent
J406 or equivalent
CA3028 or equivalent
2N3053, or 2N5190, or 2N6178, or equivalent
2N3055 or equivalent (two)
2N6178, or 2N5192, or equivalent
2N6180, or 2N5195, or equivalent

μA741, or LM301, or LM307, or equivalent (two)
μA723 or equivalent
μA78G or equivalent
MJ2955 or equivalent
LM301 or equivalent
μA555 or equivalent
2N2646, or 2N4871, or equivalent
LM566 or equivalent
LM 100, or LM105, or equivalent

NOTE: All of the discrete devices can be replaced with other devices, having equivalent NPN or PNP configurations and power ratings, with little or no change in circuit operation. This is also true for the op-amps. However, monolithic device substitutions may involve circuit and/or pin changes, unless exact substitutions are made. Also, both the CA3028 and the J406 monolithics can be replaced by discrete devices if the devices for the differential amplifier are matched.

CAPACITORS (one each except where indicated):

30 pF   0.001 pF   0.01 μF (two)   0.02 μF   0.1 μF   1 μF   100 μF (two)   1000 μF (two)

NOTE: Substitutions can be made for the capacitors if the substituted values are kept within reasonable limits. Also, all of the working voltages of the capacitors, with the exception of the 1000 μF, are below 25 volts. The 1000 μF capacitors should be about 50 volts, with 35 volts a minimum.

RESISTORS (one each except where indicated):

1 Ω (two   4.7 Ω   22 Ω   47-50 Ω   100 Ω (two)   120 Ω (two)   330 Ω   470 Ω   1 kΩ

1.2 kΩ (two)   1.5 kΩ (two)   1.8 kΩ   2.2 kΩ (two)   2.7 kΩ   3.3 kΩ   4.7 kΩ (three)

5.6 kΩ   10 kΩ (five)   22 kΩ   27 kΩ (two)   47 kΩ (two)   100 kΩ (two)   150 kΩ

470 kΩ (two)   1 MΩ   1.5 MΩ

NOTE: All of the above resistors are needed if the circuits are to be followed exactly. However, if minor resistor changes are made, about a third of the above list can be eliminated.

DIODES: 1N4001 (fourteen) or equivalents

NOTE: Fourteen diodes are needed for only one experiment. The next highest need is eight for a circuit that requires the construction of two, full-wave bridges.

ZENER DIODES (one each):   12 volt and 15 volt (one watt)

FERRITE TOROIDAL CORE (saturable)

MAGNETIC WIRE (FORVAR):   #20 to #24 gauge (approximately 20 feet)

NOTE: Appendix D gives specifications for a ferrite core that can be used in the saturable transformer construction for inverters. And the best construction to be used is normally the ferrite toroidal core form.

INDUCTOR: Most inductors are constructed to be non-saturable by using cylindrical construction, pot core construction with an air gap, or E and I construction with an air gap. Therefore, almost any non-saturable inductor of from 1 to 20 mH can be used for the switching regulator circuit.

APPENDIX C: DEVICE DATA SHEETS

 **MOTOROLA**

2N3903
2N3904

NPN SILICON ANNULAR♦ TRANSISTORS

. . . designed for general purpose switching and amplifier applications and for complementary circuitry with types 2N3905 and 2N3906.

- High Voltage Ratings — BV_{CEO} = 40 Volts (Min)
- Current Gain Specified from 100 μA to 100 mA
- Complete Switching and Amplifier Specifications
- Low Capacitance — C_{ob} = 4.0 pF (Max)

NPN SILICON
SWITCHING & AMPLIFIER
TRANSISTORS

MAXIMUM RATINGS

Rating	Symbol	Value	Unit
*Collector-Base Voltage	V_{CB}	60	Vdc
*Collector-Emitter Voltage	V_{CEO}	40	Vdc
*Emitter-Base Voltage	V_{EB}	6.0	Vdc
*Collector Current	I_C	200	mAdc
Total Power Dissipation @ T_A = 60°C	P_D	250	mW
**Total Power Dissipation @ T_A = 25°C Derate above 25°C	P_D	350 2.8	mW mW/°C
**Total Power Dissipation @ T_C = 25°C Derate above 25°C	P_D	1.0 8.0	Watts mW/°C
**Junction Operating Temperature	T_J	150	°C
**Storage Temperature Range	T_{stg}	-55 to +150	°C

THERMAL CHARACTERISTICS

Characteristic	Symbol	Max	Unit
Thermal Resistance, Junction to Ambient	$R_{\theta JA}$	357	°C/W
Thermal Resistance, Junction to Case	$R_{\theta JC}$	125	°C/W

*Indicates JEDEC Registered Data

**Motorola guarantees this data in addition to the JEDEC Registered Data.

♦Annular Semiconductors Patented by Motorola Inc.

Courtesy of Motorola Semiconductor Products Inc.

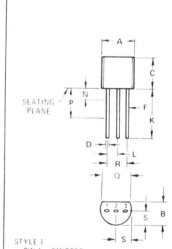

STYLE 1
PIN 1 EMITTER
    2 BASE
    3 COLLECTOR

DIM	MILLIMETERS		INCHES	
	MIN	MAX	MIN	MAX
A	4.450	5.200	0.175	0.205
B	3.180	4.190	0.125	0.165
C	4.320	5.330	0.170	0.210
D	0.407	0.533	0.016	0.021
F	0.407	0.482	0.016	0.019
K	12.700	—	0.500	—
L	1.150	1.390	0.045	0.055
N	—	1.270	—	0.050
P	6.350		0.250	—
Q	3.430		0.135	
R	2.410	2.670	0.095	0.105
S	2.030	2.670	0.080	0.105

CASE 29-02
TO-92

MOTOROLA INC 1973

DS 5127 R2

*ELECTRICAL CHARACTERISTICS (T_A = 25°C unless otherwise noted)

Characteristic	Fig. No.	Symbol	Min	Max	Unit
OFF CHARACTERISTICS					
Collector-Base Breakdown Voltage (I_C = 10 μAdc, I_E = 0)		BV_{CBO}	60	-	Vdc
Collector-Emitter Breakdown Voltage (1) (I_C = 1.0 mAdc, I_B = 0)		BV_{CEO}	40	-	Vdc
Emitter-Base Breakdown Voltage (I_E = 10 μAdc, I_C = 0)		BV_{EBO}	6.0	-	Vdc
Collector Cutoff Current (V_{CE} = 30 Vdc, $V_{EB(off)}$ = 3.0 Vdc)		I_{CEX}	-	50	nAdc
Base Cutoff Current (V_{CE} = 30 Vdc, $V_{EB(off)}$ = 3.0 Vdc)		I_{BL}	-	50	nAdc
ON CHARACTERISTICS					
DC Current Gain (1) (I_C = 0.1 mAdc, V_{CE} = 1.0 Vdc) 2N3903 / 2N3904	15	h_{FE}	20 / 40	- / -	
(I_C = 1.0 mAdc, V_{CE} = 1.0 Vdc) 2N3903 / 2N3904			35 / 70	- / -	
(I_C = 10 mAdc, V_{CE} = 1.0 Vdc) 2N3903 / 2N3904			50 / 100	150 / 300	
(I_C = 50 mAdc, V_{CE} = 1.0 Vdc) 2N3903 / 2N3904			30 / 60	- / -	
(I_C = 100 mAdc, V_{CE} = 1.0 Vdc) 2N3903 / 2N3904			15 / 30	- / -	
Collector-Emitter Saturation Voltage (1) (I_C = 10 mAdc, I_B = 1.0 mAdc) (I_C = 50 mAdc, I_B = 5.0 mAdc)	16, 17	$V_{CE(sat)}$	- / -	0.2 / 0.3	Vdc
Base-Emitter Saturation Voltage (1) (I_C = 10 mAdc, I_B = 1.0 mAdc) (I_C = 50 mAdc, I_B = 5.0 mAdc)	17	$V_{BE(sat)}$	0.65 / -	0.85 / 0.95	Vdc
SMALL-SIGNAL CHARACTERISTICS					
Current-Gain—Bandwidth Product (I_C = 10 mAdc, V_{CE} = 20 Vdc, f = 100 MHz) 2N3903 / 2N3904		f_T	250 / 300	- / -	MHz
Output Capacitance (V_{CB} = 5.0 Vdc, I_E = 0, f = 100 kHz)	3	C_{ob}	-	4.0	pF
Input Capacitance (V_{BE} = 0.5 Vdc, I_C = 0, f = 100 kHz)	3	C_{ib}	-	8.0	pF
Input Impedance (I_C = 1.0 mAdc, V_{CE} = 10 Vdc, f = 1.0 kHz) 2N3903 / 2N3904	13	h_{ie}	0.5 / 1.0	8.0 / 10	k ohms
Voltage Feedback Ratio (I_C = 1.0 mAdc, V_{CE} = 10 Vdc, f = 1.0 kHz) 2N3903 / 2N3904	14	h_{re}	0.1 / 0.5	5.0 / 8.0	X 10-4
Small-Signal Current Gain (I_C = 1.0 mAdc, V_{CE} = 10 Vdc, f = 1.0 kHz) 2N3903 / 2N3904	11	h_{fe}	50 / 100	200 / 400	-
Output Admittance (I_C = 1.0 mAdc, V_{CE} = 10 Vdc, f = 1.0 kHz)	12	h_{oe}	1.0	40	μmhos
Noise Figure (I_C = 100 μAdc, V_{CE} = 5.0 Vdc, R_3 = 1.0 k ohms, f = 10 Hz to 15.7 kHz) 2N3903 / 2N3904	9, 10	NF	- / -	6.0 / 5.0	dB
SWITCHING CHARACTERISTICS					
Delay Time (V_{CC} = 3.0 Vdc, $V_{BE(off)}$ = 0.5 Vdc, I_C = 10 mAdc, I_{B1} = 1.0 mAdc)	1, 5	t_d	-	35	ns
Rise Time	1, 5, 6	t_r	-	35	ns
Storage Time (V_{CC} = 3.0 Vdc, I_C = 10 mAdc, I_{B1} = I_{B2} = 1.0 mAdc) 2N3903 / 2N3904	2, 7	t_s	- / -	175 / 200	ns
Fall Time	2, 8	t_f	-	50	ns

(1) Pulse Test: Pulse Width = 300 μs, Duty Cycle = 2.0%.
*Indicates JEDEC Registered Data

FIGURE 1 — DELAY AND RISE TIME EQUIVALENT TEST CIRCUIT

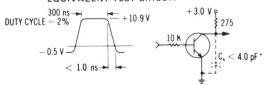

FIGURE 2 — STORAGE AND FALL TIME EQUIVALENT TEST CIRCUIT

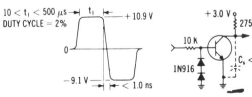

*Total shunt capacitance of test jig and connectors

Ⓜ **MOTOROLA** *Semiconductor Products Inc.*

Courtesy of Motorola Semiconductor Products Inc.

2N3905
2N3906

PNP SILICON ANNULAR♦ TRANSISTORS

.... designed for general purpose switching and amplifier applications and for complementary circuitry with types 2N3903 and 2N3904.

- High Voltage Ratings — BV_{CEO} = 40 Volts (Min)
- Current Gain Specified from 100 μA to 100 mA
- Complete Switching and Amplifier Specifications
- Low Capacitance — C_{ob} = 4.5 pF (Max)

PNP SILICON SWITCHING & AMPLIFIER TRANSISTORS

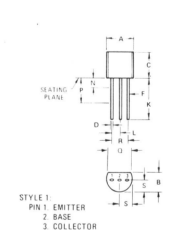

STYLE 1:
PIN 1. EMITTER
    2. BASE
    3. COLLECTOR

DIM	MILLIMETERS		INCHES	
	MIN	MAX	MIN	MAX
A	4.450	5.200	0.175	0.205
B	3.180	4.190	0.125	0.165
C	4.320	5.330	0.170	0.210
D	0.407	0.533	0.016	0.021
F	0.407	0.482	0.016	0.019
K	12.700	—	0.500	—
L	1.150	1.390	0.045	0.055
N		1.270		0.050
P	6.350		0.250	
Q	3.430		0.135	
R	2.410	2.670	0.095	0.105
S	2.030	2.670	0.080	0.105

CASE 29-02
(TO-92)

*MAXIMUM RATINGS

Rating	Symbol	Value	Unit
Collector-Base Voltage	V_{CB}	40	Vdc
Collector-Emitter Voltage	V_{CEO}	40	Vdc
Emitter-Base Voltage	V_{EB}	5.0	Vdc
Collector Current	I_C	200	mAdc
Total Power Dissipation @ T_A = 60°C	P_D	250	mW
Total Power Dissipation @ T_A = 25°C Derate above 25°C	P_D	350 2.8	mW mW/°C
Total Power Dissipation @ T_C = 25°C Derate above 25°C	P_D	1.0 8.0	Watt mW/°C
Junction Operating Temperature	T_J	+150	°C
Storage Temperature Range	T_{stg}	-55 to +150	°C

THERMAL CHARACTERISTICS

Characteristic	Symbol	Max	Unit
Thermal Resistance, Junction to Ambient	$R_{\theta JA}$	357	°C/W
Thermal Resistance, Junction to Case	$R_{\theta JC}$	125	°C/W

Courtesy of Motorola Semiconductor Products Inc.

*Indicates JEDEC Registered Data.
♦Annular semiconductors patented by Motorola Inc.

DS 5128 R2

*ELECTRICAL CHARACTERISTICS ($T_A = 25°C$ unless otherwise noted.)

Characteristic		Fig. No.	Symbol	Min	Max	Unit
OFF CHARACTERISTICS						
Collector-Base Breakdown Voltage ($I_C = 10 \mu Adc$, $I_E = 0$)			BV_{CBO}	40	—	Vdc
Collector-Emitter Breakdown Voltage (1) ($I_C = 1.0$ mAdc, $I_B = 0$)			BV_{CEO}	40	—	Vdc
Emitter-Base Breakdown Voltage ($I_E = 10 \mu Adc$, $I_C = 0$)			BV_{EBO}	5.0	—	Vdc
Collector Cutoff Current ($V_{CE} = 30$ Vdc, $V_{BE(off)} = 3.0$ Vdc)			I_{CEX}	—	50	nAdc
Base Cutoff Current ($V_{CE} = 30$ Vdc, $V_{BE(off)} = 3.0$ Vdc)			I_{BL}	—	50	nAdc
ON CHARACTERISTICS (1)						
DC Current Gain			h_{FE}			
($I_C = 0.1$ mAdc, $V_{CE} = 1.0$ Vdc)	2N3905	15		30	—	
	2N3906			60	—	
($I_C = 1.0$ mAdc, $V_{CE} = 1.0$ Vdc)	2N3905			40	—	
	2N3906			80	—	
($I_C = 10$ mAdc, $V_{CE} = 1.0$ Vdc)	2N3905			50	150	
	2N3906			100	300	
($I_C = 50$ mAdc, $V_{CE} = 1.0$ Vdc)	2N3905			30	—	
	2N3906			60	—	
($I_C = 100$ mAdc, $V_{CE} = 1.0$ Vdc)	2N3905			15	—	
	2N3906			30	—	
Collector-Emitter Saturation Voltage		16, 17	$V_{CE(sat)}$			Vdc
($I_C = 10$ mAdc, $I_B = 1.0$ mAdc)				—	0.25	
($I_C = 50$ mAdc, $I_B = 5.0$ mAdc)				—	0.4	
Base-Emitter Saturation Voltage		17	$V_{BE(sat)}$			Vdc
($I_C = 10$ mAdc, $I_B = 1.0$ mAdc)				0.65	0.85	
($I_C = 50$ mAdc, $I_B = 5.0$ mAdc)				—	0.95	
SMALL-SIGNAL CHARACTERISTICS						
Current-Gain — Bandwidth Product			f_T			MHz
($I_C = 10$ mAdc, $V_{CE} = 20$ Vdc, f = 100 MHz)	2N3905			200	—	
	2N3906			250	—	
Output Capacitance ($V_{CB} = 5.0$ Vdc, $I_E = 0$, f = 100 kHz)		3	C_{ob}	—	4.5	pF
Input Capacitance ($V_{BE} = 0.5$ Vdc, $I_C = 0$, f = 100 kHz)		3	C_{ib}	—	1.0	pF
Input Impedance		13	h_{ie}			k ohms
($I_C = 1.0$ mAdc, $V_{CE} = 10$ Vdc, f = 1.0 kHz)	2N3906			0.5	8.0	
	2N3906			2.0	12	
Voltage Feedback Ratio		14	h_{re}			X 10^{-4}
($I_C = 1.0$ mAdc, $V_{CE} = 10$ Vdc, f = 1.0 kHz)	2N3905			0.1	5.0	
	2N3906			1.0	10	
Small-Signal Current Gain		11	h_{fe}			—
($I_C = 1.0$ mAdc, $V_{CE} = 10$ Vdc, f = 1.0 kHz)	2N3905			50	200	
	2N3906			100	400	
Output Admittance		12	h_{oe}			μmhos
($I_C = 1.0$ mAdc, $V_{CE} = 10$ Vdc, f = 1.0 kHz)	2N3905			1.0	40	
	2N3906			3.0	60	
Noise Figure		9, 10	NF			dB
($I_C = 100 \mu Adc$, $V_{CE} = 5.0$ Vdc, $R_S = 1.0$ k ohm,	2N3905			—	5.0	
f = 10 Hz to 15.7 kHz)	2N3906			—	4.0	
SWITCHING CHARACTERISTICS						
Delay Time	($V_{CC} = 3.0$ Vdc, $V_{BE(off)} = 0.5$ Vdc	1, 5	t_d	—	35	ns
Rise Time	$I_C = 10$ mAdc, $I_{B1} = 1.0$ mAdc)	1, 5, 6	t_r	—	35	ns
Storage Time	2N3905	2, 7	t_s	—	200	ns
	($V_{CC} = 3.0$ Vdc, $I_C = 10$ mAdc, 2N3906			—	225	
Fall Time	$I_{B1} = I_{B2} = 1.0$ mAdc) 2N3905	2, 8	t_f	—	60	ns
	2N3906			—	75	

*Indicates JEDEC Registered Data. (1) Pulse Width = 300 μs, Duty Cycle = 2.0 %.

FIGURE 1 — DELAY AND RISE TIME EQUIVALENT TEST CIRCUIT

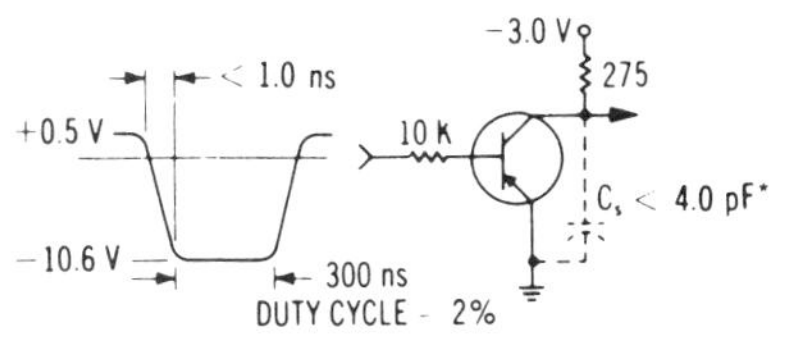

FIGURE 2 — STORAGE AND FALL TIME EQUIVALENT TEST CIRCUIT

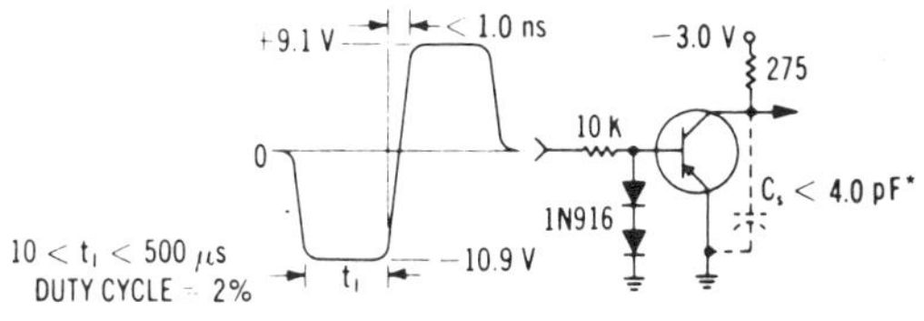

*Total shunt capacitance of test jig and connectors

TYPE 2N3819
N-CHANNEL PLANAR SILICON FIELD-EFFECT TRANSISTOR

TYPE 2N3819
BULLETIN NO. DL-S 68047, AUGUST 1965
REVISED MAY 1968

SILECT† FIELD-EFFECT TRANSISTOR
For Industrial and Consumer Small-Signal Applications
- Low C_{rss}: $\leq$ 4 pf   • High y_{fs}/C_{iss} Ratio (High-Frequency Figure of Merit)
- Cross Modulation Minimized by Square-Law Transfer Characteristics

mechanical data

This transistor is encapsulated in a plastic compound specifically designed for this purpose, using a highly mechanized process‡ developed by Texas Instruments. The case will withstand soldering temperatures without deformation. This device exhibits stable characteristics under high-humidity conditions and is capable of meeting MIL-STD-202C method 106B. The transistor is insensitive to light.

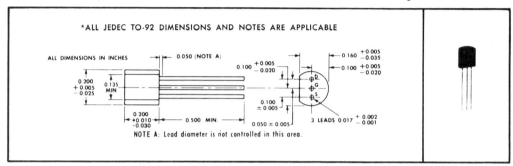

*ALL JEDEC TO-92 DIMENSIONS AND NOTES ARE APPLICABLE

ALL DIMENSIONS IN INCHES

NOTE A: Lead diameter is not controlled in this area.

*absolute maximum ratings at 25°C free-air temperature (unless otherwise noted)

Drain-Gate Voltage	25 v
Drain-Source Voltage	25 v
Reverse Gate-Source Voltage	− 25 v
Gate Current	10 ma
Continuous Device Dissipation at (or below) 25°C Free-Air Temperature (See Note 1)	360 mw
Storage Temperature Range	−65°C to 150°C
Lead Temperature 1/16 Inch from Case for 10 Seconds	260°C

*electrical characteristics at 25°C free-air temperature (unless otherwise noted)

	PARAMETER	TEST CONDITIONS	MIN	MAX	UNIT
$V_{(BR)GSS}$	Gate-Source Breakdown Voltage	$I_G = -1\,\mu a$, $V_{DS} = 0$	− 25		v
I_{GSS}	Gate Cutoff Current	$V_{GS} = -15\,v$, $V_{DS} = 0$		− 2	na
		$V_{GS} = -15\,v$, $V_{DS} = 0$, $T_A = 100°C$		− 2	μa
I_{DSS}	Zero-Gate-Voltage Drain Current	$V_{DS} = 15\,v$, $V_{GS} = 0$, See Note 2	2	20	ma
V_{GS}	Gate-Source Voltage	$V_{DS} = 15\,v$, $I_D = 200\,\mu a$	− 0.5	− 7.5	v
$V_{GS(off)}$	Gate-Source Cutoff Voltage	$V_{DS} = 15\,v$, $I_D = 2\,na$		− 8	v
$\|y_{fs}\|$	Small-Signal Common-Source Forward Transfer Admittance	$V_{DS} = 15\,v$, $V_{GS} = 0$, $f = 1\,kc$, See Note 2	2000	6500	μmho
$\|y_{os}\|$	Small-Signal Common-Source Output Admittance	$V_{DS} = 15\,v$, $V_{GS} = 0$, $f = 1\,kc$, See Note 2		50	μmho
C_{iss}	Common-Source Short-Circuit Input Capacitance	$V_{DS} = 15\,v$, $V_{GS} = 0$, $f = 1\,Mc$		8	pf
C_{rss}	Common-Source Short-Circuit Reverse Transfer Capacitance			4	pf
$\|y_{fs}\|$	Small-Signal Common-Source Forward Transfer Admittance	$V_{DS} = 15\,v$, $V_{GS} = 0$, $f = 100\,Mc$	1600		μmho

NOTES: 1. Derate linearly to 150°C free-air temperature at the rate of 2.88 mw/C°.
2. These parameters must be measured using pulse techniques. PW ≈ 100 msec, Duty Cycle ≤ 10%.
*Indicates JEDEC registered data.
†Trademark of Texas Instruments
‡Patent Pending

Courtesy of Texas Instruments Incorporated

568

TEXAS INSTRUMENTS
INCORPORATED
POST OFFICE BOX 5012 • DALLAS, TEXAS 75222

4799
(Replacement)

RCA Solid State Division

MOS Field-Effect Transistors

40673

RCA-40673 is an n-channel silicon, depletion type, dual insulated-gate field-effect transistor.

Special back-to-back diodes are diffused directly into the MOS* pellet and are electrically connected between each insulated gate and the FET's source. The diodes effectively bypass any voltage transients which exceed approximately ±10 volts. This protects the gates against damage in all normal handling and usage.

A feature of the back-to-back diode configuration is that it allows the 40673 to retain the wide input signal dynamic range inherent in the MOSFET. In addition, the low junction capacitance of these diodes adds little to the total capacitance shunting the signal gate.

The excellent overall performance characteristics of the RCA-40673 make it useful for a wide variety of rf-amplifier applications at frequencies up to 400 MHz. The two serially-connected channels with independent control gates make possible a greater dynamic range and lower cross–modulation than is normally achieved using devices having only a single control element.

The two gate arrangement of the 40673 also makes possible a desirable reduction in feedback capacitance by operating in the common-source configuration and ac-grounding Gate No. 2. The reduced capacitance allows operation at maximum gain *without neutralization;* and, of special importance in rf-amplifiers, it reduces local oscillator feedthrough to the antenna.

The 40673 is hermetically sealed in the metal JEDEC TO-72 package.

*Metal-Oxide-Semiconductor.

SILICON DUAL INSULATED-GATE FIELD-EFFECT TRANSISTOR

N-Channel Depletion Type With Integrated Gate-Protection Circuits For RF Amplifier Applications up to 400 MHz

H-1299

JEDEC TO-72

APPLICATIONS

- RF amplifier, mixer, and IF amplifier in military, industrial, and consumer communications equipment
- aircraft and marine vehicular receivers
- CATV and MATV equipment
- telemetry and multiplex equipment

PERFORMANCE FEATURES

- superior cross-modulation performance and greater dynamic range than bipolar or single-gate FETs
- wide dynamic range permits large-signal handling before overload
- dual-gate permits simplified agc circuitry
- virtually no agc power required
- greatly reduces spurious responses in fm receivers
- permits use of vacuum-tube biasing techniques
- excellent thermal stability

DEVICE FEATURES

- back-to-back diodes protect each gate against handling and in-circuit transients
- low gate leakage currents —— I_{G1SS} & I_{G2SS} = 20 nA(max.) at T_A = 25°C
- high forward transconductance —— g_{fs} = 12,000 µmho (typ.)
- high unneutralized RF power gain —— G_{ps} = 18 dB(typ.) at 200 MHz
- low VHF noise figure —— 3.5 dB(typ.) at 200 MHz

Maximum Ratings, *Absolute-Maximum Values, at* $T_A = 25°C$

DRAIN-TO-SOURCE VOLTAGE, V_{DS}	-0.2 to +20	V
GATE No.1-TO-SOURCE VOLTAGE, V_{G1S}:		
Continuous (dc)	-6 to +1	V
Peak ac	-6 to +6	V
GATE No.2-TO-SOURCE VOLTAGE, V_{G2S}:		
Continuous (dc)	-6 to 30% of V_{DS}	V
Peak ac	-6 to +6	V
DRAIN-TO-GATE VOLTAGE, V_{DG1} OR V_{DG2}	+20	V
DRAIN CURRENT, I_D	50	mA
TRANSISTOR DISSIPATION, P_T:		
At ambient } up to 25°C	330	mW
temperatures } above 25°C	derate linearly at 2.2 mW/°C	
AMBIENT TEMPERATURE RANGE:		
Storage and Operating	-65 to +175	°C
LEAD TEMPERATURE (During soldering):		
At distances ≥1/32 inch from seating surface for 10 seconds max.	265	°C

ELECTRICAL CHARACTERISTICS, at T_A = 25°C unless otherwise specified

CHARACTERISTICS	SYMBOLS	TEST CONDITIONS	LIMITS			UNITS
			Min.	Typ.	Max.	
Gate-No.1-to-Source Cutoff Voltage	$V_{G1S(off)}$	V_{DS} = +15V, I_D = 200μA V_{G2S} = +4V	–	–2	–4	V
Gate-No.2-to-Source Cutoff Voltage	$V_{G2S(off)}$	V_{DS} = +15V, I_D = 200μA V_{G1S} = 0	–	–2	–4	V
Gate-No.1-Leakage Current	I_{G1SS}	V_{G1S} = +1 or –6 V V_{DS} = 0, V_{G2S} = 0	–	–	50	nA
Gate-No.2-Leakage Current	I_{G2SS}	V_{G2S} = ±6V V_{DS} = 0, V_{G1S} = 0	–	–	50	nA
Zero-Bias Drain Current	I_{DSS}	V_{DS} = +15V V_{G2S} = +4V V_{G1S} = 0	5	15	35	mA
Forward Transconductance (Gate-No.1-to-Drain)	g_{fs}	V_{DS} = +15V, I_D = 10mA V_{G2S} = +4V, f = 1kHz	–	12,000	–	μmho
Small-Signal, Short-Circuit Input Capacitance †	C_{iss}	V_{DS} = +15V, I_D = 10mA V_{G2S} = +4V, f=1MHz	–	6	–	pF
Small-Signal, Short-Circuit, Reverse Transfer Capacitance (Drain-to-Gate No.1) ♠	C_{rss}		0.005	0.02	0.03	pF
Small-Signal, Short-Circuit Output Capacitance	C_{oss}		–	2.0	–	pF
Power Gain (see Fig. 1)	G_{PS}	V_{DS} = +15V, I_D = 10mA V_{G2S} = +4V, f = 200 MHz	14	18	–	dB
Maximum Available Power Gain	MAG		–	20	–	dB
Maximum Usable Power Gain (unneutralized)	MUG		–	20*	–	dB
Noise Figure (see Fig. 1)	NF		–	3.5	6.0	dB
Magnitude of Forward Transadmittance	$|Y_{fs}|$		–	12,000	–	μmho
Phase Angle of Forward Trans-admittance	θ		–	–35	–	degrees
Input Resistance	r_{iss}		–	1.0	–	kΩ
Output Resistance	r_{oss}		–	2.8	–	kΩ
Protective Diode Knee Voltage	V_{knee}	$I_{DIODE(REVERSE)}$=±100μA	–	±10	–	V

*Limited only by practical design considerations.

†Capacitance between Gate No. 1 and all other terminals

♠Three-terminal measurement with Gate No. 2 and Source returned to guard terminal.

OPERATING CONSIDERATIONS

The flexible leads of the 40673 are usually soldered to the circuit elements. As in the case of any high-frequency semiconductor device, the tips of soldering irons MUST be grounded.

Courtesy of Radio Corporation of America

Siliconix incorporated

**J401 J402 J403
J404 J405 J406**

GENERAL PURPOSE MONOLITHIC DUAL N-CHANNEL JFETs

LOW NOISE DUAL FETS FOR FET INPUT OP AMPS AND DIFFERENTIAL AMPS

- Low $V_{GS(off)}$ maximum 2.5 V
- Low I_G Max 100 pA
- High G_{fs} 1200 μmho minimum @ I_D = 200 μA
- Low G_{os} 2 μmho maximum @ I_D = 200 μA
- Low Noise 20 nV/$\sqrt{Hz}$ @ 10 Hz
- High CMRR 95 dB minimum (J401-404)

ABSOLUTE MAXIMUM RATINGS (25°C)

Gate-Drain or Gate-Source Voltage . . . . 50 V
Forward Gate Current . . . . 10 mA
Device Dissipation (each side)
  @ T_A = 85°C derate 7.5 mW/°C . . . . . 300 mW
Total Device Dissipation
  @ T_A = 85°C derate 11 mW/°C . . . . . . 500 mW
Storage Temperature Range . . . . . . –55 to +150°C

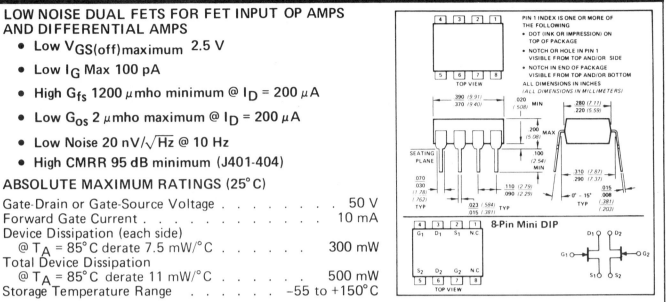

8-Pin Mini DIP

ELECTRICAL CHARACTERISTICS (@ 25°C unless otherwise noted)

		Characteristic		J401 Min	J401 Max	J402 Min	J402 Max	J403 Min	J403 Max	J404 Min	J404 Max	J405 Min	J405 Max	J406 Min	J406 Max	Unit	Test Conditions
1	STATIC	BV_{GSS}	Gate-Source Breakdown Voltage	–50		–50		–50		–50		–50		–50		V	V_{DS} = 0, I_G = –1 μA
2		I_{GSS}	Gate Reverse Current		–100		–100		–100		–100		–100		–100	pA	V_{DS} = 0, V_{GS} = –30 V
3		$V_{GS(off)}$	Gate-Source Cutoff Voltage	–.5	–2.5	–.5	–2.5	–.5	–2.5	–.5	–2.5	–.5	–2.5	–.5	–2.5	V	V_{DS} = 15 V, I_D = 1 nA
4		V_{GS}	Gate-Source Voltage (on)		–2.3		–2.3		–2.3		–2.3		–2.3		–2.3		V_{DG} = 15 V, I_D = 200 μA
5		I_{DSS}	Saturation Drain Current (Note 1)	0.5	10.0	0.5	10.0	0.5	10.0	0.5	10.0	0.5	10.0	0.5	10.0	mA	V_{DS} = 10 V, V_{GS} = 0
6		I_G	Gate Current		–100		–100		–100		–100		–100		–100	pA	V_{DG} = 15 V,
7		I_G	Gate Current		–60		–60		–60		–60		–60		–60	nA	I_D = 200 μA T_A = 125°C
8		BV_{G1-G2}	Gate-Gate Breakdown Voltage	±50		±50		±50		±50		±50		±50		V	V_{DS} = 0, V_{GS} = 0, I_G = ±1 μA
9	DYNAMIC	g_{fs}	Common-Source Forward Transconductance	2000	7000	2000	7000	2000	7000	2000	7000	2000	7000	2000	7000	μmho	V_{DS} = 10 V, V_{GS} = 0 f = 1 kHz
10		g_{os}	Common-Source Output Conductance		20		20		20		20		20		20		
11		g_{fs}	Common-Source Forward Transconductance	1200	1600	1200	1600	1200	1600	1200	1600	1200	1600	1200	1600		V_{DG} = 15 V, I_D = 200 μA
12		g_{os}	Common-Source Output Conductance		2.0		2.0		2.0		2.0		2.0		2.0		
13		C_{iss}	Common-Source Input Capacitance		8.0		8.0		8.0		8.0		8.0		8.0	pF	f = 1 MHz
14		C_{rss}	Common-Source Reverse Transfer Capacitance		3.0		3.0		3.0		3.0		3.0		3.0		
15		e_N	Equivalent Short-Circuit Input Noise Voltage		20		20		20		20		20		20	$\frac{nV}{\sqrt{Hz}}$	V_{DS} = 15 V, V_{GS} = 0 f = 10 Hz
16	MATCHING	CMRR	Common-Mode Rejection Ratio	95		95		95		95		90				dB	V_{DG} = 10 to 20 V, I_D = 200 μA
17		$\|V_{GS1} - V_{GS2}\|$	Differential Gate-Source Voltage		5		10		10		15		20		40	mV	V_{DG} = 10 V, I_D = 200 μA
18		$\frac{\Delta\|V_{GS1} - V_{GS2}\|}{\Delta T}$	Gate-Source Voltage Differential Drift		10		10		25		25		40		80	μV/°C	V_{DG} = 10 V, I_D = 200 μA T_A = –55°, T_B = +25°, T_C = +125°

NOTE 1: Pulse Test Pulse Width = 300 μsec, duty cycle ≤ 3%

Courtesy of Siliconix Incorporated **NNR**

The CA3028A and CA3028B are differential/cascode amplifiers designed for use in communications and industrial equipment operating at frequencies from dc to 120 MHz.

The CA3028B is like the CA3028A but is capable of premium performance particularly in critical dc and differential amplifier applications requiring tight controls for input offset voltage, input offset current, and input bias current.

DIFFERENTIAL/CASCODE AMPLIFIERS

For Communications and Industrial Equipment at Frequencies from DC to 120 MHz

8-Lead TO-5

FEATURES

- Controlled for Input Offset Voltage, Input Offset Current, and Input Bias Current
- Balanced Differential Amplifier Configuration with Controlled Constant-Current Source to Provide Unexcelled Versatility
- Single- and Dual-Ended Operation
- Operation from DC to 120 MHz
- Balanced-AGC Capability
- Wide Operating-Current Range

APPLICATIONS

- RF and IF Amplifiers (Differential or Cascode)
- DC, Audio, and Sense Amplifiers
- Converter in the Commercial FM Band
- Oscillator   • Mixer   • Limiter
- Companion Application Note, ICAN 5337 "Application of the RCA CA3028 Integrated Circuit Amplifier in the HF and VHF Ranges." This note covers characteristics of different operating modes, noise performance, mixer, limiter, and amplifier design considerations.

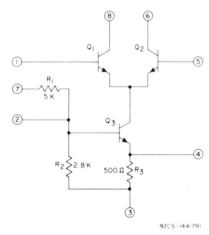

92CS-14417RI

Fig.1 - Schematic diagram for CA3028A and CA3028B.

The resistance values included on the schematic diagram have been supplied as a convenience to assist the Equipment Manufacturer in optimizing the selection of "outboard" components of his equipment designs. The values shown may vary as much as ± 30%.

RCA reserves the right to make any changes in the Resistance Values provided such changes do not adversely affect the published performance characteristics of the device.

DIMENSIONAL OUTLINE

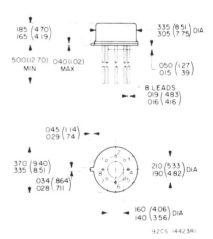

92CS-14423RI

Dimensions in Inches and Millimeters

NOTE: *Dimensions in parentheses are in millimeters and are derived from the basic inch dimensions as indicated.*

Courtesy of Radio Corporation of America

RCA Electronic Components

Supersedes File No.311 dated 11-67
Printed in U.S.A. | 3-68

CHARACTERISTIC	SYMBOL	TEST CIRCUIT Fig.	SPECIAL TEST CONDITIONS		LIMITS TYPE CA3028A Min.	Typ.	Max.	LIMITS TYPE CA3028B Min.	Typ.	Max.	UNITS	TYPICAL CHARACTERISTICS CURVE Fig.
STATIC CHARACTERISTICS												
Input Offset Voltage	V_{IO}	2	V_{CC} = +6V, V_{EE} = -6V		-	-	-	-	0.98	5	mV	4
			V_{CC} = +12V, V_{EE} = -12V		-	-	-	-	0.89	5		
Input Offset Current	I_{IO}	3	V_{CC} = +6V, V_{EE} = -6V		-	-	-	-	0.56	5	μA	4
			V_{CC} = +12V, V_{EE} = -12V		-	-	-	-	1.06	6		
Input Bias Current	I_I	3	V_{CC} = +6V, V_{EE} = -6V		-	16.6	70	-	16.6	40	μA	5
			V_{CC} = +12V, V_{EE} = -12V		-	36	106	-	36	80		
Quiescent Operating Current	I_6 or I_8	3	V_{CC} = +6V, V_{EE} = -6V		0.8	1.25	2	1	1.25	1.5	mA	6,7
			V_{CC} = +12V, V_{EE} = -12V		2	3.3	5	2.5	3.3	4		
AGC Bias Current (Into Constant-Current Source Terminal No.7)	I_{AGC}	8a	V_{AGC} = +9V, V_{CC} = +12V		-	1.28	-	-	1.28	-	mA	8b
			V_{AGC} = +12V, V_{CC} = +12V		-	1.65	-	-	1.65	-		
Input Current (Terminal No.7)	I_7	-	V_{CC} = +6V, V_{EE} = -6V		0.5	0.85	1	0.5	0.85	1	mA	-
			V_{CC} = +12V, V_{EE} = -12V		1	1.65	2.1	1	1.65	2.1		
Device Dissipation	P_T	3	V_{CC} = +6V, V_{EE} = -6V		24	36	54	24	36	42	mW	9
			V_{CC} = +12V, V_{EE} = -12V		120	175	260	120	175	220		
Power Gain	G_P	10a	f = 100 MHz, V_{CC} = +9V	Cascode	16	20	-	16	20	-	dB	10b
		11a,d		Diff.-Ampl.	14	17	-	14	17	-	dB	11b,e
		10a	f = 10.7 MHz, V_{CC} = +9V	Cascode	35	39	-	35	39	-	dB	10b
		11a		Diff.-Ampl.	28	32	-	28	32	-	dB	11b
Noise Figure	NF	10a	f = 100 MHz, V_{CC} = +9V	Cascode	-	7.2	9	-	7.2	9	dB	10c
		11a,d		Diff.-Ampl.	-	6.7	9	-	6.7	9	dB	11c,e
Input Admittance	Y_{11}	-		Cascode	-	0.6 + j1.6	-	-	0.6 + j1.6	-	mmho	12
		-		Diff.-Ampl.	-	0.5 + j0.5	-	-	0.5 + j0.5	-	mmho	13
Reverse Transfer Admittance	Y_{12}	-	f = 10.7 MHz V_{CC} = +9V	Cascode	-	0.0003 - j0	-	-	0.0003 - j0	-	mmho	14
		-		Diff.-Ampl.	-	0.01 - j0.0002	-	-	0.01 - j0.0002	-	mmho	15
Forward Transfer Admittance	Y_{21}	-		Cascode	-	99 - j18	-	-	99 - j18	-	mmho	16
		-		Diff.-Ampl.	-	-37 + j0.5	-	-	-37 + j0.5	-	mmho	17
Output Admittance	Y_{22}	-		Cascode	-	0 + j0.08	-	-	0 + j0.08	-	mmho	18
		-		Diff.-Ampl.	-	0.04 + j0.23	-	-	0.04 + j0.23	-	mmho	19
Power Output (Untuned)	P_o	20a	f = 10.7 MHz, V_{CC} = +9V	Diff.-Ampl. 50Ω Input-Output	-	5.7	-	-	5.7	-	μW	20b
AGC Range (Max. Power Gain to Full Cutoff)	AGC	21a	f = 10.7 MHz, V_{CC} = +9V	Diff.-Ampl.	-	62	-	-	62	-	dB	21b
Voltage Gain — at f = 10.7 MHz	A	22a	f = 10.7 MHz, V_{CC} = +9V, R_L = 1 kΩ	Cascode	-	98	-	-	98	-	dB	22b
		22c		Diff.-Ampl.	-	32	-	-	32	-	dB	22d
Voltage Gain — Differential at f = 1 kHz		23	V_{CC} = +6V, V_{EE} = -6V, R_L = 2 kΩ		-	-	-	35	38	42	dB	-
			V_{CC} = +12V, V_{EE} = -12V, R_L = 1.6 kΩ		-	-	-	40	42.5	45		
Max. Peak-to-Peak Output Voltage at f = 1 kHz	V_o(P-P)	23	V_{CC} = +6V, V_{EE} = -6V, R_L = 2 kΩ		-	-	-	7	11.5	-	V_{P-P}	-
			V_{CC} = +12V, V_{EE} = -12V, R_L = 1.6 kΩ		-	-	-	15	23	-		
Bandwidth at -3 dB point	BW	23	V_{CC} = +6V, V_{EE} = -6V, R_L = 2 kΩ		-	-	-	-	7.3	-	MHz	-
			V_{CC} = +12V, V_{EE} = -12V, R_L = 1.6 kΩ		-	-	-	-	8	-		
Common-Mode Input-Voltage Range	V_{CMR}	24	V_{CC} = +6V, V_{EE} = -6V		-	-	-	-2.5	(-3.2 – 4.5)	4	V	-
			V_{CC} = +12V, V_{EE} = -12V		-	-	-	-5	(-7 – 9)	7		
Common-Mode Rejection Ratio	CMR	24	V_{CC} = +6V, V_{EE} = -6V		-	-	-	60	110	-	dB	-
			V_{CC} = +12V, V_{EE} = -12V		-	-	-	60	90	-		
Input Impedance at f = 1 kHz	Z_{IN}	-	V_{CC} = +6V, V_{EE} = -6V		-	-	-	-	5.5	-	kΩ	-
			V_{CC} = +12V, V_{EE} = -12V		-	-	-	-	3	-		

Courtesy of Radio Corporation of America

RCA POWER TRANSISTORS

2N3053 2N3054
2N3055
Including
40372 40389 40392

File No. 145

RCA-2N3053, 2N3054*, and 2N3055* are silicon n-p-n transistors intended for a wide variety of medium to high-power applications.

The 2N3053 is a triple-diffused planar type useful up to 20 MHz in small-signal, medium-power applications.

The 2N3054 and 2N3055 are **Hometaxial-base**** types useful for power-switching circuits, for series- and shunt-regulator driver and output stages, and for high-fidelity amplifiers.

* Formerly Dev. Type Nos. TA2402A and TA2403A, respectively.

**"Hometaxial" was coined by RCA from "homogeneous" and "axial" to describe a single-diffused transistor with a base region of homogeneous-resistivity silicon in the axial direction (emitter-to-collector).

ALSO AVAILABLE...

Type 40372 is a 2N3054 with a factory-attached heat radiator.

Type 40389 is a 2N3053 with a factory-attached heat radiator.

Types 40372 and 40389 are intended for printed-circuit-board applications.

Type 40392 is a 2N3053 with a factory-attached diamond-shaped mounting flange.

40372 H-1470A
40389 H-1468
40392 H-1375

SILICON N-P-N GENERAL-PURPOSE TYPES FOR INDUSTRIAL AND COMMERCIAL APPLICATIONS

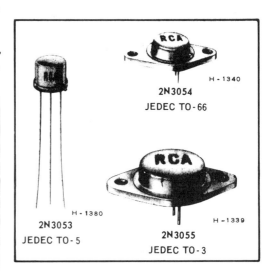

2N3054 JEDEC TO-66 H-1340

2N3053 JEDEC TO-5 H-1380

2N3055 JEDEC TO-3 H-1339

- Maximum area-of-operation curves for DC and pulse operation
- Now possible to determine maximum operating conditions for operation free from second breakdown

2N3053

- Low leakage current (I_{CBO}) and wide beta (h_{FE}) range

2N3054

- $V_{CEV}(sus) = 90$ V min.
- Low saturation voltage, $V_{CE}(sat) = 1.0$ V (at $I_C = 0.5$ A)

2N3055

- High dissipation capability -- 115 W
- $V_{CEV}(sus) = 100$ V min.
- Low saturation voltage, $V_{CE}(sat) = 1.1$ V (at $I_C = 4$ A)

MAXIMUM RATINGS

Absolute-Maximum Values:		2N3053 40389, 40392	2N3054 40372	2N3055	
COLLECTOR-TO-BASE VOLTAGE	V_{CBO}	60	90	100	V
COLLECTOR-TO-EMITTER VOLTAGE:					
With −1.5 V (V_{BE}) of reverse bias	$V_{CEV}(sus)$	60	90	100	V
With external base-to-emitter					
resistance (R_{BE}) = 10 Ω	$V_{CER}(sus)$	50	—	—	V
= 100 Ω	$V_{CER}(sus)$	—	60	70	V
With base open	$V_{CEO}(sus)$	40	55	60	V
EMITTER-TO-BASE VOLTAGE	V_{EBO}	5	7	7	V
COLLECTOR CURRENT	I_C	0.7	4	15	A
BASE CURRENT	I_B	—	2	7	A
TRANSISTOR DISSIPATION:	P_T				
At case temperatures up to 25°C		5(2N3053) 7(40392)	29(2N3054) —	115 —	W W
At free-air temperatures up to 25°C		1(2N3053) 3.5(40389)	5.8(40372) —	— —	W W
At temperatures above 25°C, See Figs.		1, 2, & 5	2, 3, & 6	2, 4, & 6	
TEMPERATURE RANGE:					
Storage & Operating (Junction)		←——— −65 to 200 ———→			°C
LEAD OR PIN TEMPERATURE					
(During soldering):					
At distance ⟩ 1/32" from seating plane for 10 s max.		255	235	235	°C

RADIO CORPORATION OF AMERICA
ELECTRONIC COMPONENTS AND DEVICES, HARRISON, N.J.

Trademark(s) ® Registered
Marca(s) Registrada(s)

Printed in U.S.A.
2N3053, 40389, 40392,
2N3054, 40372, 2N3055   8/66
Supersedes issue dated 10/64

Courtesy of Radio Corporatin of America

ELECTRICAL CHARACTERISTICS

Case Temperature (T_C) = 25°C, Unless Otherwise Specified

Characteristics	Symbol	V_{CB}	V_{CE}	V_{EB}	V_{BE}	I_C	I_E	I_B	2N3053 40389 40392 Min.	Max.	2N3054 40372 Min.	Max.	2N3055 Min.	Max.	Units
Collector-Cutoff Current	I_{CBO}	30						0	—	0.25	—	—	—	—	μA
	I_{CEV}		90		−1.5				—	—	—	1.0	—	—	mA
			100		−1.5				—	—	—	—	—	5.0	
At T_C = 150°C	I_{CEV}		30		−1.5				—	—	—	5.0	—	—	mA
			60		−1.5				—	—	—	—	—	10.0	
Emitter-Cutoff Current	I_{EBO}			4		0			—	0.25	—	—	—	—	μA
				7		0			—	—	—	1.0	—	—	mA
				7		0			—	—	—	—	—	5.0	mA
DC Forward-Current Transfer Ratio	h_{FE}		10			150[a]			50	250	—	—	—	—	
			4			500			—	—	25	100	—	—	
			4			4 A[a]			—	—	—	—	20	70	
Collector-to-Base Breakdown Voltage	BV_{CBO}					0.1	0		60	—	—	—	—	—	V
Emitter-to-Base Breakdown Voltage	BV_{EBO}					0	0.1		5	—	—	—	—	—	V
						0	1		—	—	7	—	—	—	
						0	5		—	—	—	—	7	—	
Collector-to-Emitter Sustaining Voltage: With base open	V_{CEO}(sus)					100[a]		0	40	—	—	—	—	—	V
						100		0	—	—	55	—	—	—	
						200		0	—	—	—	—	60	—	
With base-emitter junction reverse biased	V_{CEV}(sus)				−1.5	100			—	—	—	—	100	—	V
With external base-to-emitter resistance (R_{BE}) = 10 Ω	V_{CER}(sus)					100[a]			50	—	—	—	—	—	V
= 100 Ω						100			—	—	60	—	—	—	
= 100 Ω						200			—	—	—	—	70	—	
Base-to-Emitter Voltage	V_{BE}		4			500			—	—	—	1.7	—	—	V
			4			4 A[a]			—	—	—	—	—	1.8	
Base-to-Emitter Saturation Voltage						150		15	—	1.7	—	—	—	—	V
Collector-to-Emitter Saturation Voltage	V_{CE}(sat)					150		15	—	1.4	—	—	—	—	V
						500		50	—	—	—	1.0	—	—	
						4 A[a]		400	—	—	—	—	—	1.1	
Small-Signal, Forward Current Transfer Ratio (At 20 MHz)	h_{fe}		10			50			5	—	—	—	—	—	
Gain-Bandwidth Product	f_T					200			—	—	800	—	—	—	kHz
						1 A			—	—	—	—	800	—	
Output Capacitance	C_{ob}	10					0		—	15	—	—	—	—	pF
Input Capacitance	C_{ib}			0.5		0			—	80	—	—	—	—	pF
Power Rating Test	PRT		39			3 A			—	—	—	—	—	1[b]	s
Thermal Resistance: Junction-to-Case	θ_{J-C}								35(max.) 2N3053		6(max.) 2N3054		—	1.5	°C/W
									25(max.) 40392				—		°C/W
Junction-to-Free Air	θ_{J-FA}								175(max.) 2N3053		30(max.) 40372		—	—	°C/W
									50(max.) 40389				—	—	°C/W

[a] Pulsed; pulse duration = 300 μs, duty factor = 1.8%. [b] At 115 W.

Courtesy of Radio Corporation of America

SILICON NPN POWER TRANSISTORS

. . . for use in power amplifier and switching circuits, — excellent safe area limits. Complement to PNP 2N5193, 2N5194, 2N5195 and MJE5193, MJE5194, MJE5195.

2N5190 thru 2N5192
MJE5190 thru MJE5192

4 AMPERE
POWER TRANSISTORS
SILICON NPN

40-80 VOLTS
40 and 60 WATTS

NOVEMBER 1971 – DS 3106 R1

*MAXIMUM RATINGS

Rating	Symbol	2N5190 MJE5190	2N5191 MJE5191	2N5192 MJE5192	Unit
Collector-Emitter Voltage	V_{CEO}	40	60	80	Vdc
Collector-Base Voltage	V_{CB}	40	60	80	Vdc
Emitter-Base Voltage	V_{EB}	← 5 0 →			Vdc
Collector Current	I_C	← 4 0 →			Adc
Base Current	I_B	← 1 0 →			Adc
		2N5190 Series	MJE5190 Series		
Total Device Dissipation @ $T_C = 25°C$	P_D	40	60		Watts
Derate above 25°C		320	480		mW/°C
Operating and Storage Junction Temperature Range	T_J, T_{stg}	← -65 to +150 →			°C

THERMAL CHARACTERISTICS

Characteristic	Symbol	2N5190 Series	MJE5190 Series	Unit
Thermal Resistance, Junction to Case	θ_{JC}	3.12	2.08	°C/W

*ELECTRICAL CHARACTERISTICS ($T_C = 25°C$ unless otherwise noted)

Characteristic	Symbol	Min	Max	Unit
OFF CHARACTERISTICS				
Collector-Emitter Sustaining Voltage (1)	$V_{CEO(sus)}$		–	Vdc
($I_C = 0.1$ Adc, $I_B = 0$) 2N5190,MJE5190		40		
2N5191,MJE5191		60		
2N5192,MJE5192		80		
Collector Cutoff Current	I_{CEO}			mAdc
($V_{CE} = 40$ Vdc, $I_B = 0$) 2N5190,MJE5190		–	1.0	
($V_{CE} = 60$ Vdc, $I_B = 0$) 2N5191,MJE5191		–	1.0	
($V_{CE} = 80$ Vdc, $I_B = 0$) 2N5192,MJE5192		–	1.0	
Collector Cutoff Current	I_{CEX}			mAdc
($V_{CE} = 40$ Vdc, $V_{EB(off)} = 1.5$ Vdc) 2N5190,MJE5190		–	0.1	
($V_{CE} = 60$ Vdc, $V_{EB(off)} = 1.5$ Vdc) 2N5191,MJE5191		–	0.1	
($V_{CE} = 80$ Vdc, $V_{EB(off)} = 1.5$ Vdc) 2N5192,MJE5192		–	0.1	
($V_{CE} = 40$ Vdc, $V_{EB(off)} = 1.5$ Vdc, $T_C = 125°C$) 2N5190,MJE5190		–	2.0	
($V_{CE} = 60$ Vdc, $V_{EB(off)} = 1.5$ Vdc, $T_C = 125°C$) 2N5191,MJE5191		–	2.0	
($V_{CE} = 80$ Vdc, $V_{EB(off)} = 1.5$ Vdc, $T_C = 125°C$) 2N5192,MJE5192		–	2.0	
Collector Cutoff Current	I_{CBO}			mAdc
($V_{CB} = 40$ Vdc, $I_E = 0$) 2N5190,MJE5190		–	0.1	
($V_{CB} = 60$ Vdc, $I_E = 0$) 2N5191,MJE5191		–	0.1	
($V_{CB} = 80$ Vdc, $I_E = 0$) 2N5192,MJE5192		–	0.1	
Emitter Cutoff Current	I_{EBO}			mAdc
($V_{BE} = 5.0$ Vdc, $I_C = 0$)		–	1.0	
ON CHARACTERISTICS				
DC Current Gain(1)	h_{FE}			
($I_C = 1.5$ Adc, $V_{CE} = 2.0$ Vdc) 2N5190,MJE5190		25	100	
2N5191,MJE5191		25	100	
2N5192,MJE5192		20	80	
($I_C = 4.0$ Adc, $V_{CE} = 2.0$ Vdc) 2N5190,MJE5190		10	–	
2N5191,MJE5191		10	–	
2N5192,MJE5192		7.0	–	
Collector-Emitter Saturation Voltage(1)	$V_{CE(sat)}$			Vdc
($I_C = 1.5$ Adc, $I_B = 0.15$ Adc)		–	0.6	
($I_C = 4.0$ Adc, $I_B = 1.0$ Adc)		–	1.4	
Base-Emitter On Voltage (1)	$V_{BE(on)}$			Vdc
($I_C = 1.5$ Adc, $V_{CE} = 2.0$ Vdc)		–	1.2	
DYNAMIC CHARACTERISTICS				
Current-Gain-Bandwidth Product	f_T			MHz
($I_C = 1.0$ Adc, $V_{CE} = 10$ Vdc, f = 1.0 MHz)		2.0		

(1) Pulse Test: Pulse Width ≤ 300 μs, Duty Cycle ≤ 2.0%.
*Indicates JEDEC Registered Data for 2N5190 Series.

Courtesy of Motorola Semiconductor Products Inc.

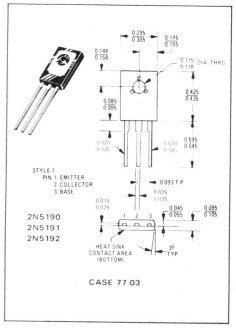

STYLE 1
PIN 1. EMITTER
2. COLLECTOR
3. BASE

2N5190
2N5191
2N5192

HEAT SINK
CONTACT AREA
(BOTTOM)

CASE 77-03

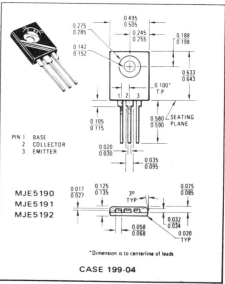

PIN 1. BASE
2. COLLECTOR
3. EMITTER

MJE5190
MJE5191
MJE5192

*Dimension is to centerline of leads

CASE 199-04

© MOTOROLA INC., 1971

310

MOTOROLA

2N5193 thru 2N5195
MJE5193 thru MJE5195

SILICON PNP POWER TRANSISTORS

... for use in power amplifier and switching circuits, − excellent safe area limits. Complement to NPN 2N5190, 2N5191, 2N5192 and MJE5190, MJE5191, MJE5192.

4 AMPERE
POWER TRANSISTORS
SILICON PNP

40-80 VOLTS
40 and 60 WATTS

NOVEMBER 1971 − DS 3109 R1

*MAXIMUM RATINGS

Rating	Symbol	2N5193 MJE5193	2N5194 MJE5194	2N5195 MJE5195	Unit
Collector-Emitter Voltage	V_{CEO}	40	60	80	Vdc
Collector-Base Voltage	V_{CB}	40	60	80	Vdc
Emitter-Base Voltage	V_{EB}		5.0		Vdc
Collector Current	I_C		4.0		Adc
Base Current	I_B		1.0		Adc

		2N5193 Series	MJE5193 Series	
Total Device Dissipation @ T_C = 25°C Derate above 25°C	P_D	40 320	60 480	Watts mW/°C
Operating and Storage Junction Temperature Range	T_J, T_{stg}		− 65 to +150	°C/W

THERMAL CHARACTERISTICS

Characteristic	Symbol	2N5193 Series	MJE5193 Series	Unit
Thermal Resistance, Junction to Case	θ_{JC}	3.12	2.08	°C/W

*ELECTRICAL CHARACTERISTICS (T_C = 25°C unless otherwise noted)

Characteristic	Symbol	Min	Max	Unit
OFF CHARACTERISTICS				
Collector-Emitter Sustaining Voltage (1) (I_C = 0.1 Adc, I_B = 0) 2N5193, MJE5193 2N5194, MJE5194 2N5195, MJE5195	$V_{CEO(sus)}$	40 60 80	− − −	Vdc
Collector Cutoff Current (V_{CE} = 40 Vdc, I_B = 0) 2N5193, MJE5193 (V_{CE} = 60 Vdc, I_B = 0) 2N5194, MJE5194 (V_{CE} = 80 Vdc, I_B = 0) 2N5195, MJE5195	I_{CEO}	− − −	1.0 1.0 1.0	mAdc

Collector Cutoff Current
 (V_{CE} = 40 Vdc, $V_{BE(off)}$ = 1.5 Vdc) 2N5193, MJE5193
 (V_{CE} = 60 Vdc, $V_{BE(off)}$ = 1.5 Vdc) 2N5194, MJE5194
 (V_{CE} = 80 Vdc, $V_{BE(off)}$ = 1.5 Vdc) 2N5195, MJE5195
 (V_{CE} = 40 Vdc, $V_{BE(off)}$ = 1.5 Vdc, 2N5193, MJE5193
    T_C = 125°C)
 (V_{CE} = 60 Vdc, $V_{BE(off)}$ = 1.5 Vdc, 2N5194, MJE5194
    T_C = 125°C)
 (V_{CE} = 80 Vdc, $V_{BE(off)}$ = 1.5 Vdc, 2N5195, MJE5195
    T_C = 125°C)
	I_{CEX}	−
		 −
		 −
		 −
		
		 −
		
		 −
			0.1
			 0.1
			 0.1
			 2.0
			
			 2.0
			
			 2.0
				mAdc

Collector Cutoff Current (V_{CB} = 40 Vdc, I_E = 0) 2N5193, MJE5193 (V_{CB} = 60 Vdc, I_E = 0) 2N5194, MJE5194 (V_{CB} = 80 Vdc, I_E = 0) 2N5195, MJE5195	I_{CBO}	− − −	0.1 0.1 0.1	mAdc
Emitter Cutoff Current (V_{BE} = 5.0 Vdc, I_C = 0)	I_{EBO}	−	1.0	mAdc
ON CHARACTERISTICS				
DC Current Gain (1) (I_C = 1.5 Adc, V_{CE} = 2.0 Vdc) 2N5193, MJE5193, 2N5194, MJE5194, 2N5195, MJE5195 (I_C = 4.0 Adc, V_{CE} = 2.0 Vdc) 2N5193, MJE5193, 2N5194, MJE5194, 2N5195, MJE5195	h_{FE}	25 20 10 7.0	100 80 − −	
Collector-Emitter Saturation Voltage (1) (I_C = 1.5 Adc, I_B = 0.15 Adc) (I_C = 4.0 Adc, I_B = 1.0 Adc)	$V_{CE(sat)}$	− −	0.6 1.2	Vdc
Base-Emitter On Voltage (1) (I_C = 1.5 Adc, V_{CE} = 2.0 Vdc)	$V_{BE(on)}$	−	1.2	Vdc
DYNAMIC CHARACTERISTICS				
Current-Gain-Bandwidth Product (I_C = 1.0 Adc, V_{CE} = 10 Vdc, f = 1.0 MHz)	f_T	2.0		MHz

*Indicates JEDEC Registered Data for 2N5193 Series.

(1) Pulse Test: Pulse Width ≤ 300 μs, Duty Cycle ≤ 2.0%.

Courtesy of Motorola Semiconductor Products Inc.

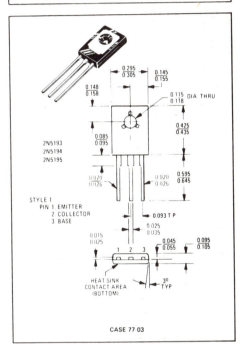

2N5193
2N5194
2N5195

STYLE 1
PIN 1. EMITTER
2. COLLECTOR
3. BASE

CASE 77-03

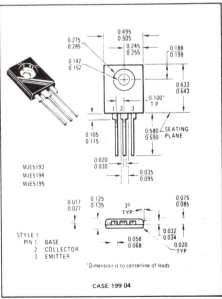

MJE5193
MJE5194
MJE5195

STYLE 1
PIN 1. BASE
2. COLLECTOR
3. EMITTER

*Dimension is to centerline of leads

CASE 199-04

© MOTOROLA INC., 1971

311

µA741
FREQUENCY-COMPENSATED OPERATIONAL AMPLIFIER
FAIRCHILD LINEAR INTEGRATED CIRCUITS

GENERAL DESCRIPTION — The µA741 is a high performance monolithic Operational Amplifier constructed using the Fairchild Planar* epitaxial process. It is intended for a wide range of analog applications. High common mode voltage range and absence of "latch-up" tendencies make the µA741 ideal for use as a voltage follower. The high gain and wide range of operating voltage provides superior performance in integrator, summing amplifier, and general feedback applications.

- NO FREQUENCY COMPENSATION REQUIRED
- SHORT CIRCUIT PROTECTION
- OFFSET VOLTAGE NULL CAPABILITY
- LARGE COMMON-MODE AND DIFFERENTIAL VOLTAGE RANGES
- LOW POWER CONSUMPTION
- NO LATCH UP

ABSOLUTE MAXIMUM RATINGS

Supply Voltage	
Military (741)	±22 V
Commercial (741C)	±18 V
Internal Power Dissipation (Note 1)	
Metal Can	500 mW
DIP	670 mW
Mini DIP	310 mW
Flatpak	570 mW
Differential Input Voltage	±30 V
Input Voltage (Note 2)	±15 V
Storage Temperature Range	
Metal Can, DIP, and Flatpak	−65°C to +150°C
Mini DIP	−55°C to +125°C
Operating Temperature Range	
Military (741)	−55°C to +125°C
Commercial (741C)	0°C to +70°C
Lead Temperature (Soldering)	
Metal Can, DIP, and Flatpak (60 seconds)	300°C
Mini DIP (10 seconds)	260°C
Output Short Circuit Duration (Note 3)	Indefinite

EQUIVALENT CIRCUIT

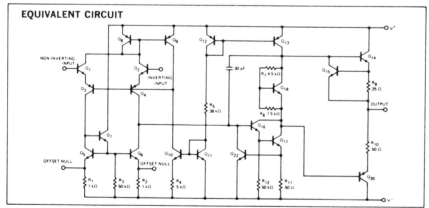

Notes on following pages.

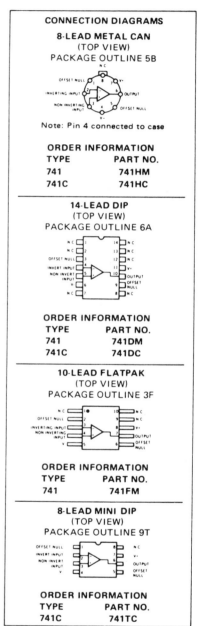

CONNECTION DIAGRAMS

8-LEAD METAL CAN
(TOP VIEW)
PACKAGE OUTLINE 5B

Note: Pin 4 connected to case

ORDER INFORMATION

TYPE	PART NO.
741	741HM
741C	741HC

14-LEAD DIP
(TOP VIEW)
PACKAGE OUTLINE 6A

ORDER INFORMATION

TYPE	PART NO.
741	741DM
741C	741DC

10-LEAD FLATPAK
(TOP VIEW)
PACKAGE OUTLINE 3F

ORDER INFORMATION

TYPE	PART NO.
741	741FM

8-LEAD MINI DIP
(TOP VIEW)
PACKAGE OUTLINE 9T

ORDER INFORMATION

TYPE	PART NO.
741C	741TC

*Planar is a patented Fairchild process.

Fairchild cannot assume responsibility for use of any circuitry described other than circuitry entirely embodied in a Fairchild product. No other circuit patent licenses are implied.

Courtesy of Fairchild Semiconductor

741

ELECTRICAL CHARACTERISTICS ($V_S = \pm 15$ V, $T_A = 25°C$ unless otherwise specified)

PARAMETERS (see definitions)	CONDITIONS		MIN.	TYP.	MAX.	UNITS
Input Offset Voltage	$R_S \leqslant 10$ kΩ			1.0	5.0	mV
Input Offset Current				20	200	nA
Input Bias Current				80	500	nA
Input Resistance			0.3	2.0		MΩ
Input Capacitance				1.4		pF
Offset Voltage Adjustment Range				±15		mV
Large Signal Voltage Gain	$R_L \geqslant 2$ kΩ, $V_{OUT} = \pm 10$ V		50,000	200,000		
Output Resistance				75		Ω
Output Short Circuit Current				25		mA
Supply Current				1.7	2.8	mA
Power Consumption				50	85	mW
Transient Response (Unity Gain)	Risetime	$V_{IN} = 20$ mV, $R_L = 2$ kΩ, $C_L \leqslant 100$ pF		0.3		μs
	Overshoot			5.0		%
Slew Rate	$R_L \geqslant 2$ kΩ			0.5		V/μs

The following specifications apply for $-55°C \leqslant T_A \leqslant +125°C$:

		MIN.	TYP.	MAX.	UNITS
Input Offset Voltage	$R_S \leqslant 10$ kΩ		1.0	6.0	mV
Input Offset Current	$T_A = +125°C$		7.0	200	nA
	$T_A = -55°C$		85	500	nA
Input Bias Current	$T_A = +125°C$		0.03	0.5	μA
	$T_A = -55°C$		0.3	1.5	μA
Input Voltage Range		±12	±13		V
Common Mode Rejection Ratio	$R_S \leqslant 10$ kΩ	70	90		dB
Supply Voltage Rejection Ratio	$R_S \leqslant 10$ kΩ		30	150	μV/V
Large Signal Voltage Gain	$R_L \geqslant 2$ kΩ, $V_{OUT} = \pm 10$ V	25,000			
Output Voltage Swing	$R_L \geqslant 10$ kΩ	±12	±14		V
	$R_L \geqslant 2$ kΩ	±10	±13		V
Supply Current	$T_A = +125°C$		1.5	2.5	mA
	$T_A = -55°C$		2.0	3.3	mA
Power Consumption	$T_A = +125°C$		45	75	mW
	$T_A = -55°C$		60	100	mW

TYPICAL PERFORMANCE CURVES FOR 741

OPEN LOOP VOLTAGE GAIN AS A FUNCTION OF SUPPLY VOLTAGE

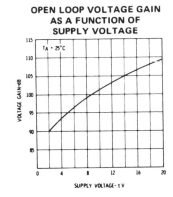

OUTPUT VOLTAGE SWING AS A FUNCTION OF SUPPLY VOLTAGE

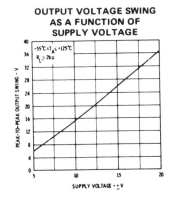

INPUT COMMON MODE VOLTAGE RANGE AS A FUNCTION OF SUPPLY VOLTAGE

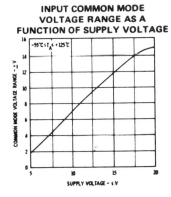

Courtesy of Fairchild Semiconductor

LINEAR INTEGRATED CIRCUITS

DESCRIPTION

The µA723 is a Monolithic Precision Voltage Regulator capable of operation in positive or negative supplies as a series, shunt, switching or floating regulator. The µA723 contains a temperature compensated reference amplifier, error amplifier, series pass transistor, and current limiter, with access to remote shutdown.

FEATURES

- POSITIVE OR NEGATIVE SUPPLY OPERATION
- SERIES, SHUNT, SWITCHING OR FLOATING OPERATION
- .01% LINE AND LOAD REGULATION
- OUTPUT VOLTAGE ADJUSTABLE FROM 2 TO 37 VOLTS
- OUTPUT CURRENT TO 150mA WITHOUT EXTERNAL PASS TRANSISTOR

EQUIVALENT CIRCUIT

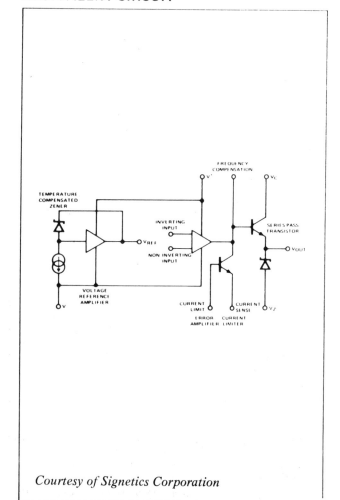

Courtesy of Signetics Corporation

PIN CONFIGURATION

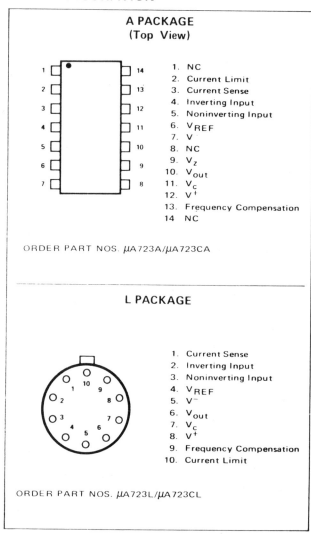

A PACKAGE (Top View)

1. NC
2. Current Limit
3. Current Sense
4. Inverting Input
5. Noninverting Input
6. V_{REF}
7. V^-
8. NC
9. V_Z
10. V_{out}
11. V_C
12. V^+
13. Frequency Compensation
14. NC

ORDER PART NOS. µA723A/µA723CA

L PACKAGE

1. Current Sense
2. Inverting Input
3. Noninverting Input
4. V_{REF}
5. V^-
6. V_{out}
7. V_C
8. V^+
9. Frequency Compensation
10. Current Limit

ORDER PART NOS. µA723L/µA723CL

ABSOLUTE MAXIMUM RATINGS

	µA723	µA723C
Pulse Voltage from V^+ to V^- (50ms)	50V	
Continuous Voltage from V^+ to V^-	40V	40V
Input-Output Voltage Differential	40V	40V
Maximum Output Current	150mA	150mA
Current from V_{REF}	15mA	
Current from V_Z		25mA
Internal Power Dissipation (Note 1)	800mW	800mW
Operating Temperature Range	−55 to +125°C	0 to 70°C
Storage Temperature Range	−65°C to +150°C	−65°C to +150°C
Lead Temperature	300°C	300°C

SIGNETICS ■ μA723/723C — PRECISION VOLTAGE REGULATOR

ELECTRICAL CHARACTERISTICS (T_A = 25°C unless otherwise specified – Note 1)

PARAMETER (See definitions)	MIN	TYP	MAX	UNITS	CONDITIONS
μA723C					
Line Regulation (Note 2)		0.01	0.1	% V_{out}	V_{in} = 12V to V_{in} = 15V
		0.1	0.5	% V_{out}	V_{in} = 12V to V_{in} = 40V
Load Regulation (Note 2)		0.03	0.2	% V_{out}	I_L = 1mA to I_L = 50mA
Ripple Rejection		74		dB	f = 50 Hz to 10 kHz, C_{REF} = 0
		86		dB	f = 50 Hz to 10 kHz, C_{REF} = 5μF
Short Circuit Current Limit		65		mA	R_{sc} = 10Ω, V_{out} = 0
Reference Voltage	6.80	7.15	7.50	V	
Output Noise Voltage		20		μV rms	BW = 100 Hz to 10 kHz, C_{REF} = 0
		2.5		μV rms	BW = 100 Hz to 10 kHz, C_{REF} = 5μF
Long Term Stability			0.1	%/1000 hrs.	
Standby Current Drain		2.3	4.0	mA	I_L = 0, V_{in} = 30V
Input Voltage Range	9.5		40	V	
Output Voltage Range	2.0		37	V	
Input-Output Voltage Differential	3.0		38	V	
The Following Specifications Apply Over the Operating Temperature Ranges					
Line Regulation			0.3	% V_{out}	
Load Regulation			0.6	% V_{out}	
Average Temperature Coefficient of Output Voltage		0.003	0.015	%/°C	V_{in} = 12V to V_{in} = 15V / I_L = 1mA to I_L = 50mA
μA723					
Line Regulation (Note 2)		0.01	0.1	%V_{out}	V_{in} = 12V to V_{in} = 15V
		0.02	0.2	%V_{out}	V_{in} = 12V to V_{in} =40V
Load Regulation (Note 2)		0.03	0.15	%V_{out}	I_L = 1mA to I_L = 50mA
Ripple Rejection		74		dB	f = 50 Hz to 10 kHz, C_{REF} = 0
		86		dB	f = 50 Hz to 10 kHz, C_{REF} = 5μF
Short Circuit Current Limit		65		mA	R_{SC} = 10Ω, V_{out} = 0
Reference Voltage	6.95	7.15	7.35	V	
Output Noise Voltage		20		μV rms	BW = 100 Hz to 10 kHz, C_{REF} = 0
		2.5		μV rms	RW = 100 Hz to 10 kHz, C_{REF} = 5μF
Long Term Stability		0.1		%/1000 hrs	
Standby Current Drain		2.3	3.5	mA	I_L = 0, V_{in} = 30V
Input Voltage Range	9.5		40	V	
Output Voltage Range	2.0		37	V	
Input-Output Voltage Differential	3.0		38	V	
The Following Specifications Apply Over the Operating Temperature Ranges					
Line Regulation			0.3	% V_{out}	
Load Regulation			0.6	% V_{out}	
Average Temperature Coefficient of Output Voltage		0.002	0.015	%/°C	V_{in} = 12V to V_{in} = 15V / I_L = 1mA to I_L = 50mA

NOTES

1. Unless otherwise specified, T_A = 25°C, V_{in} = V+ = V_c = 12V, V– = 0V, V_{out} = 5V, I_L = 1mA, R_{sc} = 0, C_1 = 100pF, C_{REF} = 0 and divider impedance as seen by error amplifier ⩽ 10kΩ when connected as shown in Figure 3.

2. The load and line regulation specifications are for constant junction temperature. Temperature drift effects must be taken into account separately when the unit is operating under conditions of high dissipation.

Courtesy of Signetics Corporation

315

μA78G · μA79G
4-TERMINAL POSITIVE AND NEGATIVE ADJUSTABLE VOLTAGE REGULATORS
FAIRCHILD LINEAR INTEGRATED CIRCUITS

GENERAL DESCRIPTION — The μA78G and μA79G are 4-Terminal Adjustable Voltage Regulators. They are designed to deliver continuous load currents of up to 1.0 A with a maximum input voltage of 40 V for the positive regulator 78G and −40 V for the negative regulator 79G. Output current capability can be increased to greater than 1.0 A through use of one or more external transistors. The output voltage range of the 78G positive voltage regulator is 5 V to 30 V and the output voltage range of the negative 79G is −30 V to −2.2 V. For systems requiring both a positive and negative, the 78G and 79G are excellent for use as a dual tracking regulator with appropriate external circuitry. These 4-terminal voltage regulators are constructed using the Fairchild Planar* process.

- OUTPUT CURRENT IN EXCESS OF 1.0 A
- μA78G POSITIVE OUTPUT VOLTAGE 5 TO 30 V
- μA79G NEGATIVE OUTPUT VOLTAGE −30 TO −2.2 V
- INTERNAL THERMAL OVERLOAD PROTECTION
- INTERNAL SHORT CIRCUIT CURRENT PROTECTION
- OUTPUT TRANSISTOR SAFE AREA PROTECTION
- MILITARY AND COMMERCIAL VERSIONS AVAILABLE
- AVAILABLE IN 4-PIN TO-202 TYPE AND 4-PIN TO-3

ABSOLUTE MAXIMUM RATINGS

Input Voltage
μA78G, μA78GC .... 40 V
μA79G, μA79GC .... −40 V

Control Pin Voltage
μA78G, μA78GC .... $0 \leqslant V \leqslant V_{OUT}$
μA79G, μA79GC .... $-V_{OUT} \leqslant -V \leqslant 0$

Power Dissipation .... Internally Limited

Operating Junction Temperature Range
Military (μA78G, μA79G) .... −55°C to 150°C
Commercial (μA78GC, μA79GC) .... 0°C to 150°C

Storage Temperature Range
4-Pin Power TAB (U1) .... −55°C to +150°C
4-Pin TO-3 (K) .... −65°C to +150°C

Lead Temperature
4-Pin Power TAB (U1) (Soldering, 10 s) .... 230°C
4-Pin TO-3 (K) (Soldering, 60 s) .... 300°C

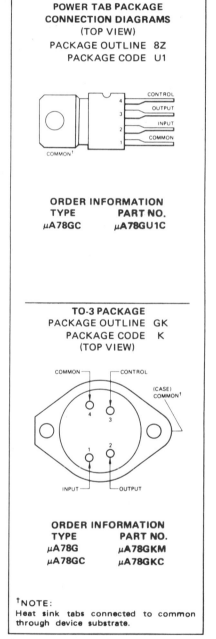

Courtesy of Fairchild Semiconductor

316

μA78G EQUIVALENT CIRCUIT

μA78G, μA78GC

ELECTRICAL CHARACTERISTICS Unless otherwise specified, the following applies: $0°C \leq T_J \leq 125°C$ for 78GC and $-55°C \leq T_J \leq 150°C$ for 78G, $V_{IN} = 10\ V$, $I_{OUT} = 500\ mA$, Test Circuit 1.

PARAMETER	CONDITION (Note 1)		MIN	TYP	MAX	UNITS
Input Voltage Range	$T_J = 25°C$		7.5		40	V
Output Voltage Range	$V_{IN} = V_{OUT} + 5\ V$		5.0		30	V
Output Voltage Tolerance	$V_{OUT} + 3\ V \leq V_{IN} \leq V_{OUT} + 15\ V$, $5\ mA \leq I_{OUT} \leq 1.0\ A$	$T_J = 25°C$			4.0	%(V_{OUT})
	$P_D \leq 15\ W$, $V_{IN(MAX)} = 38\ V$				5.0	%(V_{OUT})
Line Regulation	$T_J = 25°C$, $V_{OUT} \leq 10\ V$ $(V_{OUT} + 2.5\ V) \leq V_{IN} \leq (V_{OUT} + 20\ V)$				1.0	%(V_{OUT})
	$T_J = 25°C$, $V_{OUT} \geq 10\ V$ $(V_{OUT} + 3\ V) \leq V_{IN} \leq (V_{OUT} + 15\ V)$				0.75	%(V_{OUT})
	$(V_{OUT} + 3\ V) \leq V_{IN} \leq (V_{OUT} + 7\ V)$				0.67	
Load Regulation	$T_J = 25°C$	$250\ mA \leq I_{OUT} \leq 750\ mA$			1.0	%(V_{OUT})
	$V_{IN} = V_{OUT} + 5\ V$	$5\ mA \leq I_{OUT} \leq 1.5\ A$			2.0	
Control Pin Current	$T_J = 25°C$			1.0	5.0	μA
					8.0	μA
Quiescent Current	$T_J = 25°C$			3.2	5.0	mA
					6.0	mA
Ripple Rejection	$8\ V \leq V_{IN} \leq 18\ V$, $f = 120\ Hz$	μA78G	68	78		dB
	$V_{OUT} = 5\ V$	μA78GC	62	78		dB
Output Noise Voltage	$T_J = 25°C$, $10\ Hz < f < 100\ kHz$, $V_{OUT} = 5\ V$			40		μV
Dropout Voltage	Note 2	μA78G			3.0	V
		μA78GC			2.5	V
Short Circuit Current	$T_J = 25°C$, $V_{IN} = 30\ V$			750		mA
Peak Output Current	$T_J = 25°C$			2.2		A
Average Temperature Coefficient of Output Voltage	$V_{OUT} = 5\ V$, $I_{OUT} = 5\ mA$			-1.1		mV/°C
Control Pin Voltage (Reference)	$T_J = 25°C$		4.8	5.0	5.2	V
			4.75		5.25	V

NOTES:
1. V_{OUT} is defined for the 78GC as $V_{OUT} = \frac{R1 + R2}{R2}(5.0)$; The 79GC as $V_{OUT} = \frac{R1 + R2}{R2}(-2.23)$.
2. Dropout voltage is defined as that input-output voltage differential which causes the output voltage to decrease by 5% of its initial value.

Courtesy of Fairchild Semiconductor

79G EQUIVALENT CIRCUIT

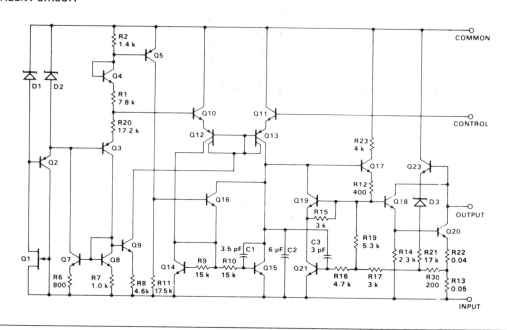

μA79G, μA79GC

ELECTRICAL CHARACTERISTICS Unless otherwise specified, the following applies: $0°C \leq T_J \leq 125°C$ for 79GC and $-55°C \leq T_J \leq 150°C$ for 79G, $V_{IN} = -10$ V, $I_{OUT} = 500$ mA, Test Circuit 2 and Note 3.

PARAMETER	CONDITION (Note 1)		MIN	TYP	MAX	UNITS
Input Voltage Range	$T_J = 25°C$		-40		-7.0	V
Nominal Output Voltage Range	$V_{IN} = V_{OUT} - 5$ V		-30		-2.23	V
Output Voltage Tolerance	$V_{OUT} - 15$ V $\leq V_{IN} \leq V_{OUT} - 3$ V, 5 mA $\leq I_{OUT} \leq 1.0$ A	$T_J = 25°C$			4.0	%(V_{OUT})
	$P_D \leq 15$ W, $V_{IN(MAX)} = -38$ V				5.0	%(V_{OUT})
Line Regulation	$T_J = 25°C$, $V_{OUT} \geq -10$ V ($V_{OUT} - 20$ V) $\leq V_{IN} \leq$ ($V_{OUT} - 2.5$ V)				1.0	%(V_{OUT})
	$T_J = 25°C$, $V_{OUT} \leq -10$ V ($V_{OUT} - 15$ V) $\leq V_{IN} \leq$ ($V_{OUT} - 3$ V)				0.75	%(V_{OUT})
	($V_{OUT} - 7$ V) $\leq V_{IN} \leq$ ($V_{OUT} - 3$ V)				0.67	
Load Regulation	$T_J = 25°C$	250 mA $\leq I_{OUT} \leq 750$ mA			1.0	%(V_{OUT})
	$V_{IN} = V_{OUT} - 5$ V	5 mA $\leq I_{OUT} \leq 1.5$ A			2.0	%(V_{OUT})
Control Pin Current	$T_J = 25°C$			0.4	2.0	μA
					3.0	μA
Quiescent Current	$T_J = 25°C$			0.5	1.5	mA
					2.0	mA
Ripple Rejection	-18 V $\leq V_{IN} \leq -8$ V	μA79G	50	60		dB
	$V_{OUT} = -5$ V, $f = 120$ Hz	μA79GC	50	60		dB
Output Noise Voltage	$T_J = 25°C$, 10 Hz $\leq f \leq 100$ kHz, $V_{OUT} = -5$ V			125		μV
Dropout Voltage	Note 2	μA79G			2.8	V
		μA79GC			2.3	V
Short Circuit Current	$T_J = 25°C$, $V_{IN} = -30$ V			250		mA
Peak Output Current	$T_J = 25°C$			2.2		A
Average Temperature Coefficient of Output Voltage	$V_{OUT} = -5$ V, $I_{OUT} = 5$ mA			-0.4		mV/°C
Control Pin Voltage (Reference)	$T_J = 25°C$		-2.32	-2.23	-2.14	V
			-2.35		-2.11	V

Note 3.

The convention for negative regulators is the algebraic value, thus -15 is less than -10 V.

MJ2955

15 AMPERE POWER TRANSISTOR

PNP SILICON

60 VOLTS
150 WATTS

DS 3215

PNP SILICON POWER TRANSISTOR

. . . designed for general-purpose switching and amplifier applications.

- DC Current Gain —
  h_{FE} = 20-70 @ I_C = 4.0 Adc
- Collector-Emitter Saturation Voltage —
  $V_{CE(sat)}$ = 1.1 Vdc (Max) @ I_C = 4.0 Adc
- Excellent Safe Operating Area
- Complement to Motorola's "Epi-Base" Transistor, 2N3055

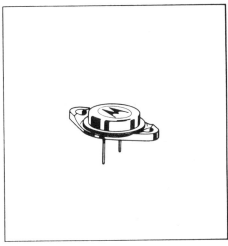

MAXIMUM RATINGS

Rating	Symbol	Value	Unit
Collector-Emitter Voltage	V_{CEO}	60	Vdc
Collector-Emitter Voltage	V_{CER}	70	Vdc
Collector-Base Voltage	V_{CB}	100	Vdc
Emitter-Base Voltage	V_{EB}	7.0	Vdc
Collector Current — Continuous	I_C	15	Adc
Base Current	I_B	7.0	Adc
Total Device Dissipation @ T_C = 25°C Derate above 25°C	P_D	150 0.86	Watts W/°C
Operating and Storage Junction Temperature Range	T_J, T_{stg}	-65 to +200	°C

THERMAL CHARACTERISTICS

Characteristic	Symbol	Max	Unit
Thermal Resistance, Junction to Case	θ_{JC}	1.17	°C/W

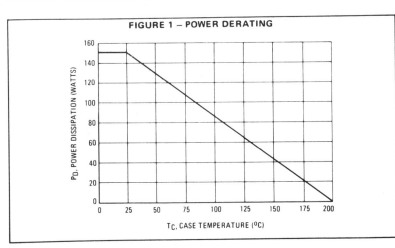

FIGURE 1 — POWER DERATING

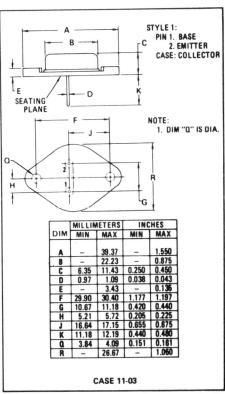

STYLE 1:
PIN 1. BASE
2. EMITTER
CASE: COLLECTOR

NOTE:
1. DIM "Q" IS DIA.

DIM	MILLIMETERS		INCHES	
	MIN	MAX	MIN	MAX
A	–	39.37	–	1.550
B	–	22.23	–	0.875
C	6.35	11.43	0.250	0.450
D	0.97	1.09	0.038	0.043
E	–	3.43	–	0.135
F	29.90	30.40	1.177	1.197
G	10.67	11.18	0.420	0.440
H	5.21	5.72	0.205	0.225
J	16.64	17.15	0.655	0.675
K	11.18	12.19	0.440	0.480
Q	3.84	4.09	0.151	0.161
R	–	26.67	–	1.050

CASE 11-03

Courtesy of Motorola Semiconductor Products Inc.

ELECTRICAL CHARACTERISTICS ($T_C = 25^{\circ}C$ unless otherwise noted)

Characteristic	Symbol	Min	Max	Unit
OFF CHARACTERISTICS				
Collector-Emitter Sustaining Voltage (1) (I_C = 200 mAdc, I_B = 0)	$V_{CEO(sus)}$	60	—	Vdc
Collector-Emitter Breakdown Voltage (1) (I_C = 200 mAdc, R_{BE} = 100 Ohms)	BV_{CER}	70	—	Vdc
Collector Cutoff Current (V_{CE} = 30 Vdc, I_B = 0)	I_{CEO}	—	0.7	mAdc
Collector Cutoff Current (V_{CE} = 100 Vdc, $V_{BE(off)}$ = 1.5 Vdc) (V_{CE} = 100 Vdc, $V_{BE(off)}$ = 1.5 Vdc, T_C = 150°C)	I_{CEX}	— —	1.0 5.0	mAdc
Emitter Cutoff Current (V_{BE} = 7.0 Vdc, I_C = 0)	I_{EBO}	—	5.0	mAdc
ON CHARACTERISTICS (1)				
DC Current Gain (I_C = 4.0 Adc, V_{CE} = 4.0 Vdc) (I_C = 10 Adc, V_{CE} = 4.0 Vdc)	h_{FE}	20 5.0	70 —	—
Collector-Emitter Saturation Voltage (I_C = 4.0 Adc, I_B = 400 mAdc) (I_C = 10 Adc, I_B = 3.3 Adc)	$V_{CE(sat)}$	— —	1.1 3.0	Vdc
Base-Emitter On Voltage (I_C = 4.0 Adc, V_{CE} = 4.0 Vdc)	$V_{BE(on)}$	—	1.8	Vdc
DYNAMIC CHARACTERISTICS				
Current Gain — Bandwidth Product (I_C = 0.5 Adc, V_{CE} = 10 Vdc, f = 1.0 MHz)	f_T	4.0	—	MHz
Small-Signal Current Gain (I_C = 1.0 Adc, V_{CE} = 4.0 Vdc, f = 1.0 kHz)	h_{fe}	15		—
Small-Signal Current Gain Cutoff Frequency (V_{CE} = 4.0 Vdc, I_C = 1.0 Adc, f = 1.0 kHz)	$f_{\propto e}$	10	—	kHz

(1) Pulse Test: Pulse Width $\leq$ 300 µs, Duty Cycle $\leq$ 2.0%.

FIGURE 2 — SWITCHING TIME TEST CIRCUIT

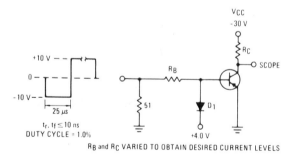

R_B and R_C VARIED TO OBTAIN DESIRED CURRENT LEVELS

D_1 MUST BE FAST RECOVERY TYPE, eg:
MBD5300 USED ABOVE $I_B \sim$ 100 mA
MSD6100 USED BELOW $I_B \sim$ 100 mA

FIGURE 3 — TURN-ON TIME

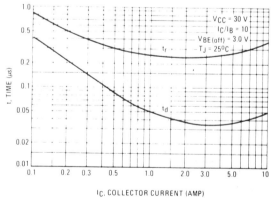

MOTOROLA *Semiconductor Products Inc.*

Courtesy of Motorola Semiconductor Products Inc.

Operational Amplifiers

LM301A operational amplifier
general description

The LM301A is a general-purpose operational amplifier which features improved performance over the 709C and other popular amplifiers. Advanced processing techniques make possible an order of magnitude reduction in input currents, and a redesign of the biasing circuitry reduces the temperature drift of input current.

This amplifier offers many features which make its application nearly foolproof: overload protection on the input and output, no latch-up when the common mode range is exceeded, freedom from oscillations and compensation with a single 30 pF capacitor. It has advantages over internally compensated amplifiers in that the compensation can be tailored to the particular application. For

example, as a summing amplifier, slew rates of 10 V/μs and bandwidths of 10 MHz can be realized. In addition, the circuit can be used as a comparator with differential inputs up to ±30V; and the output can be clamped at any desired level to make it compatible with logic circuits.

The LM301A provides better accuracy and lower noise than its predecessors in high impedance circuitry. The low input currents also make it particularly well suited for long interval integrators or timers, sample and hold circuits and low frequency waveform generators. Further, replacing circuits where matched transistor pairs buffer the inputs of conventional IC op amps, it can give lower offset voltage and drift at reduced cost.

schematic** and connection diagrams

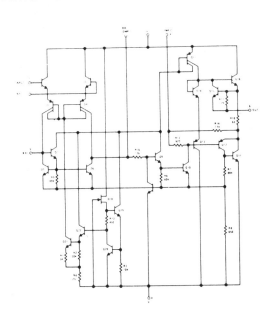

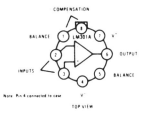

Order Number LM301AH
See Package 11

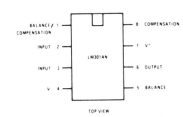

Order Number LM301AN
See Package 20

typical applications **

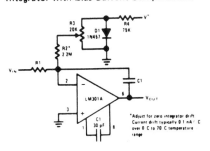

Integrator with Bias Current Compensation

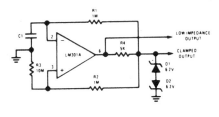

Low Frequency Square Wave Generator

Voltage Comparator for Driving
DTL or TTL Integrated Circuits

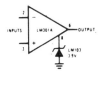

**Pin connections shown are for metal can.

absolute maximum ratings   LM301A

Supply Voltage	±18V
Power Dissipation (Note 1)	500 mW
Differential Input Voltage	±30V
Input Voltage (Note 2)	±15V
Output Short-Circuit Duration (Note 3)	Indefinite
Operating Temperature Range	0°C to 70°C
Storage Temperature Range	−65°C to 150°C
Lead Temperature (Soldering, 10 sec)	300°C

electrical characteristics (Note 4)

PARAMETER	CONDITIONS	MIN	TYP	MAX	UNITS
Input Offset Voltage	$T_A = 25°C$, $R_S < 50\,k\Omega$		2.0	7.5	mV
Input Offset Current	$T_A = 25°C$		3	50	nA
Input Bias Current	$T_A = 25°C$		70	250	nA
Input Resistance	$T_A = 25°C$	0.5	2		MΩ
Supply Current	$T_A = 25°C$, $V_S = ±15V$		1.8	3.0	mA
Large Signal Voltage Gain	$T_A = 25°C$, $V_S = ±15V$ $V_{OUT} = ±10V$, $R_L \geq 2\,k\Omega$	25	160		V/mV
Input Offset Voltage	$R_S < 50\,k\Omega$			10	mV
Average Temperature Coefficient of Input Offset Voltage			6.0	30	µV/°C
Input Offset Current				70	nA
Average Temperature Coefficient of Input Offset Current	$25°C \leq T_A \leq 70°C$ $0°C \leq T_A \leq 25°C$		0.01 0.02	0.3 0.6	nA/°C nA/°C
Input Bias Current				300	nA
Large Signal Voltage Gain	$V_S = ±15V$, $V_{OUT} = ±10V$ $R_L \geq 2\,k\Omega$	15			V/mV
Output Voltage Swing	$V_S = ±15V$, $R_L = 10\,k\Omega$ $R_L = 2\,k\Omega$	±12 ±10	±14 ±13		V V
Input Voltage Range	$V_S = ±15V$	±12			V
Common Mode Rejection Ratio	$R_S < 50\,k\Omega$	70	90		dB
Supply Voltage Rejection Ratio	$R_S < 50\,k\Omega$	70	96		dB

Note 1: For operating at elevated temperatures, the device must be derated based on a 100°C maximum junction temperature and a thermal resistance of 150°C/W junction to ambient or 45°C/W junction to case.

Note 2: For supply voltages less than ±15V, the absolute maximum input voltage is equal to the supply voltage.

Note 3: Continuous short circuit is allowed for case temperatures to 70°C and ambient temperatures to 55°C.

Note 4: These specifications apply for $0°C \leq T_A < 70°C$, $±5V \leq V_S \leq ±15V$ and C1 = 30 pF unless otherwise specified.

Courtesy of National Semiconductor Corporation

µA555
SINGLE TIMING CIRCUIT
FAIRCHILD LINEAR INTEGRATED CIRCUIT

GENERAL DESCRIPTION — The µA555 Timing Circuit is a very stable controller for producing accurate time delays or oscillations. In the time delay mode, the delay time is precisely controlled by one external resistor and one capacitor; in the oscillator mode, the frequency and duty cycle are both accurately controlled with two external resistors and one capacitor. By applying a trigger signal, the timing cycle is started and an internal flip-flop is set, immunizing the circuit from any further trigger signals. To interrupt the timing cycle a reset signal is applied ending the time-out.

The output, which is capable of sinking or sourcing 200 mA, is compatible with TTL circuits and can drive relays or indicator lamps.

- **MICROSECONDS THROUGH HOURS TIMING CONTROL**
- **ASTABLE OR MONOSTABLE OPERATING MODES**
- **ADJUSTABLE DUTY CYCLE**
- **200 mA SINK OR SOURCE OUTPUT CURRENT CAPABILITY**
- **TTL OUTPUT DRIVE CAPABILITY**
- **TEMPERATURE STABILITY OF 0.005% PER °C**
- **NORMALLY ON OR NORMALLY OFF OUTPUT**
- **DIRECT REPLACEMENT FOR SE555/NE555**

ABSOLUTE MAXIMUM RATINGS

Supply Voltage	+18 V
Power Dissipation (Note 1)	600 mW
Operating Temperature Ranges	
µA555TC/HC	0°C to +70°C
µA555HM	−55°C to +125°C
Storage Temperature Range	−65°C to +150°C
Lead Temperature	
Plastic Mini DIP(9T) (Soldering, 10 s)	260°C
Metal Can (5T) (Soldering, 60 s)	300°C

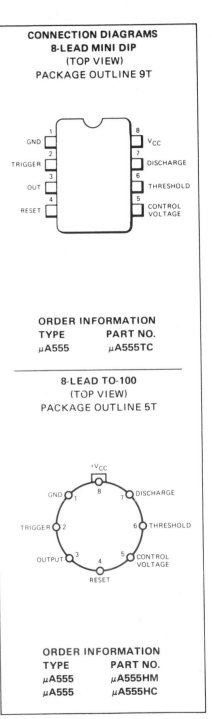

CONNECTION DIAGRAMS
8-LEAD MINI DIP
(TOP VIEW)
PACKAGE OUTLINE 9T

ORDER INFORMATION

TYPE	PART NO.
µA555	µA555TC

8-LEAD TO-100
(TOP VIEW)
PACKAGE OUTLINE 5T

ORDER INFORMATION

TYPE	PART NO.
µA555	µA555HM
µA555	µA555HC

BLOCK DIAGRAM

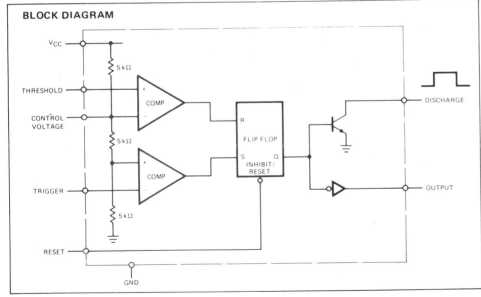

©1977 Fairchild Camera and Instrument Corporation   Printed in U.S.A.   232-11-0009-037 2M

Manufactured under one or more of the following U.S. Patents: 2981877, 3015048, 3025589, 3064167, 3108359, 3117260; other patents pending.

Courtesy of Fairchild Semiconductor

ELECTRICAL CHARACTERISTICS ($T_A = 25°C$, V_{CC} = +5.0 V to +15 V, unless otherwise specified)

PARAMETER	TEST CONDITIONS	μA555HM			μA555TC/HC			UNITS
		MIN	TYP	MAX	MIN	TYP	MAX	
Supply Voltage		4.5		18	4.5		16	V
Supply Current	V_{CC} = 5.0 V, $R_L = \infty$		3.0	5.0		3.0	6.0	mA
	V_{CC} = 15 V, $R_L = \infty$ LOW State (Note 1)		10	12		10	15	mA
Timing Error								
Initial Accuracy	R_A, R_B = 1 kΩ to 100 kΩ		0.5	2.0		1.0		%
Drift with Temperature	C = 0.1 μF (Note 2)		30	100		50		ppm/°C
Drift with Supply Voltage			0.05	0.2		0.1		%V
Threshold Voltage			2/3			2/3		X V_{CC}
Trigger Voltage	V_{CC} = 15 V	4.8	5.0	5.2		5.0		V
	V_{CC} = 5.0 V	1.45	1.67	1.9		1.67		V
Trigger Current			0.5			0.5		μA
Reset Voltage		0.4	0.7	1.0	0.4	0.7	1.0	V
Reset Current			0.1			0.1		mA
Threshold Current	Note 3		0.1	0.25		0.1	0.25	μA
Control Voltage Level	V_{CC} = 15 V	9.6	10	10.4	9.0	10	11	V
	V_{CC} = 5.0 V	2.9	3.33	3.8	2.6	3.33	4.0	V
Output Voltage Drop (LOW)	V_{CC} = 15 V							
	I_{SINK} = 10 mA		0.1	0.15		0.1	0.25	V
	I_{SINK} = 50 mA		0.4	0.5		0.4	0.75	V
	I_{SINK} = 100 mA		2.0	2.2		2.0	2.5	V
	I_{SINK} = 200 mA		2.5			2.5		V
	V_{CC} = 5.0 V							
	I_{SINK} = 8.0 mA		0.1	0.25				V
	I_{SINK} = 5.0 mA					0.25	0.35	V
Output Voltage Drop (HIGH)	I_{SOURCE} = 200 mA V_{CC} = 15 V		12.5			12.5		V
	I_{SOURCE} = 100 mA V_{CC} = 15 V	13	13.3		12.75	13.3		V
	V_{CC} = 5.0 V	3.0	3.3		2.75	3.3		V
Rise Time of Output			100			100		ns
Fall Time of Output			100			100		ns

NOTES:
1. Supply Current is typically 1.0 mA less when output is HIGH.
2. Tested at V_{CC} = 5.0 V and V_{CC} = 15 V.
3. This will determine the maximum value of $R_A + R_B$. For 15 V operation, the max total R = 20 MΩ.
4. For operating at elevated temperatures the device must be derated based on a +125°C maximum junction temperature and a thermal resistance of +45°C/W junction to case for TO-5 and +150°C/W junction to ambient for both packages.

Courtesy of Fairchild Semiconductor

The General Electric 2N2646 and 2N2647 Silicon Unijunction Transistors have an entirely new structure resulting in lower saturation voltage, peak-point current and valley current as well as a much higher base-one peak pulse voltage. In addition, these devices are much faster switches.

The 2N2646 is intended for general purpose industrial applications where circuit economy is of primary importance, and is ideal for use in firing circuits for Silicon Controlled Rectifiers and other applications where a guaranteed minimum pulse amplitude is required. The 2N2647 is intended for applications where a low emitter leakage current and a low peak point emitter current (trigger current) are required (i.e. long timing applications), and also for triggering high power SCR's.

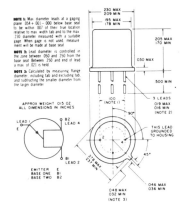

Power Dissipation (Note 1)	300 mw
RMS Emitter Current	50 ma
Peak Emitter Current (Note 2)	2 amperes
Emitter Reverse Voltage	30 volts
Interbase Voltage	35 volts
Operating Temperature Range	$-65°C$ to $+125°C$
Storage Temperature Range	$-65°C$ to $+150°C$

ELECTRICAL CHARACTERISTICS at $T_A = 25°C$

PARAMETER		2N2646			2N2647			
		Min.	Typ.	Max.	Min.	Typ.	Max.	
Intrinsic Standoff Ratio ($V_{BB} = 10V$)	η	0.56	0.65	0.75	0.68	0.75	0.82	
Interbase Resistance ($V_{BB} = 3V$, $I_E = 0$)	R_{BBO}	4.7	7	9.1	4.7	7	9.1	$K\Omega$
Emitter Saturation Voltage ($V_{BB} = 10V$, $I_E = 50$ ma)	$V_{E(SAT)}$		2			2		volts
Modulated Interbase Current ($V_{BB} = 10V$, $I_E = 50$ ma)	$I_{B2(MOD)}$		12			12		ma
Emitter Reverse Current ($V_{B2E} = 30V$, $I_{B1} = 0$)	I_{EO}		0.05	12		0.01	0.2	μa
Peak Point Emitter Current ($V_{BB} = 25V$)	I_P		0.4	5		0.4	2	μa
Valley Point Current ($V_{BB} = 20V$, $R_{B2} = 100\Omega$)	I_V	4	6		8	11	18	ma
Base-One Peak Pulse Voltage (Note 3)	V_{OB1}	3.0	6.5		6.0	7.5		volts
SCR Firing Conditions (See Figure 26, back page)								

1. Derate 3.0 MW/°C increase in ambient temperature. The total power dissipation (available power to Emitter and Base-Two) must be limited by the external circuitry.
2. Capacitor discharge—10µfd or less, 30 volts or less.
3. The Base-One Peak Pulse Voltage is measured in the circuit below. This specification on the 2N2646 and 2N2647 is used to ensure a minimum pulse amplitude for applications in SCR firing circuits and other types of pulse circuits.

4. The intrinsic standoff ratio, η, is essentially constant with temperature and interbase voltage. η is defined by the equation:

$$V_P = \eta \, V_{BB} + V_D$$

Where V_P = Peak Point Emitter Voltage
V_{BB} = Interbase Voltage

V_D = Junction Diode Drop (Approx. .5V)

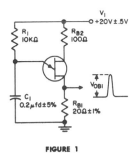

FIGURE 1

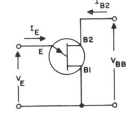

FIGURE 2
Unijunction Transistor Symbol with Nomenclature used for voltage and currents.

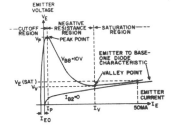

FIGURE 3
Static Emitter Characteristics curves showing important parameters and measurement points (exaggerated to show details).

GENERAL ELECTRIC

Courtesy of General Electric

Consumer Circuits

LM566/LM566C voltage controlled oscillators

general description

The LM566/LM566C are general purpose voltage controlled oscillators which may be used to generate square and triangular waves, the frequency of which is a very linear function of a control voltage. The frequency is also a function of an external resistor and capacitor.

The LM566 is specified for operation over the $-55°C$ to $+125°C$ military temperature range. The LM566C is specified for operation over the $0°C$ to $+70°C$ temperature range.

features

■ Wide supply voltage range: 10 to 24 volts
■ Very linear modulation characteristics

■ High temperature stability
■ Excellent supply voltage rejection
■ 10 to 1 frequency range with fixed capacitor
■ Frequency programmable by means of current, voltage, resistor or capacitor.

applications

■ FM modulation
■ Signal generation
■ Function generation
■ Frequency shift keying
■ Tone generation

schematic and connection diagrams

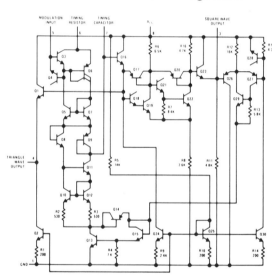

Metal Can

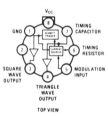

TOP VIEW

**Order Number LM566H or LM566CH
See Package 11**

Dual-In-Line Package

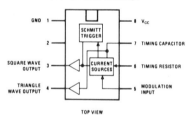

TOP VIEW

**Order Number LM566CN
See Package 20**

typical application

**1 kHz and 10 kHz TTL Compatible
Voltage Controlled Oscillator**

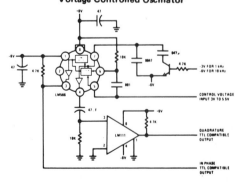

applications information

The LM566 may be operated from either a single supply as shown in this test circuit, or from a split (±) power supply. When operating from a split supply, the square wave output (pin 4) is TTL compatible (2 mA current sink) with the addition of a 4.7 kΩ resistor from pin 3 to ground.

A .001 μF capacitor is connected between pins 5 and 6 to prevent parasitic oscillations that may occur during VCO switching.

Coutesy of National Semiconductor Corporation

absolute maximum ratings

Power Supply Voltage   26V
Power Dissipation (Note 1)   300 mW
Operating Temperature Range   LM566 -55°C to $+125^\circ$C
                               LM566C 0°C to 70°C
Lead Temperature (Soldering, 10 sec) 300°C

electrical characteristics V_{CC} = 12V, T_A = 25°C, AC Test Circuit

PARAMETER	CONDITIONS	LM566			LM566C			UNITS
		MIN	TYP	MAX	MIN	TYP	MAX	
Maximum Operating Frequency	R_0 = 2k C_0 = 2.7 pF		1			1		MHz
Input Voltage Range Pin 7		3/4 V_{CC}		V_{CC}	3/4 V_{CC}		V_{CC}	
Average Temperature Coefficient of Operating Frequency			100			200		ppm/$^\circ$C
Supply Voltage Rejection			1			2		%/V
Input Impedance Pin 5			1			1		MΩ
VCO Sensitivity	f_0 = 10 kHz		6600			6600		Hz/V
FM Distortion	±10% Deviation		.2	.75		.2	1.5	%
Maximum Sweep Rate			1			1		MHz
Sweep Range			10:1			10:1		
Output Impedance Pin 3			50			50		Ω
Pin 4			50			50		Ω
Square Wave Output Level	R_{L1} = 10k	5.0	5.4		5.0	5.4		V p-p
Triangle Wave Output Level	R_{L2} = 10k	2.0	2.4					V p-p
Square Wave Duty Cycle		45		55	40		60	%
Square Wave Rise Time			20			20		ns
Square Wave Fall Time			50			50		ns
Triangle Wave Linearity			.2			.5		%

Note 1: The maximum junction temperature of the LM566 is 150°C, while that of the LM566C is 100°C. For operating at elevated junction temperatures, devices in the TO-5 package must be derated based on a thermal resistance of 150°C/W. The thermal resistance of the dual-in-line package is 100°C/W.

Courtesy of National Semiconductor Corporation

Voltage Regulators

LM100/LM200/LM300 voltage regulator

general description

The LM100, LM200 and LM300 are integrated voltage regulators designed for a wide range of applications from digital power supplies to precision regulators for analog circuitry. Built on a single silicon chip, these devices are encapsulated in either an 8-lead, low profile TO-5 header or a 1/4 x 1/4 metal flat package. Outstanding characteristics are:

- Output voltage adjustable from 2V to 30V (LM300 adjustable from 2V to 20V)
- Better than one percent load and line regulation
- One percent temperature stability
- Adjustable short-circuit limiting
- Output currents in excess of 5A possible by adding external transistors

- Can be used as either a linear or high-efficiency switching regulator.

Additional features are fast response to both load and line transients, small standby power dissipation, freedom from oscillations with varying resistive and reactive loads, and the ability to start reliably on any load within rating.

The LM100 is specified for operation over the -55°C to $+125^\circ$C military temperature range. The LM200 and LM300 are low cost, commercial-industrial versions of the LM100. They are identical to the LM100 except that they are specified for operation from -25°C to 85°C and from 0°C to 70°C respectively.

schematic and connection diagrams

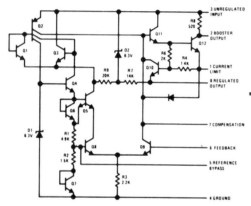

Metal Can

Order Number LM100H
or LM200H or LM300H
See Package 11

Flat Package

Order Number LM100F
or LM200F or LM300F
See Package 3

Pin connections shown are for TO-5 package

typical applications

Basic Regulator Circuit

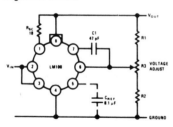

200 mA Regulator

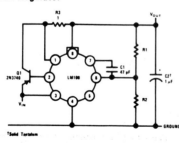

2A Regulator With Foldback Current Limiting

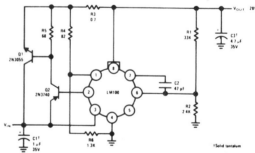

4A Switching Regulator

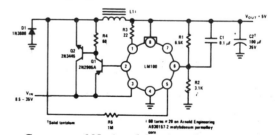

Courtesy of National Semiconductor Corporations

absolute maximum ratings

Input Voltage

    LM100, LM200   40V

    LM300   35V

Input-Output Voltage Differential

    LM100, LM200   40V

    LM300   30V

Power Dissipation (Note 1)

    LM100, LM200   800 mW

    LM300   500 mW

Operating Temperature Range

    LM100, LM200 -55°C to $+150^\circ$C

    LM300 0°C to 70°C

Storage Temperature Range -65°C to 150°C

Lead Temperature (soldering, 10 sec) 300°C

electrical characteristics (Note 2)

PARAMETER	CONDITIONS	MIN	TYP	MAX	UNITS
Input Voltage Range					
LM100/LM200		8.5		40	V
LM300		8.5		30	
Output Voltage Range					
LM100/LM200				30	V
LM300		2.0		20	
Output-Input Voltage Differential					
LM100/LM200				30	V
LM300		3.0		20	
Load Regulation (Note 3)	$R_{SC} = 0$, $I_O < 12$ mA		0.1	0.5	%
Line Regulation	$V_{IN} - V_{OUT} \leq 5V$		0.1	0.2	%/V
	$V_{IN} - V_{OUT} \geq 5V$		0.05	0.1	%/V
Temperature Stability					
LM100	$-55^\circ C < T_A < +125^\circ C$		0.3	1.0	
LM200	$-25^\circ C < T_A < 85^\circ C$		0.3	1.0	%
LM300	$0^\circ C < T_A < 70^\circ C$		0.3	2.0	
Feedback Sense Voltage		1.63	1.7	1.81	V
Output Noise Voltage	$10\ Hz \leq f \leq 10\ kHz$				
	$C_{REF} = 0$		0.005		%
	$C_{REF} = 0.1\ \mu F$		0.002		%
Long Term Stability			0.1	1.0	%
Standby Current Drain					
LM100/LM200	$V_{IN} = 40V$				
LM300	$V_{IN} = 30V$		1.0	3.0	mA
Minimum Load Current					
LM100/LM200	$V_{IN} - V_{OUT} = 30V$				
LM300	$V_{IN} - V_{OUT} = 20V$		1.5	3.0	mA

Note 1: The maximum junction temperature of the LM100 is 150°C, while that of the LM200 is 100°C, and the LM300 is 85°C. For operating at elevated temperatures, devices in the TO-5 package must be derated based on a thermal resistance of 150°C/W junction to ambient or 45°C/W, junction to case. For the flat package, the derating is based on a thermal resistance of 185°C/W when mounted on a 1/16-inch-thick, epoxy-glass board with ten, 0.03-inch-wide, 2-ounce copper conductors. Peak dissipations to 1.0W are allowable providing the dissipation rating is not exceeded with the power averaged over a five second interval for the LM100 and LM200, and a two second interval for the LM300.

Note 2: These specifications apply for an operating temperature between -55°C to $+125^\circ$C for the LM100, between -25°C to 85°C for the LM200 and between 0°C to 70°C for the LM300 devices for input and output voltages within the ranges given, and for a divider impedance seen by the feedback terminal of 2 kΩ, unless otherwise specified. The load and line regulation specifications are for constant junction temperature. Temperature drift effects must be taken into account separately when the unit is operating under conditions of high dissipation.

Note 3: The output currents given, as well as the load regulation, can be increased by the addition of external transistors. The improvement factor will be roughly equal to the composite current gain of the added transistors.

Courtesy of National semiconductor Corporation

APPENDIX D: TORIOD CORE AND POT CORE SPECIFICATION SHEETS

SERIES K300 TOROID CORES

TOROID CORES

1. These toroid core types are
   - manufactured in 3E ferrite material.
   - available as lacquer coated core.

2. Electrical parameters are expressed in the MKS System.

3. All terms and symbols used are defined in the Glossary at the rear of the catalog.

4. Characteristics of the ferrite material used are described on the next page.

MECHANICAL CHARACTERISTICS & DIMENSIONS

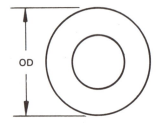
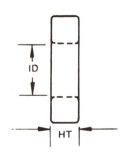

All dimensions are in inches and are nominal, measured with laquer coating on the core.

MECHANICAL CHARACTERISTICS

		K300502	K300500	K300501
OUTSIDE DIA.	OD	1.142	1.417	1.417
INSIDE DIA.	ID	.748	.905	.905
THICKNESS	HT	.295	.394	.591
MAGNETIC PATH LENGTH	ℓ_e	2.95 in. 7.50 CM	3.63 in. 9.20 CM	3.63 in. 9.20 CM
CORE CONSTANT	$\Sigma \frac{\ell_e}{A_e}$	51.10 in.$^{-1}$ 20.10 CM^{-1}	36.10 in.$^{-1}$ 14.2 CM^{-1}	24.0 in.$^{-1}$ 9.42 CM^{-1}
EFFECTIVE CORE AREA	A_e	.058 in.^{2} .373 CM^2	.100 in.^{2} .647 CM^2	.147 in^2 .977 CM^2
EFFECTIVE CORE VOLUME	V_e	.170 in.^{3} 2.58 CM^3	.362 in.^{3} 5.60 CM^3	.520 in.^{3} 9.00 CM^3
WEIGHT		.455 oz. 13 grams	1.02 oz. 29 grams	1.54 oz. 44 grams

ELECTRICAL CHARACTERISTICS

CORE PART NUMBER	CORE MAT'L TYPE	A_L mH PER 1000 TURNS (±20%)	μ_o (REF.)
K300502	3E	1688	2700
K300500	3E	2422	2700
K300501	3E	3639	2700

DIVISION OF AMPEREX ELECTRONIC CORP.

5083 Kings Highway, Saugerties, New York 12477

Telephone (914) 246-2811

A North American Philips Company

Courtesy of Ferroxcube

TOROID CORES

3E MATERIAL

A versatile material with a medium permeability. Losses under high excitation are low, making this MnZn ferrite suitable for power-handling transformers and inductors.

Used in Series K300 toroid cores described on page 3-9.

3E CHARACTERISTICS

Parameters shown are typical values, based upon measurements of a 1″ toroid.

Initial Permeability at 25°C	μ_0	2700 (±20%)
Saturation Flux Density at 25°C., H = 5 oersteds	B_s	3000 gauss*
Coercive Force	H_c	0.1 oersted*
Loss Factor at B ≤ 1 gauss $\frac{\tan \delta}{\mu_0}$ 100 KHz 500 KHz		$\leqslant 15 \times 10^{-6}$ 90×10^{-6} *
Temperature Factor (+20° to +50°C)	TF	$+4 \times 10^{-6}$ *
Disaccommodation Factor (10-100 minutes)	DF	$< 5 \times 10^{-6}$ *
Hysteresis Loss Constant at 4 KHz	η_B	$\leqslant 1.8 \times 10^{-3}$ tesla^{-1}
Curie Temperature	T_c	$\geqslant 125°C$

*Typical values

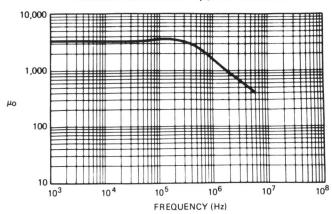

INITIAL PERMEABILITY (μ_0) vs. FREQUENCY

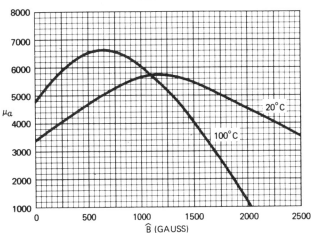

PERMEABILITY (μ_a) vs. FLUX DENSITY (B)

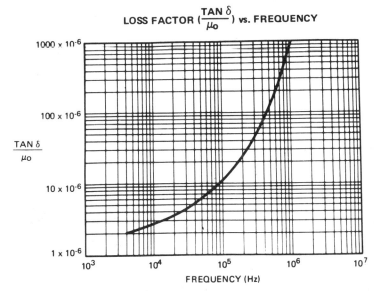

LOSS FACTOR ($\frac{TAN \delta}{\mu_0}$) vs. FREQUENCY

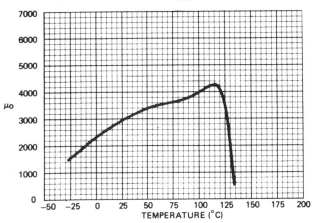

INITIAL PERMEABILITY (μ_0) vs. TEMPERATURE

Courtesy of Ferroxcube

SERIES 905 POT CORES

1 **This core type is**

- manufactured in the following ferrite materials:
     3B7   3B9   3D3   4C4

- available in the following types:
     ungapped;   fixed gap;   adjustable gap.

- available with optional Single-Section Standard Bobbin only.

2 Electrical parameters are expressed in the MKS System.

3 All terms and symbols used are defined in the Glossary at the rear of the catalog.

4 Characteristics of the ferrite materials used are described in the Materials section at the front of the catalog.

TWICE ACTUAL SIZE

MECHANICAL CHARACTERISTICS & DIMENSIONS

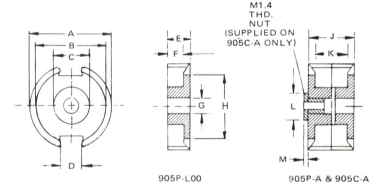

M1.4 THD. NUT (SUPPLIED ON 905C-A ONLY)

905P-L00

905P-A & 905C-A

MECHANICAL CHARACTERISTICS

NOTE: Values given apply to a core **set.**

MAGNETIC PATH LENGTH	ℓ_e	.493 in. 1.25 CM
CORE CONSTANT	$\Sigma \dfrac{\ell_e}{A_e}$	31.5 in.$^{-1}$ 12.4 CM^{-1}
EFFECTIVE CORE AREA	A_e	.0156 in.^{2} .101 CM^2
EFFECTIVE CORE VOLUME	V_e	.0075 in.^{3} 0.126 CM^3
WEIGHT		.046 oz. 1.3 Grams

NOTE: MINIMUM CORE AREA .0729 CM^2

OUTLINE DIMENSIONS
All dimensions are in inches.

	MINIMUM	MAXIMUM		MINIMUM	MAXIMUM
A	.354	.366	G	.079	.083
B	.295	.305	H	.246	.266
C	.146	.154	J	.204	.212
D	.071	.087	K	.142	.154
E	.102	.106	L	——	.114
F	.071	.077	M	——	.028

Courtesy of Ferroxcube

ELECTRICAL CHARACTERISTICS

UNGAPPED POT CORES

CORE PART NUMBER*	CORE MATERIAL	A_L† (mH PER 1000 TURNS) (±25%)	μ_e† (REF.)
905P-L00-3B7	3B7	1229	1217
905P-L00-3B9	3B9	1150	1135
905P-L00-3D3	3D3	665	650
905P-L00-4C4	4C4	125	124

*Part number for a core half. †Per pair of cores.

GAPPED POT CORES

NON-ADJUSTABLE GAPPED POT CORE PART NO.*	ADJUSTABLE-GAP POT CORE ASSEMBLY PART NO.**	CORE MATERIAL	A_L† VALUE	μ_e (REF.)	APPROX. GAP LENGTH (IN.)	TEMPERATURE COEFFICIENT PPM/°C	
						MIN-MAX	TEMP. RANGE
905P-A60-3B7	905C-A60-3B7 Green Adjustor	3B7 (To 300 kHz)	60 ±1.5%	55.0	.009	−33 to +33	+20° to +70°C
905P-A100-3B7	905C-A100-3B7 Yellow Adjustor		100 ±2%	91.5	.0045	−55 to +55	
905P-A160-3B7	905C-A160-3B7 Brown Adjustor		160 ±2%	147.0	.0024	−88 to +88	
905P-A60-3B9	905C-A60-3B9 Green Adjustor	3B9 (To 300 kHz)	60 ±1.5%	55.0	.010	+49 to +105	−30° to +70°C
905P-A100-3B9	905C-A100-3B9 Yellow Adjustor		100 ±2%	91.5	.0047	+82 to +174	
905P-A160-3B9	905C-A160-3B9 Brown Adjustor		160 ±2%	147.0	.0025	+132 to +279	
905P-A25-3D3	905C-A25-3D3 Green Adjustor	3D3 (200 kHz to 2.5 MHz)	25 ±1.5%	23.0	.028	+23 to +69	
905P-A40-3D3	905C-A40-3D3 Green Adjustor		40 ±1.5%	36.7	.015	+36 to +108	
905P-A60-3D3	905C-A60-3D3 Yellow Adjustor		60 ±1.5%	55.0	.0065	+55 to +165	
905P-A16-4C4	905C-A16-4C4 Green Adjustor	4C4 (1 MHz to 20 MHz)	16 ±1.5%	14.7	.031	−88 to +88	+5° to +55°C
905P-A25-4C4	905C-A25-4C4 Green Adjustor		25 ±1.5%	23.0	.026	−138 to +138	
905P-A40-4C4	905C-A40-4C4 Green Adjustor		40 ±1.5%	36.7	.013	−220 to +220	

*Part number is for a core set (2 cores).   **Part number is for a core set (2 cores), nut, and specified adjustor.

†The A_L values are based on a fully wound bobbin without adjustor: mH/1000 turns.

Courtesy of Ferroxcube

APPENDIX E: REFERENCES

Brazee, J.G, *SEMICONDUCTORS AND TUBE ELECTRONICS.* New York, Holt, Rinehart, and Winston, Inc., 1969.

Fairchild Semiconductor, *THE VOLTAGE REGULATOR HANDBOOK.* Mountain View, California, 1974.

Ferroxcube, *LINEAR FERRITE MATERIALS AND COMPONENTS.* Saugerties, New York, 1975

MacDonald, Lorne, *PRACTICAL ANALYSIS OF AMPLIFIER CIRCUITS THROUGH EXPERIMENTATION.* Seal Beach, The Technical Education Press, 1975.

MIL-HDBK-215, *MILITARY HANDBOOK ON ELECTRONIC CIRCUITS, PART 9.* Washington, D.C., Government Printing Press, June 15, 1960.

Motorola Semiconductor Products Inc., *THEORY AND CHARACTERISTICS OF THE UNIJUNCTION TRANSISTOR.* Phoenix, Arizona, 1975.

National Semiconductor Corporation, *VOLTAGE REGULATOR HANDBOOK.* Santa Clara, California, May 1975.

Pierce, J.F., and Paulus, T.J., *APPLIED ELECTRONICS.* Columbus, Ohio, Charles Merrill Publishing Company, 1972.

Siliconix Incorporated, *AN INTRODUCTION TO FIELD EFFECT TRANSISTORS.* Santa Clara, California, 1970.

SPECIFICATION SHEETS. Appendix C.

RCA, *SILCON POWER CIRCUITS MANUAL.* Harrison, New Jersey, 1969.

INDEX